RECURSIVE FUNCTION THEORY AND LOGIC

Computer Science and Applied Mathematics

A SERIES OF MONOGRAPHS AND TEXTBOOKS

Hans P. Künzi, H. G. Tzschach, and C. A. Zehnder
NUMERICAL METHODS OF MATHEMATICAL OPTIMIZATION: WITH ALGOL AND FORTRAN PROGRAMS, CORRECTED AND AUGMENTED EDITION, 1971

Azriel Rosenfeld
PICTURE PROCESSING BY COMPUTER, 1969

James Ortega and Werner Rheinboldt
ITERATIVE SOLUTION OF NONLINEAR EQUATIONS IN SEVERAL VARIABLES, 1970

A. T. Berztiss
DATA STRUCTURES: THEORY AND PRACTICE, 1971

Azaria Paz
INTRODUCTION TO PROBABILISTIC AUTOMATA, 1971

David Young
ITERATIVE SOLUTION OF LARGE LINEAR SYSTEMS, 1971

Ann Yasuhara
RECURSIVE FUNCTION THEORY AND LOGIC, 1971

In preparation

James M. Ortega
NUMERICAL ANALYSIS: AN INTERMEDIATE COURSE

RECURSIVE FUNCTION THEORY AND LOGIC

Ann Yasuhara
New York University

ACADEMIC PRESS
NEW YORK and LONDON

ACADEMIC PRESS, INC.
111 Fifth Avenue, New York, New York 10003

United Kingdom Edition published by
ACADEMIC PRESS, INC. (LONDON) LTD.
24/28 Oval Road, London NW1 7DD

Library of Congress Catalog Card Number: 74-154379

AMS (MOS) 1970 Subject Classification: 0201, 02B05, 02B10, 02F05, 02F10, 02F15, 02F20, 02F25, 02F35, 02F47, 02G05, 02G10, 02G15, 02G20, 02H05, 02H10, 02H13

PRINTED IN THE UNITED STATES OF AMERICA

IN GRATITUDE TO THOSE WHO HAVE TAUGHT ME,
BEGINNING WITH MY PARENTS AND INCLUDING MY HUSBAND

Ce n'est point de l'espace que je dois chercher ma dignité, mais c'est du réglement de ma pensée. Je n'aurai pas davantage en posedant des terres. Par l'espace l'univers me comprend et m'engloutit comme un point; par la pensée je le comprends.

Blaise Pascal, 1623–1662, *Pensées*

The original question, "Can machines think?" I believe to be too meaningless to deserve discussion. Nevertheless I believe that at the end of the century the use of words and general educated opinion will have altered so much that one will be able to speak of machines thinking without expecting to be contradicted.

Alan M. Turing, Computing Machinery and Intelligence, *Mind* **LIX**, 1950

CONTENTS

PREFACE

The purpose of this book is to present the basic facts of recursive function theory and mathematical logic in such a way that the reader understands it well, begins to appreciate its significance, learns how to think about and use it, and is prepared to go on in several different directions: automata theory, formal languages, complexity of computation, mechanical theorem-proving, the use of finite automata with respect to logic, more advanced recursive function theory, and more advanced mathematical logic.

The general scheme of the book is to study the notion of effective computability and how it relates to sets of natural numbers and to mathematical theories. In Part I formal definitions of effective computability are developed (Turing machines and partial recursive functions), and results about sets of numbers that can or cannot be obtained by these methods of computation are studied (e.g., the complete set, Friedberg's theorem). Part II is devoted to setting up the proper framework for showing the relationship between effective computability and mathematical theories and to studying the results of that relationship. This is done by studying formalized theories and their interpretations (via the Tarski definition of truth), as well as the extent and limitations of these theories (e.g., Gödel's completeness and incompleteness theorems).

Although the book includes that material about recursive function theory and mathematical logic which would be included in any logic text, the arrangement and presentation of the material and the extra topics are intended for computer scientists. These remarks are elucidated as follows:

1. *The standard recursive function and logic material.* This is covered in chapters I, III–V, VII, VIII (sections 1–3), IX, X, XI, XII, and XIV (sections 1–4). In Part II considerable attention is given to the semantic (meaning) as well as the syntactic (symbols) approaches to logic. Mathematics is not just a collection of axioms and formal deductions, as the reader learns in this book. Thus, even though a computer itself can only operate on a set of symbols

according to a specified set of rules (i.e., as a syntactic system) a theoretically minded computer scientist who uses a computer in mathematical investigations cannot restrict his attention merely to the syntactic aspects of logic. He must also study that part of logic which is concerned with syntax, semantics, and their relationship in the context of mathematics.

2. *The arrangement and presentation of the material.* Although logic books usually have the material of Part II first, followed by some discussion of the material of Part I, the order is reversed in this book for two main reasons. (i) The material of Part I is much less abstract than that of Part II and is therefore easier to grasp for the reader whose mathematical background may not be very strong. In particular, the notion of a Turing machine is very accessible, especially to those familiar with programming. Thus with ideas that are fairly familiar the reader can get very quickly to one of the most basic theorems—the unsolvability of the halting problem for Turing machines —and can immediately appreciate some of its significance. It is a great advantage for the student to be able, from the beginning, not only to do something but also to understand it. (ii) The basic topics of Part II are standard for any study of logic. However it is the computable and non-computable aspects of these topics which are particularly important to the computer scientist and so these are used as a framework for presenting the material in Part II.

3. *Extra topics.* Chapter II studies the basic ideas about semi-Thue systems, thus giving an introduction to word problems, and shows how these ideas lead into the study of formal grammars and associated languages and into certain kinds of algebraic problems. Chapter VI takes up several computational topics and, in particular, offers an introduction to some of the views of computational complexity. Chapter XIII presents Herbrand's theorem and some of the recent work in mechanical theorem proving. In chapter XIV, section 5, the method of elimination of quantifiers is introduced in order to show the decidability of the first-order theory of addition (Presburger's theorem). In section 6 there is an introduction to second-order languages and to finite automata; these are then combined to prove the decidability of the weak second-order theory of successor (Elgot's theorem).

The book was written for a one-year course, although the inclusion of all the material in the book in one year would probably be difficult. It was developed from mimeographed notes that I prepared for a graduate course I gave for three years at New York University. Although the majority of the students had studied mathematics as undergraduates, there were quite a few who had majored in engineering. Consequently the manner of presentation is based on the assumption that the reader is fairly capable in terms of reading and studying but may not be very well trained in sophisticated mathematical thinking. With a few exceptions, everything in the book is defined and examples are given. The exceptions are (i) very elementary theory of equivalence relations and equivalence classes, (ii) very elementary set theory, (iii) very elementary cardinal arithmetic, and (iv) knowledge of the definition of a group and a field. That the student be familiar with (i) and (ii) is important. As for (iii) what is really important is given in the book; the rest plays a minor role. Groups and fields are used here and there as examples.

Some parts of the book can be omitted and still others, particularly the computations of primitive recursive and recursive functions, should be left to the reader to work through. The natural flow of the book is as follows: chapter I, chapter III, chapter IV, chapter V (sections 1–4), chapter VII, chapter II (section 1), chapter VIII (sections 1–3), chapter IX (sections 1–5, and 7), chapter X, chapter XI, chapter XII, chapter XIII (sections 1–5), and chapter XIV (sections 1–3). (The course as I gave it included this sequence.) The material mentioned here that belongs in Part II (perhaps without chapter XIII) could be given as a one-semester logic course provided that either chapter I or some intuitive notions of effective procedures and decidable problems are taught first.

The exercises are a very important part of the book. They are an integral part of the flow of thought and the development of the ideas and methods. Most of them must be read in order to go on in the book. The position of an exercise in the text is usually determined by the result (i.e., fact) which the problem states and the method of proof to be employed. Most of the exercises are used immediately and many are used again later. There are several kinds of exercises to serve several purposes but they are all directed toward helping the reader to think about the material and to understand it. Once certain methods of proof have been introduced further lemmas and theorems whose proofs can be done by those methods are given as exercises. At that point it is more interesting, as well as more instructive, to do the proof yourself than to read it. A proof is rarely given that the reader should be able to supply for himself. Sometimes an exercise is so trivial as to be merely an observation; others are fairly substantial. Most are in between.

Princeton, New Jersey

ACKNOWLEDGMENTS

My general indebtedness to the books of Church (1956), Kleene (1952), Davis (1958), Mendelson (1964), Minsky (1967), Rogers (1967), and Shoenfield (1967) will be obvious to anyone who knows those books and reads this one. When I have used specific information from a particular book or paper I have indicated that source in the text. I wish to thank Professor Martin Davis very much for reading the first draft of Part I and for his comments and suggestions. I am grateful to the students who had to suffer through the roughness of the mimeographed notes. They usually did so with patience and sometimes with enthusiasm. Among them, I am particularly grateful to Mrs. Frances Gustavson and Mr. Jeffrey Landau for reading the first draft with, so to speak, a fine tooth comb, for pointing out mistakes, and for making many helpful suggestions. My thanks again to Mrs. Gustavson for her careful reading of section 13.6; it is much better than it would have been without her comments. I appreciate very much the helpful comments of many friends who were kind enough to read various drafts of the Introduction. The difficulties of turning a manuscript into a book have been almost pleasant thanks to the staff of Academic Press. It is impossible to express my thanks to my husband, Mitsuru Yasuhara, for hours and hours of discussion, many suggestions, and answers to many questions, to say nothing of general encouragement and sustained good humor. However, lest his reputation be besmirched by whatever may be bad about the book, let it be known that he has not read all of the manuscript and I have not always followed his suggestions.

RECURSIVE
FUNCTION
THEORY
AND
LOGIC

INTRODUCTION

PLAYING TAG

Suppose you are waiting someplace and have nothing to do but you do have a paper and pencil. Here is something with which to amuse yourself.† Write

abccac.

Looking at the left-most symbol, *a*, write *bab* at the right end of *abccac* and erase the first three symbols on the left, thus obtaining

cacbab.

This time the left-most symbol is *c*. Write *cab* at the right end of *cacbab* and erase the first three letters on the left to obtain

babcab.

Now the left-most symbol is *b*. Write *abba* at the right and erase the first three symbols to obtain

cababba.

By always appending *bab* on the right if *a* is on the left, appending *abba* on the right if *b* is on the left, appending *cab* on the right if *c* is on the left and each time erasing the first three symbols, we can continue to play for awhile—continuing from above

abbacab
acabbab
bbabbab
bbababba
babbaabba
.
.
.

† The symbols that are used here, {*a*, *b*, *c*}, are chosen in a completely arbitrary manner as are the pairings, *a* with *bab*, *b* with *abba*, and *c* with *cab*.

So the question is, by continuing to play will you find that (1) eventually one of the new sequences of symbols is the same as a previous one, or (2) eventually the new sequence is completely erased, or (3) you can play forever? (The first two possibilities are illustrated in examples below.)

The game described here is an example of a Tag system.† A Tag system is given by: a finite set‡ of symbols, $\{a_1, \ldots, a_m\}$; a finite set of pairs $\{(a_1, A_1), \ldots, (a_m, A_m)\}$, where each A_i is some specific nonempty finite sequence‡ of symbols from the given set; a positive integer n; and a particular sequence of those symbols, called the "initial word," A_0. The rules for playing Tag, i.e., generating sequences of symbols by a Tag system, are:

Instruction 0: Set W equal to A_0. Go to Instruction 1.
Instruction 1: Put a copy of W in a list called "the list."
Look at the left-most symbol of W.
If it is a_i, write A_i to the right of W and erase the first n symbols from the left of W with A_i written to the right.
Now set W equal to that result.
Go to Instruction 2.
Instruction 2: If W equals nothing, i.e., the erasing process erased everything, go to Instruction 3.
Otherwise, check to see if W equals any entry in the list.
If so, go to Instruction 3, if not go to Instruction 1.
Instruction 3: Stop.

If Instruction 3 is ever obeyed we say that the Tag system halts. The set of sequences of symbols generated by the Tag system is just the words in the list.

Here is another Tag system: $\{a, b, c\}$, $\{(a, bab), (b, abba), (c, ca)\}$, $n = 3$, $A_0 = abccba$. If we play awhile we get:

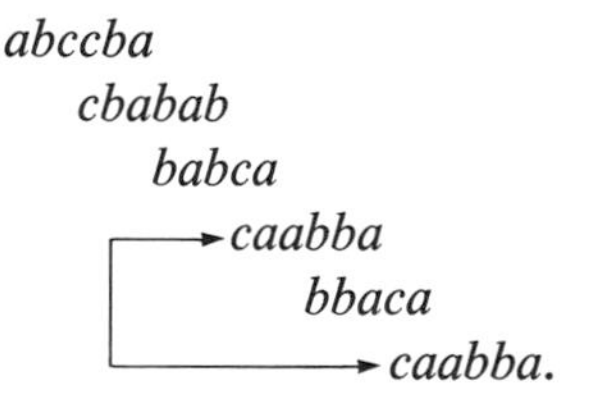

In this case the words start repeating and would, of course, continue to do

† Tag systems were first invented by Emil Post and were first written about by him (1943, pp. 203–204).

‡ In the Introduction all sets of symbols are finite and all sequences of symbols are finite so we will no longer specifically indicate that they are to be finite. Sets of sequences of symbols are not necessarily finite.

so forever were we not instructed to halt. In this situation we will say that the Tag system cycles. Here is another Tag system: $\{a, b, c\}$, $\{(a, cac), (b, bab), (c, ca)\}$, $n = 3$ and $A_0 = cabcba$. We have

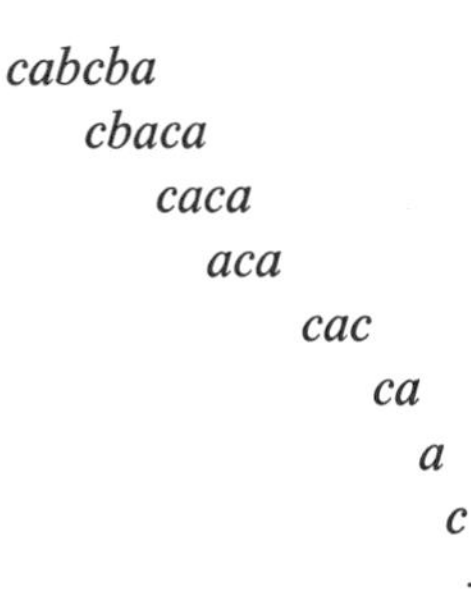

In this situation we say that the Tag system tags itself. The reader is invited to play with this Tag system: $\{0, 1\}$, $\{(0, 00), (1, 1101)\}$, $n = 3$, and $A_0 =$ 100100100100100100100. Although he is invited to play, it is only fair also to warn the reader that this is one of the first Tag systems that Post started playing with and after much playing by both human beings and computers, it is not yet known whether it cycles, tags itself or goes on forever.

Given a Tag system it is obvious that machines, children, and adults can play tag with it for hours or weeks in an attempt to see if it cycles or tags itself. But some people might become interested in a little more sophisticated problem. Given a Tag system, to figure out by whatever means whether or not it halts. For example, the reader can easily prove for himself that if the positive integer, n, of a Tag system is smaller than the length (total number of symbols) of each A_i, the Tag system does not tag itself; also that if n is greater than each of the lengths of the A_i the Tag system does tag itself. But what about Tag systems which do not belong to those two cases? (Notice that none of the four examples given belongs to one of those cases.) Instead of trying to figure out if a given Tag system halts, an even more sophisticated problem would be to try to find a procedure or method which one could use to decide of *any* Tag system whether or not it halts. Could we not write a program for a computing machine so that we could give the machine the set of symbols, the set of pairs, the integer n, and the initial word A_0 for any Tag system and the machine would give as output the words "yes" or "no" according to whether or not the Tag system used as input does or does not halt? The most obvious approach to try is to give a machine programmed to play tag a Tag system and wait to see if it stops. While it is true that if, at some point (maybe after a few thousand years), the Tag system does halt, the computer can write "yes" on its output tape, it is not true that it is able, by this method, truthfully to write "no." That this approach does not work does not necessarily mean that there is no approach that does work. With such a simple thing as a Tag system, we are inclined to say that surely there is some approach that works. But, however simple all this may seem, it has been proved that there can be no rule, no machine (which exists or which in principle might exist) that can tell of an arbitrary Tag system whether or not it halts.† Thus we see that there are apparently simple things that machines cannot do. One of the main purposes of this book is to explore the theoretical limits of what computing machines can do.

EFFECTIVE PROCEDURES

We will say that a finite set of instructions, each instruction being to perform some simple operation, is an effective procedure. An example is the set of instructions for playing tag. Notice that the definition of effective procedure is not a precise definition because "instruction," "simple," and "operation" are not precisely defined. But probably we all have some notion as to what would or would not be an effective procedure even if there is no precise way of making it absolutely clear. Certainly an ordinary digital computing machine with a program is an effective procedure.

We know that every positive integer is represented by a unique sequence of elements in the set of symbols $\{0, 1, 2, 3, 4, 5, 6, 7, 8, 9\}$ and that no two positive integers are represented by the same sequence. We will say that a set S is finitely representable if there is a finite set of symbols, Σ, such that every element of S can be represented by a unique finite sequence of symbols of Σ, no two sequences of S are represented by the same sequence and if there is an effective procedure for determining of any sequence of symbols of Σ whether or not that sequence represents an element of S. Certainly the set of positive integers is finitely representable. Since a Tag system on $\{a, b, c\}$ is given by a sequence of symbols of the form

$$\{a, b, c\}; \quad \{(a, A), (b, B), (c, C)\}; \quad n; \quad A_0$$

where A, B, and C are sequences of symbols from $\{a, b, c\}$, we see that the set of all Tag systems on $\{a, b, c\}$ is a finitely representable set. (Notice that the set of irrational numbers is not finitely representable.)

By definition, a function is a rule which associates to each element of one set (the domain of the function) a unique element of another (perhaps the same) set. The word "rule" as used in this definition does not mean anything precise. Certainly, an effective procedure is such a rule, but as we soon will see there are rules which are not effective procedures. We will say that a function is a total effectively computable function if its domain is finitely

† The proof of this fact will be easy for the reader to understand by the time, if not before, he has reached the end of chapter II. The references will be given there.

representable and if there exists an effective procedure by means of which one can obtain the value of the function for any element in the domain. As a matter of fact, most of the functions on the integers that the reader is likely to think of could reasonably be classified as total effectively computable functions. However, not all functions are total effectively computable functions. Any given Tag system does or does not halt; so we can define the function, *d*, from Tag systems to {"yes," "no"} by the rule:

> the value of *d* for a given Tag system is "yes"
> if that Tag system halts and is "no" if it does not.

Call this function "the deciding function for Tag systems." The implication of the end of the previous section is that the deciding function for Tag systems is not a total effectively computable function. As we noticed above, if a particular Tag system does halt, a human being or a machine can truthfully assert that the Tag system halts by simply obeying the instructions for playing tag with that system. Thinking once more of the deciding function for Tag systems, we see that there is an effective procedure for obtaining the values of that function for some of the elements of its domain, namely for just those Tag systems which do halt. We will say that a function (with finitely representable domain) is a partial effectively computable function if there is a subset of its domain and an effective procedure by means of which one can obtain the value of the function for any element in that subset. Thus, the deciding function for Tag systems is a partial effectively computable function though not a total effectively computable function.

In the early 1930's several logicians were thinking about the general notion of effective procedure and wished to define precisely procedures which definitely were effective and at the same time as broad as possible. That is, the precisely defined procedure should be effective and one should feel that any effective procedure could be realized by one of these precise procedures. Several such precise procedures have been defined and investigated (by Gödel, Herbrand, Church, Turing, Post, Markov, and Kleene). Most of these have been shown to be equivalent in the sense that whatever can be done by means of one kind of procedure can be done by means of any other kind. In this book (chapters I, IV, and V) we will study carefully two of these precise procedures —they have come to be known as Turing machines and partial (or μ) recursive functions of Kleene. It will be clear from our study of Turing machines that any kind of computation that can be done by an ordinary computing machine can be done by a Turing machine. Also as a result of our investigations, we will see that it is unlikely that the informal notion of "effective procedure" means anything stronger than "Turing machine" and that it is unlikely that there is any function which is a partial effectively computable function which is not also a partial recursive function.

WHAT CAN MACHINES DO FOR MATHEMATICIANS?

An immediate answer to this question would be, they can perform computations. But then, in view of the previous discussion, we should probably ask: What kinds of computations? Or even, what do we mean by "computation"? Is everything that appears at first sight to be computable in some naive sense actually computable in a mechanical sense? What kinds of machines are we allowed to use? Since Turing machines encompass all possible machines, in Part I we will study them and try to answer the other questions in terms of them.

Among many other possible answers to the question "What can machines do for mathematicians?" a less immediate, but nevertheless natural answer is that perhaps machines can be used to get answers to mathematical questions. That is, maybe they can be used to perform operations which do not appear to be computations in the usual sense. For example, can a machine be used to decide of certain sets of mathematical statements which are true and which are false? (From the previous discussion we know that a machine cannot, in general, be used to tell which statements are true and which false in the set {The Tag system given by $\{a_1, \ldots, a_m\}$, $\{(a_1, A_1), \ldots, (a_m, A_m)\}$, n, $A_0 \mid$ where $A_0, A_1, \ldots, A_m$ vary over all possible sequences of symbols from the given set and n varies over all positive integers}.) As we know, machines are very precise creatures, so if this is an example of something that a machine might actually do, the machine must have a precise definition of "truth." Furthermore, there must be some way of representing a mathematical "statement" so that it can be given to the machine to operate on. It is one of the tasks of logicians to try to provide solutions to these problems. In Part II formal (precise) languages (sets of symbols) are defined which are to be used in representing mathematical ideas, the relationship between such a language and the ideas it is meant to represent is studied, and finally the truth of a statement of a formal language is defined in terms of the mathematical idea it expresses.

It is well known that one of the main activities of mathematicians is proving mathematical statements. Perhaps computers can be of assistance. If so, there must also be a precise definition of "proof." But, one might ask, don't truth and provability mean the same thing in mathematics? The preciseness of the notion of mathematical proof varies according to what branch of mathematics you are dealing with. The most precise use that the reader is likely to be familiar with is the grade school version of Euclidean geometry. As he may recall, a few specific geometric statements (statements about geometry) are designated as the axioms because they are accepted as being obviously true. Certain rules for operating on the axioms are given as the only ones to be used and those are chosen because a statement resulting from such an operation would be true if the statements operated on were true. By means of these rules other geometric statements are deduced from the axioms and because of the choice of axioms and rules of deduction the statements so obtained are also accepted as being true. (If a statement can be obtained from the axioms by means of the rules then we say that it is provable.) It was thought that each true geometrical statement would be either an axiom or provable by this method. The difficulty is that the truth of one of the geometrical statements designated as axioms, namely the parallel line axiom, is not really so obviously true. The whole system, i.e., axioms and provable statements, would be more securely anchored in truth, so to speak, if that axiom could be dropped. But, most people (certainly including Aristotle and Euclid) believe that for any given statement, *S*, either *S* is true or not-*S* is true. Thus, the price of dropping the suspicious axiom would be that the statements provable from the remaining axioms would not include all true geometric statements—*unless* the parallel line axiom itself could be proved from the remaining axioms. Thus great effort was devoted over the years to trying to show that either the parallel line axiom or its negation is provable from the rest of Euclid's axioms. However, in the 19th century it was shown that it cannot be done. Although one of them, the parallel line axiom or its negation, *is* true (at least in this galaxy) neither is provable from the other axioms of Euclid. In Part II of this book a precise definition of proof of a statement in a formal language from a set of statements (axioms) in that formal language will be given. We will see that a proof is, in fact, an effective procedure. There is thus the possibility of using a machine to prove statements from a set of axioms. (This is somewhat analogous to using a machine to play tag.) One might also hope to use a machine to decide which formal statements in some set are (and which are not) provable from a particular set of formal statements taken as the axioms. [This is somewhat analogous to using a machine, given a particular Tag system, to decide of an arbitrary set of sequences of symbols (from the set of symbols of the given Tag system) which are generated by that Tag system and which are not.] These possibilities will be explored in Part II. Finally certain systems used by mathematicians will be considered from several points of view—e.g., what kind of formal language is necessary to express the ideas of that system; is there a set of axioms (statements of the formal language) from which all or at least many statements true in terms of the system can be proved; is there a machine that can decide which statements are and which are not provable from those axioms; is there a machine that can decide which statements of the formal language are true in terms of that system and which are not, and so forth.

If the above discussion is not all immediately clear to the reader, please let him remember that it takes a whole book to explain it carefully.

SYNTAX AND SEMANTICS

Common to Parts I and II of the book is the problem of symbolic representation of mathematical ideas. The words on these pages are simply symbols (ink marks) and in so far as the reader can make sense of it all (assuming there is some sense) it is because in his mind these symbols—singly and in combination—have come to refer to certain ideas; or putting it another way, they have become attached to certain meanings.

Language is a human activity which enables us to communicate with each other about our understanding of the universe. A careful discussion of language would be long and difficult and is not suitable for this book.† We do consider briefly two points of view of natural language, say English, because they are useful for the purposes of this book. (However, no claim is made that either of these is *the correct* view.) The first is that a natural language is a collection of symbols ‡ and sequences of symbols which can be thought of as existing independently of any meaning. The second is that the symbols and sequences of symbols of a natural language represent our ideas. Since we know whatever we know about the objects of the universe and the relations (or our concepts of relations) among them by means of our senses and mental processes (perception, imagination, abstraction, deduction, etc.), this second point of view is not as strange as it may first seem. Thus, we will assume that one uses a word to represent an idea, where the idea may or may not have been directly influenced by objects exterior to the mind. Similarly, we assume that one uses sentences, paragraphs, etc., to represent in detail complex ideas about relations, states of affairs, events, processes, etc. In this way, a great deal of language can be thought of as representing our ideas about the complexities of the universe.

As a written language, English has the alphabet as a basic set of symbols, certain (but not all) sequences of these are words and, again, certain (but not all) sequences of words are grammatically correct English sentences. By including a blank space with the alphabet we can think of both words and sentences as being sequences of symbols on the enlarged alphabet; thus we will refer to both words and sentences as sequences. In this way most ideas (some people would say all) are representable by sequences of symbols. Somehow or other we more or less learn how to form (generate) words from

† The reader who is not already aware of it should be warned that what language really is, what its symbols really stand for is a much discussed and very controversial philosophical subject. *Some* of the interesting discussion can be found in Frege (1892), Wittgenstein (1961), Russell (1940), Carnap (1947), Quine (1953), and Church (1956, chapter 0).

‡ Often the word "symbol" is used with the idea that it signifies something. Logicians use the word "symbol" to mean just some ink marks or sounds without any reference to what they may mean. (We usually have some meaning in mind, but it is part of the job to pretend that we do not.)

the letters of the alphabet and sentences from the words so as to express our ideas and we also learn how to recognize which sequences of letters are words and which sequences of words are sentences and what ideas they represent. How and why it all works out as it does is studied by linguists. The study of rules for generating and recognizing correct sequences of a natural languages is known as syntax. The theory of the relationship between the correct sequences and the ideas they represent is known as semantics. What ideas are expressed by the language is the semantic interpretation of the language. The mere existence of a collection of symbols and sequences of symbols does not automatically carry with it a semantic interpretation. Certain, or all, sequences may have no semantic interpretation or may have several semantic interpretations. For example, although "colorless purple dreams struggle listlessly" is a grammatically correct sentence according to English syntax, it has no semantic interpretation with respect to English.† On the other hand, the word "mean" and the phrase "those creatures are flying machines" have more than one semantic interpretation.

Since there are various peculiarities in any natural language, one might wish for a language that is free of such peculiarities and that can be used to represent our ideas clearly. For this reason logicians try to construct precise languages (sets of symbols and sequences of those symbols) in such a way that the formulation of correct sequences does not depend on any intended semantic interpretation. But at the same time, they require that this precise language be constructed in such a way that they can justify that it (the set of all correct sequences) can be used to represent the ideas represented by natural languages. In order to satisfy that requirement, the different ways in which natural language is used must be studied, and, indeed, our ideas themselves must be examined and made more precise. Thus the ultimate aim of these endeavors is that we make our ideas precise and that we can represent them precisely. Mathematical logicians try to construct precise languages for those of our ideas known as mathematics. Since mathematical ideas are quite precise, constructing such a language is a relatively simple matter. Some examples of how this has been done are discussed in detail in Part II of this book.

SOME GOALS IN ESTABLISHING A LANGUAGE FOR MATHEMATICS

Among the most basic of the mathematical ideas are the positive integers. Any language for mathematics must include symbols for them. There are many well-known ways of representing them.

† For a discussion of this sort of problem see Chomsky (1968, p, 15).

Examples 1. They can be represented as sequences on the set of symbols $\{a, b, c, \ldots, x, y, z\}$ in several ways such as

{one two, three, ...}
{un, deux, trois, ...}.

2. They can be represented on the set of symbols {1} by

{1, 11, 111, 1111, ...}.

3. They can be represented as sequences on the set {0, 1} by

{1, 10, 11, 100, ...}.

4. They can be represented as sequences on the set of symbols {0, 1, 2, 3, 4, 5, 6, 7, 8, 9} by

{1, 2, ..., 9, 10, 11, ...}.

It follows that any subset of the positive integers is also representable by any of these sets. Notice that in only one case (the second) does each sequence of the symbols represent a positive integer according to our standard conventions. To show that the positive integers are finitely representable on these sets we must have an effective procedure for determining of a given set of symbols which sequences of those symbols represent the positive integers and which do not. For the binary and decimal (and any in between) cases we have an easy way to decide:

> look at the left-most symbol of the sequence, if it is "0" the sequence does not represent a positive integer, otherwise it does represent a positive integer.

This rule is simply a convention which has been established some time in the past. Suppose we consider the decimal representations of the positive integers, can we recognize if such a sequence of symbols represents an even positive integer? Two possible procedures are as follows:

(a) divide the number represented by two and see if there is a remainder
(b) look at the rightmost symbol and see if it is "0," "2," "4," "6" or "8."

In order to use procedure (a) one must know the semantic interpretation of "divide" and "remainder" as well as a bit of arithmetic. In order to use procedure (b) one simply checks the rightmost symbol to see if it matches one of a given set of five symbols. Certainly (b) is an effective procedure and so the set of even positive integers is finitely representable. Notice further that the instruction refers only to the *form* of the sequence of symbols, not to anything about its semantic interpretation. However, in order to justify that by following instruction (b) one will in fact pick out the decimal representations of the even positive integers one must know what an even positive integer is and refer to something like instruction (a). If all the instructions of an effective procedure are to perform simple operations on symbols or sequences of symbols without any reference to any semantic interpretation of them, then we will say that the effective procedure is a "good procedure." As examples of good procedures we have (b) just above and the rules for playing tag. Given a set of symbols we will say that a subset of the set of all sequences of those symbols is recognizable if there is a good procedure for recognizing which sequences belong to the subset and which do not. We have just seen that the set of decimal representations of both the positive integers and the even positive integers are recognizable sets. Of course, a machine can be used to pick out a set of sequences only if that set is recognizable.

We will be interested not only in recognizing sets of sequences of symbols, but also in generating them. The reader is asked to think about what procedure he is following when he starts writing down the decimal representation of the positive integers in order of the size of the integers. Is he thinking about the size of the numbers or a certain ordering of the symbols 0, 1, . . . , 9? He is probably following the instructions of a good procedure which was instilled in him in first grade. We will say that a set of sequences of symbols is generable if there is a good procedure for combining the symbols of a given set so as to obtain exactly that set of sequences. Since the set of rules for playing tag is a good procedure, the set of sequences of symbols generated by a Tag system is a generable set. Notice that we can generate subsets of the decimal representation of the positive integers by playing tag with Tag systems on the set of symbols {1, 2, 3, 4, 5, 6, 7, 8, 9}. We mentioned earlier that the native speaker of a natural language can, somehow, both generate and recognize grammatically correct sentences. It is natural to wonder whether a set (of sequences) which is generable is also recognizable, and conversely, if a set which is recognizable is also generable. The reader is invited to prove for himself that the latter is true. He could also see why the former is false by showing that the set of all Tag systems on $\{a, b, c\}$ that halt is generable whereas it is not recognizable, as we can conclude from the discussion in the first section. In this book there are two questions which will often be asked about a set of ideas: is it finitely representable in such a way that the set of sequences representing it is generable, and is it finitely representable in such a way that the set of sequences representing it is recognizable?

Among the sets of ideas that are of interest are sets of questions. Assuming that such a set is finitely representable and that the set of sequences representing it is recognizable, we should also like to find a way of producing the sequences of symbols that represent the answers. Consider for a moment what one does when presented with a sequence of words representing a question in terms of English plus mathematics. In order to write down the sequence of words representing the answer one expects to go through various mental and physical operations, reach some mental conclusion, find the symbols which represent that conclusion and write them down. However, there are many sets of questions which are conventionally represented by sequences of symbols in such a way that one can produce the sequences of symbols representing the answers without paying much attention to the semantic interpretation. For familiar examples one need only think of adding, or at a little more sophisticated level, finding the derivatives of polynomials. Thus one might expect that there are sets of questions which are finitely representable in such a way that the sets of sequences are recognizable and that for such a set there is a good procedure for obtaining the sequences which represent the answers.

In essence the major problems the book will study are:

1. Which sets of questions are finitely representable in such a way that the set of sequences representing them is recognizable, and
2. For which of those sets of sequences is there a good procedure for obtaining the sequences which represent the answers?

We can see that this pair of questions is closely related to the discussion in the second section as follows. Every question, by virtue of being a question, has an answer.† Therefore, for each set of questions there is a function whose domain is that set and whose value for each question is the answer. Recall that a function is a total effectively computable function if its domain is finitely representable and there is an effective procedure for obtaining the function values. We will say that a total effectively computable function is a mechanical function‡ if the set of sequences representing the domain is recognizable and if the effective procedure for obtaining the function values is good. Thus the pair of questions above is equivalent to "For which sets of questions is there a mechanical function whose values represent the answers?" or, roughly, "For which sets of questions can there be a machine that can provide the answers?"

† This statement is arguable, philosophically. As examples of sentences which *may* be thought of as unanswerable questions, consider: "Is the present king of France bald?" and, at another level, "What is the meaning of life?" For the purposes of this discussion we assume that a sentence in question form is a question if it implies the possibility of an answer.

‡ In the rest of the book "total effectively computable function" is used also for "mechanical function."

PART I

Recursive Function Theory

Chapter I

TURING MACHINES

1.1 THE DEFINITION OF A TURING MACHINE†

If we were to give physical embodiment to a Turing machine (Tm), trying to make it resemble the computing machines we know, we would say that it is a machine with an infinite tape marked off in equal sections (squares of tape) and a tapehead which can move to the right or left along the tape and read and print a symbol in the square of tape at which it is positioned. We would also say that a Tm operates at discrete intervals of time and that at each interval the Tm is in an internal state. At a particular time t, what symbol is to be printed and whether the tapehead is to move to the right or to the left depend on the combination of the internal state of the Tm and the symbol being read at the time t; this combination also determines the next internal state. Thus, at each interval of time, the tapehead reads the symbol on the square of tape where it is positioned, prints a symbol there (i.e., erases the old symbol and prints a new one, which may be the same as before and may also be the blank), moves to the right or left, and the Tm goes into some particular internal state (possibly the same). A given Tm can have only a finite number of internal states and can read and print only a finite number of tape symbols. We shall let $q_H, q_0, q_1, \ldots$ be internal states and $*, s_1, s_2, \ldots$ be tape symbols, where, in particular, $*$ is the blank or empty symbol. If a tape has $s_2 s_4 s_5 s_1 s_1 s_3$ printed on it and all the rest blank and if the tapehead is positioned over the s_5 and is in internal state q_1, we can indicate this symbolically by

$$\cdots \mid * \mid * \mid * \mid s_2 \mid s_4 \mid s_5 \mid s_1 \mid s_1 \mid s_3 \mid * \mid * \mid * \mid \cdots$$
$$\uparrow q_1$$

† In his paper, Turing (1936–1937) defined and investigated an idealized machine. Such machines have subsequently come to be known as Turing machines and the main theorem of that paper has come to be known as the "unsolvability of the halting problem," which will be Theorem 1.1. The definition given here differs slightly from the original. At the same time, a very similar definition was proposed by Post (1936).

Suppose that if this particular Tm is in internal state q_1 while reading tape symbol s_5, it is to print s_1, go into state q_4, and move to the right. Then we have, at the next interval of time,

···	*	*	*	s_2	s_4	s_1	s_1	s_1	s_3	*	*	*	···
							$\uparrow q_4$						

Of course, the fact that if the Tm is in internal state q_1 and reading tape symbol s_5 it is to print s_1, go into q_4, and move to the right is quite independent of the tape and so we can indicate this simply by the quintuple $q_1 s_5 q_4 s_1 R$.

Although we can pretend to give a physical description of a Tm, it is not really a physical entity but an abstract one. A *Tm is defined* to be (given by) a *finite* set of quintuples such that each quintuple is either of the form $q_i s_j q_k s_t R$ or $q_i s_j q_k s_t L$, where q_i and q_k are any of the internal states except that q_i may not be q_H (because q_H is the state that stops the Tm; the *halt state*) and s_j and s_t are any of the tape symbols, and, further, where no two quintuples begin with the same pair $q_i s_j$. The finite set of tape symbols appearing in the finite set of quintuples is the *alphabet of the Tm* and the finite set of internal states appearing in the set of quintuples is the *set of internal states of the Tm*. For every pair $q_i s_j$, there is to be a quintuple beginning with that pair. The condition that no two quintuples begin with the same $q_i s_j$ ensures that the Tm is never confused, that is, when in a certain state and reading a certain symbol, it always does the same thing. (When this requirement is not imposed, the resulting Tm is said to be nondeterministic. Our concern will be with *deterministic* Tm's.) This kind of property should be familiar to all; it is the property of being a function. In this case, the domain of the function is the set of ordered pairs of internal states and tape symbols of the Tm and the range is the set of ordered triples consisting of an internal state, a tape symbol, and a direction either right or left.

Example 1 Let the Tm be

$$\begin{matrix} q_0 & s_1 & q_0 & * & R \\ q_0 & s_2 & q_0 & * & R \\ q_0 & * & q_H & * & R \end{matrix}$$

If this Tm is given a tape with $s_1 s_2 s_1$ written on it so that the tapehead is over the leftmost nonblank symbol, we can follow the operation (computation) of the Tm thus:

	···	*	s_1	s_2	s_1	*	*	*	···
$t = 0$			$\uparrow q_0$						

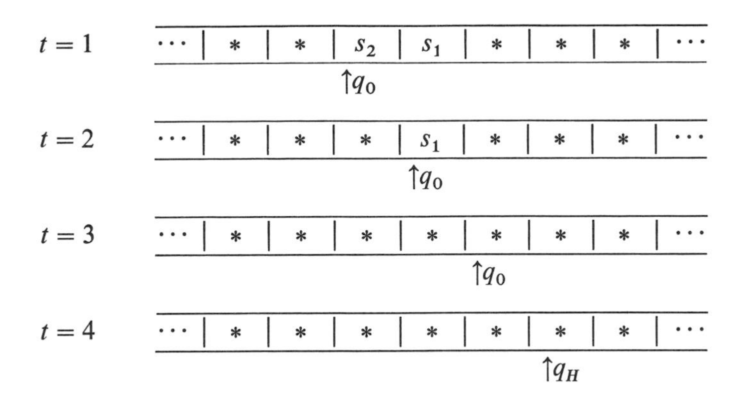

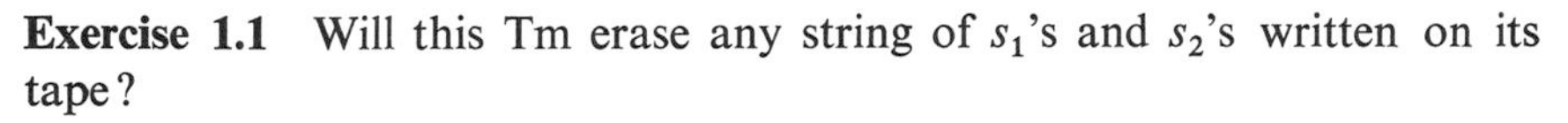

Exercise 1.1 Will this Tm erase any string of s_1's and s_2's written on its tape?

Example 2 Let the Tm be

$$\begin{array}{lllll} q_0 & 1 & q_1 & 2 & R \\ q_1 & 1 & q_0 & 1 & R \\ q_0 & * & q_2 & * & R \\ q_1 & * & q_2 & * & R \\ q_2 & * & q_H & * & L \end{array}$$

If this Tm is given a tape with a string of 1's (say $n+1$ of them) written on it and the tapehead is positioned over the leftmost nonblank symbol, we can follow the computation thus.

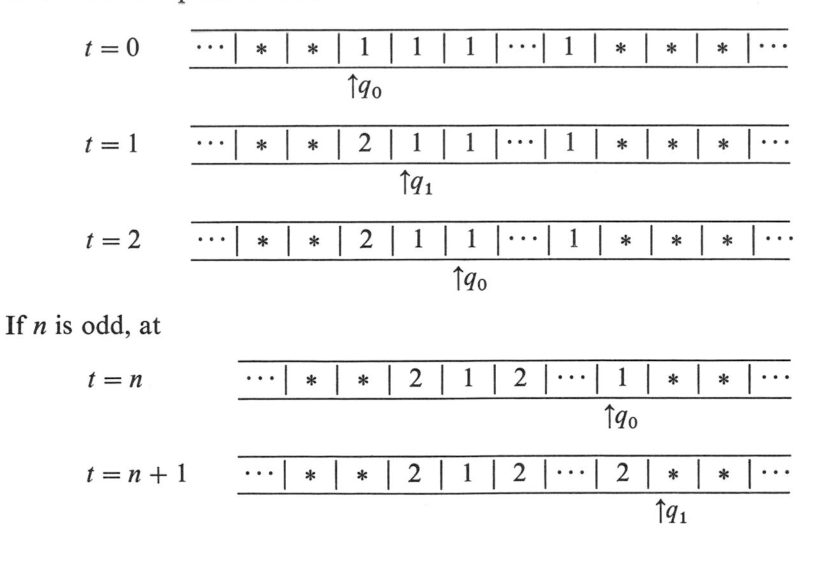

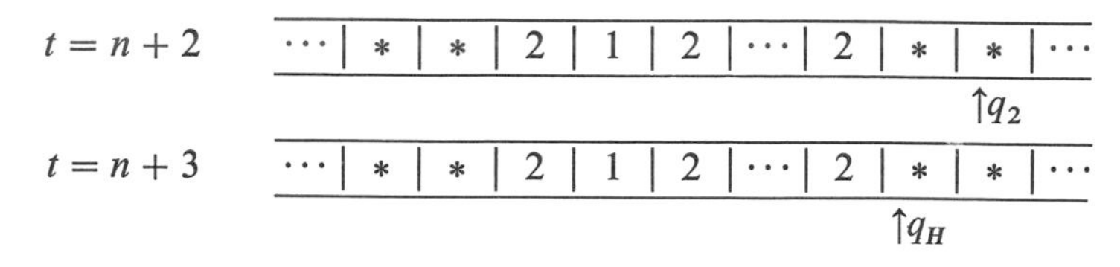

Exercise 1.2 Design a Tm which writes on its tape forever.

In general, since the tape is infinite, there is some question as to where to set the tapehead before "turning on" the Tm. We adopt the convention that the Tm will start with the tapehead positioned over the leftmost nonblank square if the tape is not entirely blank, and otherwise, anywhere. At the moment a Tm is turned on, tape in place, it is always in the same internal state: the *initial state*. What is written on the tape before the computation begins is called the *initial tape inscription*. A Tm will, hopefully, compute something, so we must agree upon what an answer is and where to find it on the tape. A standard convention is that, unless otherwise specified, the string of tape symbols to the left of the tapehead when the Tm halts (enters internal state q_H) is the *answer computed* by the Tm. We could specify that each machine must erase all the data and scratchwork before halting, but we will allow cases where the Tm does not "clean up," on the assumption that where the answer begins will be made clear somehow. When it is necessary, we will assume that the machine has "cleaned up." An important assumption about Tm's is that only a finite number of its tape squares are ever nonblank.

Now that we know what a Tm is and how it operates, the obvious question is, What can they do? This question may be divided into two questions: (1) Given a problem in a mathematical sense, when can I design a Tm to get the answer? (2) Given a Tm, how can I find out what mathematical problems it can get answers to? Another question of interest is, of course, How do Tm's compare to the machines we use today or even those we may use 50 years from today? The general aim of this chapter is to begin to answer questions 1 and 2. The answers are in some sense disappointing and as the reader begins to appreciate the tremendous power of Tm's he will find the answers to those questions asked in terms of real machines even more disappointing.

1.2 SOME SIMPLE PROBLEMS TM'S CAN DO

To become quite at home with Tm's, we shall formulate some simple mathematical problems (in line with question 1 of the previous paragraph) and design machines to compute the answers. When numbers are involved, the problems will be posed in terms of the nonnegative integers.

First, we should think a little about what kinds of tape symbols are convenient and the reader is encouraged to play around with decimal, binary, unary, or any other kind of notation. We shall usually use unary with a few other symbols (A, B, or X or whatever) which will be used like parentheses or commas or placekeepers. Symbols used for these purposes are usually called *markers*. By unary notation, we mean to represent the nonnegative integer n by $n+1$ consecutive 1's. Thus, 111111 is the unary representation of the number that is represented in decimal notation by the digit 5. We will denote the unary representation of n by $\bar{n}$. For example, $\bar{5}$ denotes 111111, $\bar{0}$ denotes 1, $\bar{1}$ denotes 11, etc. The "extra" 1 is included to enable us to distinguish between the number 0 and the blank square.

The $n+1$ or Successor Machine

Perhaps the simplest example of a Tm is the machine which for any n, given the number n, computes the number $n+1$. Imagine the tape as

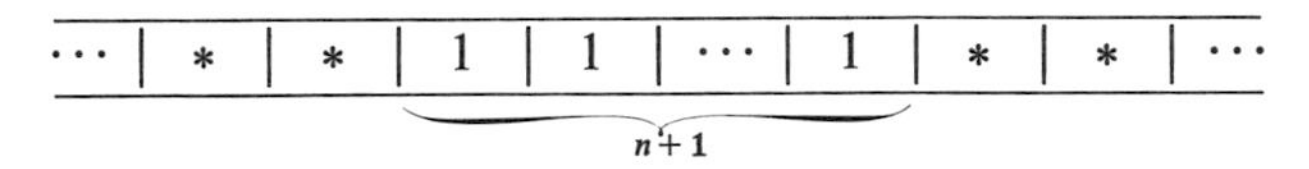

that is, $\bar{n}$ is the initial tape inscription. A plan for this computation would be: (1) start at leftmost 1 and keep reading rightward leaving the 1's unchanged; (2) upon encountering the first blank square on the right of $\bar{n}$, print 1 and move right; (3) halt. A Tm to do this is

q_0	1	q_0	1	R	Step 1
q_0	*	q_1	1	R	Step 2
q_1	*	q_2	*	R	Step 3
q_2	*	q_H	*	L	

Exercise 1.3 Design a Tm (i.e., write out the quintuples) which for any $n > 1$, computes $n - 1$.

The Copy Machine

This machine is simple and extremely useful. Suppose that we have a tape inscribed

···	*	*	s_1	s_2	···	s_k	X	*	*	···

where the s_j are various nonblank tape symbols, not necessarily distinct from one another but all distinct from the symbol X, which is used here as a marker. Suppose further that we want the machine to do whatever is necessary to end with a tape inscribed

···	*	*	X	s_1	s_2	···	s_k	*	*	···

where X is the original X of the initial tape inscription. The plan is that the machine must: (1) start by reading s_1, (2) run right until it has read X and print s_1 just to the right of X, (3) go back leftward until it finds s_2, read it, and (4) run rightward past Xs_1 to print s_2, etc. until all the original inscription except X has been copied. Further, either after all is copied, or during the process, it must erase (i.e., print blanks in place of) the original $s_1 s_2 \cdots s_k$. The reader should stop reading here and try to work out his own scheme even though, as a precaution, one will be given. In designing his own Tm and reading the Tm about to be given, the reader will realize that there are many possible Tm's which can compute the same answer from a given initial tape inscription. Some are more efficient, some more intuitive, some have few instructions, and so forth. For the time being, we aim at clarity. Suppose that there are m distinct nonblank symbols occurring in $s_1 s_2 \cdots s_k$. Then the Tm we are designing will have those m symbols, $s_{i_1}, s_{i_2}, \ldots, s_{i_m}$, as well as $*$ and X for its alphabet. We think of the machine as beginning with the tapehead on s_1 and for a start we give an outline of the set of quintuples in the order in which the operations would actually take place:

q_0	s_1	q_1	$*$	R	—
q_1	Any symbol except $*$	q_1	Same	R	(Read rightward remembering s_1, printing blank in its place, and leaving other symbols unchanged)
q_1	$*$	q_2	s_1	L	(At first blank, print s_1 and start left)
q_2	Any symbol except $*$	q_2	Same	L	(Search for first blank to the left leaving all symbols unchanged)
q_2	$*$	q_0	$*$	R	(Having found the blank, go right one step prepared to read and remember the next symbol)
q_0	s_2	q_3	$*$	R	—

At this point, we are in a situation similar to that of the first step and this should continue in a similar way until the tape looks like

···	*	*	X	s_1	s_2	···	s_k	*	*	···

Suppose the machine has just printed s_k and is starting leftward as usual, not knowing that its job is done. It goes leftward past X, reads $*$, and, as before, in q_0 moves right, but this time to read X. As yet, there is no quintuple of the form beginning q_0 X, and so we may have quintuples

q_0	X	q_j	X	R
q_j	Any symbol except $*$	q_j	Same	R
q_j	$*$	q_H	$*$	R

where q_j is an internal state not used previously. Before we fill in the details of the outline, we should notice that for each distinct symbol s_i that is neither $*$ nor X, we need one corresponding internal state to remember s_i while moving rightward so as to print it in the first available blank square. So we shall number the internal states a little differently than in the outline. q_0 will be as in the outline, $q_1, q_2, \ldots, q_m$ to go with $s_{i_1}, s_{i_2}, \ldots, s_{i_m}$, respectively, and then q_{m+1} for the leftward search and q_{m+2} where q_j was written in the outline. We emphasize that the number of tape symbols is finite and that the number of internal states is finite, as is required by the definition of a Tm. The Tm given here has exactly four more internal states than tape symbols. Notice that the number of internal states depends *not* on the length of the initial tape inscription, but on the number of distinct symbols appearing in it. Finally, we have the set of quintuples:

$$
\begin{array}{lllll}
q_0 & s_1 & q_1 & * & R \\
q_0 & s_2 & q_2 & * & R \\
\vdots & & & & \\
q_0 & s_m & q_m & * & R \\
q_0 & X & q_{m+2} & X & R \\
q_1 & * & q_{m+1} & s_1 & L \\
q_1 & s_1 & q_1 & s_1 & R \\
q_1 & s_2 & q_1 & s_2 & R \\
\vdots & & & & \\
q_1 & s_m & q_1 & s_m & R \\
q_1 & X & q_1 & X & R \\
q_2 & * & q_{m+1} & s_2 & L \\
q_2 & s_1 & q_2 & s_1 & R \\
\vdots & & & & \\
q_m & * & q_{m+1} & s_m & L \\
q_m & s_1 & q_m & s_1 & R \\
\vdots & & & & \\
q_m & s_m & q_m & s_m & R \\
q_{m+1} & * & q_0 & * & R \\
q_{m+1} & s_1 & q_{m+1} & s_1 & L \\
\vdots & & & & \\
q_{m+1} & s_m & q_{m+1} & s_m & L \\
q_{m+1} & X & q_{m+1} & X & L \\
q_{m+2} & * & q_H & * & R \\
q_{m+2} & s_1 & q_{m+2} & s_1 & R \\
\vdots & & & & \\
q_{m+2} & s_m & q_{m+2} & s_m & R
\end{array}
$$

Before going on to another kind of Tm, the reader should make up examples of copy machines and see if his design and the one given really work. For example, suppose the tape to be copied is given as

⋯	*	*	3	3	5	7	7	7	5	X	*	*	⋯

The reader should be sure that he can design a Tm which will copy that sequence of symbols to the right of X and that that *same* Tm will copy any sequence of 3's, 5's, and 7's, *no matter how long.*

Exercise 1.4 Design a machine which when given the above tape will first erase all of the sequence 335775 and *then* write it out to the right of X. Can you construct a Tm which erases *any* sequences of 3's, 5's, and 7's and then writes it out to the right of X? If not all such sequences, then which ones?

Exercise 1.5 The copy machine given assumed that $s_1 s_2 \cdots s_k$ was a sequence of nonblank symbols. Design a Tm where that restriction is lifted, that is, s_1 and s_k are not blank but any of $s_2, s_3, \ldots, s_{k-1}$ may be.

Exercise 1.6 Design a Tm which when given the tape

⋯	*	*	X	1	1	⋯	1	Y	1	1	⋯	1	Z	*	*	⋯
					$n+1$					$m+1$						

will compute n times m and write $\overline{nm}$ to the right of Z. Hint: We may think of nm as being n copies of m strung out next to each other.

1.3 SOME TERMINOLOGY AND NOTATION

So as to cut down on the chattiness and increase the precision of the discussion, we introduce some terminology and notation. As we have been doing, we continue to let $*, s_1, s_2, \ldots$ be tape symbols and $q_H, q_0, q_1, \ldots$ be internal states and $T, T_1, T_2, \ldots$ denote Tm's. Further, let $\Sigma_1, \Sigma_2, \ldots$ be alphabets and $Z_1, Z_2, \ldots$ be finite sets of internal states. A *word* on an alphabet Σ_1 is just what one would imagine—some or all of those symbols written next to each other in any order with any number of occurrences of each symbol, the only provision being that the total number of occurrences is finite. That is, giving an inductive definition, if $s_i \in \Sigma_1$, then s_i is a word on Σ_1, and if τ is a word on Σ_1 and $s_i \in \Sigma_1$, then τs_i is a word on Σ_1. (Words are also known as *strings* of symbols or *finite concatenations* of symbols.) Let $\tau_1, \tau_2, \ldots$ be variables for words. The *length of a word* τ is the total number of symbols occurring in it, counting repetitions; it is denoted by $|\tau|$. Examples are $|ab| = 2$, $|aaa| = 3$, $|\text{computer}| = 8$. Two words τ_1 and τ_2 are *equal* if they are identical, symbol by symbol. τ_1 equal to τ_2 is written $\tau_1 \equiv \tau_2$.

Exercise 1.7 Given an alphabet, how many distinct words are there on that alphabet? (For those who have difficulty with this exercise, the proof will be given in chapter III.)

The Pick-Out (Projection) Machine

Suppose that for a specific i and n ($i \leq n$), we want to be able to pick out the ith component (coordinate) of any n-tuple of words on some alphabet. For example, let the alphabet be $\{a, b\}$ and $i = 2$ and $n = 4$. One 4-tuple of words would be (*aa*, *aba*, *bbb*, *aabb*). In this case, the second component is *aba*. With this idea in mind, for each i and n, $i \leq n$, define the function $U_i^n(x_1, \ldots, x_i, \ldots, x_n)$ to have the value x_i. We want to design a Tm to compute this function. That is, if the Tm is given the tape†

···	*	A		A		A	···	A		A	···	A		A	*	···
			x_1		x_2				x_i				x_n			

we want it to write x_i in the blanks immediately to the right of the rightmost A. Thus, the Tm, starting at the leftmost A, must: (1) search for x_i, (2) copy x_i to the right of the rightmost A, and (3) restore the original x_i in case the data were mutilated during the computation. Let the machine have alphabet $*$, A, those symbols occurring in the representation of $x_1, x_2, \ldots, x_n$, and for each of those symbols, another one. (For example, if the x_j are given in decimal representation, the alphabet would be $*$, A, 0, 1, 2, 3, 4, 5, 6, 7, 8, 9, $B_0, B_1, \ldots, B_9$.) As we did for the copy machine, we will begin in outline form.

1. The machine begins reading the leftmost A and must move rightward until it encounters the ith A. So,

q_0	A	q_1	A	R
q_1	Any symbol but A	q_1	Same	R
q_1	A	q_2	A	R
q_2	Any symbol but A	q_2	Same	R
⋮				
q_{i-1}	A	q_i	A	R

2. Now the Tm is in a position to copy x_i. This will be just like the copy machine except that instead of erasing x_i as the machine copies it, it will

† When we want to say that something is written on a tape to represent x but we do not care how, we indicate this by

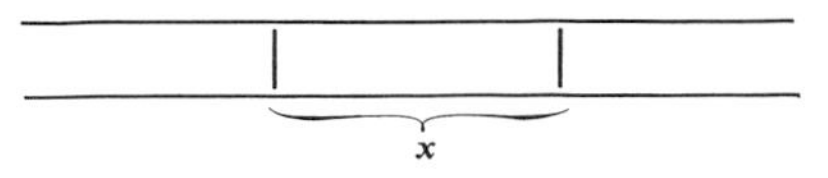

replace each tape symbol by its corresponding extra symbol. (Thus, in the example, if x_i is 369, as the machine copies the 3, it prints B_3 in place of 3, etc., so that by the time all x_i is copied, in its place the tape reads $B_3 B_6 B_9$.)

3. Once the x_i is copied, the Tm goes back to the original x_i position and restores the original representation of x_i.

Exercise 1.8 (a) Write out the quintuples. (b) On what does the number of internal states depend? (3) Suppose the Tm were designed for $n = 67$ and $i = 51$. Would it also work for $n = 67$ and $i = 23$? For $n = 60$ and $i = 51$? For $n = 10003$ and $i = 1001$?

Exercise 1.9 Define a new function $V(i, n, x_1, \ldots, x_n) = x_i$ for $0 < i \leq n$. Design a Tm to do this computation. Consider the differences between the function (machines) U_i^n and V.

Exercise 1.10 Suppose that all x_i's are to be represented in unary notation. Then consider the truth or falsity of the following statements and how these statements relate to one another:

(a) For all n and i, $0 < i \leq n$, there exists a Tm $T_{U_i^n}$ such that $T_{U_i^n}$ computes $U_i^n(x_1, \ldots, x_n)$.

(b) There exists a Tm T such that for all i and n, $0 < i \leq n$, T computes $U_i^n(x_i, \ldots, x_n)$. Notice that T computes from a tape containing only a tape representation of $x_1, \ldots, x_n$.

(c) For all n and i, $0 < i \leq n$, there exists a Tm, T_V such that T_V computes $V(i, n, x_1, \ldots, x_n)$.

(d) There exists a Tm T_V such that for all i and n, $0 < i \leq n$, T_V computes $V(i, n, x_1, \ldots, x_n)$.

Exercise 1.11 Suppose you have Tm's T and T', both of which use the tape symbols 0, 1, A, and $*$. Unfortunately, however, T' is in need of repairs and only moves rightward. Suppose also that you have a stockroom containing infinitely many tapes, one for each positive integer. On each is written, in binary, a positive integer with an A at each end. We may assume that the tapes are stored in numerical order starting with

···	*	*	A	1	A	*	*	···

Suppose you tried the first 5383 tapes, one at a time, first in T and, after having restored the tape to its original form, in T', and you found that, in each case, the number on the tape was copied onto the tape just to the right of the rightmost A. Would you be correct in concluding that both T and T' could copy any tape in your stockroom. If not, what do you think is the case?

Do you think that for any tape in your stockroom you could build a right-moving Tm that will copy it? If so, what other tapes in the stockroom will such a machine copy? A Tm like T' which can move only to the right is sometimes called a *sequential machine* or a *finite transducer*.

Exercise 1.12 Suppose that a tape has exactly one nonblank square. Design a Tm which, when given the tape anywhere, will stop on the nonblank square and the tape will have only that square nonblank when the machine stops.

Exercise 1.13 Recalling the discussion of "cleaning up" we should be sure that such an idea can be realised. (a) Suppose the initial tape inscription τ contains no blanks. Suppose also that T is a Tm such that T computing from τ halts with τ' written to the left of the tapehead, but may have other things written to the right. Design a Tm T' so that T' computing from τ halts with τ' written to the left of the tapehead and the rest of the tape blank. (b) Is it essential that τ contain no blanks? Is this a serious restriction?

Exercise 1.14 (The Comparing Machine) It is often the case that we have a finite set of words on an alphabet Σ and wish to know whether or not an arbitrary word τ on Σ is in the list. The problem is to design a Tm to do the job. Specifically, let the Tm have as its alphabet $\Sigma \cup \{M_1, \ldots, M_k\}$, where $k \geq 4$, and $\Sigma \cap \{M_1, \ldots, M_k\} = \varnothing$. The Tm is to compute from the following tape:

⋯	*	M_1	τ_1	M_2	τ_2	M_2	⋯	τ_n	M_2	M_3	τ	M_4	*	⋯

Design the Tm so that it will start with the tapehead over the leftmost M_1, go rightward to M_3, read τ, search through $\tau_1, \ldots, \tau_n$, and if there is an i such that $\tau \equiv \tau_i$, the Tm is to stop on the M_2 just to the right of τ_i, but if there is no such i, then the tapehead should go to M_4 and halt.

Exercise 1.15 Let T_1 and T_2 be two Tm's and $\Sigma_1 = \{*, A, B, C\}$ and $\Sigma_2 = \{*, D, E, F\}$ be their respective alphabets. If

$$T_1 = \{q_0\, A q_0\, AR,\quad q_0\, B q_0\, BR,\quad q_0\, C q_0\, CR,\quad q_0 * q_1 CR,\quad q_1 * q_H * R\}$$

and

$$T_2 = \{q_3\, D q_3\, DR,\quad q_3\, E q_3\, ER,\quad q_3\, F q_3\, FR,\quad q_3 * q_4\, FR,\quad q_4 * q_H * R\}$$

In what ways to T_1 and T_2 differ?

Let T_1 and T_2 be two Tm's such that T_i, $i = 1, 2$, has alphabet Σ_i and set of internal states Z_i. Define *T_1 and T_2 to be similar* if there exist one-to-one functions f, g, and h for $f\colon \Sigma_1 \to \Sigma_2$, $g : Z_1 \to Z_2$, and $h : T_1 \to T_2$ with the properties that g takes the initial state and the halt state of T_1 into the initial state and the halt state of T_2, respectively, and if $(q_i s_j q_k s_t R) \in T_1$, then $h(q_i s_j q_k s_t R) = (g(q_i)f(s_j)g(q_k)f(s_t)R)$.†

Exercise 1.16 Show that the relation of being similar for Tm's is an equivalence relation.

Exercise 1.17 For any given n and m show that there are only finitely many nonsimilar Tm's having n tape symbols and m internal states.

Exercise 1.18 Show that there are not more than countably many nonsimilar Tm's. (This exercise is worked out in chapter III for those who are not familiar with counting procedures.)

From the considerations of the previous paragraph, we see that we can account for all nonsimilar Tm's by choosing their tape symbols from a given, fixed, countable set. Let that set be: $H, R, L, M_1, M_2, M_3, M_4, M_5, M_6, B, Q, S, *, 1, a_2, a_3, \ldots,$ where $*$ is still the blank (empty) symbol. We will, in general, further limit this list of available tape symbols to: $*, 1, a_1, a_2, \ldots,$ since, as we have seen, for any Tm with alphabet from the first list, there is a Tm similar to it with alphabet from the second list. Hence, if a Tm has n tape symbols, we will assume that they are $*, 1, a_2, \ldots, a_{n-1}$. By similar considerations, we will assume, unless otherwise specified, that if a machine has m internal states, they are to be $q_H, q_0, q_1, \ldots, q_{m-2}$, where q_H is the halt state and q_0 the initial state. We can now think of a Tm as being given by a certain finite set of words (the quintuples) on a certain finite alphabet $\{*, 1, a_2, \ldots, a_{n-1}\} \cup \{q_H, q_0, \ldots, q_{m-2}\} \cup \{R, L\}$. Thus, to "give a Tm" is equivalent to "giving a certain set of words."

Using the language of functions, we denote by $T(\tau)$ the answer, if any, computed by machine T when given a tape with the word τ as initial tape inscription. Since, as we know from exercise 1.2, there are Tm's which, given certain tapes, never halt, we also know that $T(\tau)$ is not necessarily defined. Thus, if Tm T halts on initial tape inscription τ, we say $T(\tau)$ *is defined* (T *converges* on τ), if T does not halt on τ, $T(\tau)$ *is undefined* (T does *not converge* on τ). For a given Tm T, let $S = \{\tau \mid T(\tau) \text{ is defined}\}$. Then we say that T *accepts* S. A set of words S on an alphabet Σ is *Turing acceptable* if there is a Tm T such that $T(\tau)$ is defined if and only if $\tau \in S$.

Exercise 1.19 Let $\Sigma = \{a, b\}$. Let $S = \{a^n ba^n \mid \text{for all } n \in N\}$. (a) Is S a Turing-acceptable set? (b) Do you think there can be a right-moving Turing

† We understand from exercise 1.15 that two Tm's are similar if they differ only in the choice of symbols to represent them.

machine which accepts S? (Remark: a^n is an abbreviation for $\underbrace{aaa\ldots a}_{n}$.)

Exercise 1.20 Consider the following Tm T:

$$\begin{array}{lllll@{\quad}lllll@{\quad}lllll}
q_0 & 0 & q_0 & 0 & R & q_1 & 0 & q_0 & 0 & R & q_0 & * & q_0 & * & R \\
q_0 & 1 & q_1 & 1 & R & q_2 & 0 & q_0 & 0 & R & q_1 & * & q_1 & * & R \\
q_1 & 1 & q_2 & 1 & R & q_3 & 1 & q_3 & 1 & R & q_2 & * & q_2 & * & R \\
q_2 & 1 & q_3 & 1 & R & q_3 & 0 & q_3 & 0 & R & & & & & \\
 & & & & & q_3 & * & q_H & * & R & & & & &
\end{array}$$

Notice that it is not only a right-moving Tm, but also that it always prints just what it reads. That is, in essence, it does not print. (Tm's with the two properties of being right-moving and nonprinting are called *finite automata*.) Let $S = \{\tau \mid \tau$ is a word on $\{0, 1\}$ and τ contains at least three consecutive 1's$\}$. Prove that $T(\tau)$ is defined if and only if $\tau \in S$; that is, T accepts the set S.

From our understanding of functions, we know what it means for two functions f and g to be equal. That is, they have the same domain D and range, and for every $x \in D$, $f(x) = g(x)$. How f "gets its answer" $f(x)$ and how g "gets its answer" $g(x)$ are unimportant provided the answers are the same. We extend this notion to Tm's, saying that two *Tm's* T_1 *and* T_2 *are equivalent* if $\Sigma_1 = \Sigma_2$ and if for all words τ on Σ_1 as initial inscriptions, $T_1(\tau) \equiv T_2(\tau)$. That is, either they are both defined and equal to the same thing or they are both undefined.

Exercise 1.21 Show that for every Tm T there are infinitely many nonsimilar Tm's equivalent to T.

Exercise 1.22 Let T be a Tm and τ an initial tape inscription containing no blank squares. Show that there is a Tm T' computing from a half-infinite tape such that $T(\tau) \equiv T'(\tau)$. That is, for T' we assume that the initial tape inscription is written at the left end of the tape and that the tape extends infinitely to the right.

In the next section, we will want to represent a Tm on a tape. Thus, we must be able to write out, in tape symbols, the quintuples of a Tm on a tape. There are various ways of representing $q_i s_j q_k s_t R$, the most obvious being to use q_i, s_j, q_k, s_t, and R themselves as tape symbols. But we must be careful, for our problem will concern a machine that is to read the representation of *any* Tm and so we must avoid a machine requiring an infinite set of tape symbols. To get around this, we just use the symbols B, Q, S, R, L, H, $M_1, \ldots, M_6$, as follows. Let q_i, $i \geq 0$, be represented by

$$\underbrace{QQQ\cdots QQ}_{i+1}$$

and let this string be denoted by ρ_{q_i}; in the same fashion, let a_j be represented by $\underbrace{SSS \cdots SS}_{j+1}$ and let this be denoted by ρ_{a_j}. Let q_H be represented by H and let this be denoted by ρ_{q_H}. Of course, ρ_* is just S and ρ_1 is SS. Suppose that Tm T we wish to represent has m internal states and n tape symbols. The quintuple of T, $q_i s_j q_k s_t R$, is then represented by

$$\underbrace{B \cdots B}_{m-i}\underbrace{Q \cdots Q}_{i+1}\underbrace{B \cdots B}_{n-j}\underbrace{S \cdots S}_{j+1}\underbrace{B \cdots B}_{m-k}\underbrace{Q \cdots Q}_{k+1}\underbrace{B \cdots B}_{n-t}\underbrace{S \cdots S}_{t+1}R$$

To represent the whole Tm T, the representations of the quintuples will have M_1 on the left end, M_2 on the right end, and M_3 between each pair of quintuples. Thus, in the future we will refer to this representation of a Tm T as *the standard tape representation of* T and will denote it by ρ_T. A *standard tape representation of* τ, ρ_τ, for a word τ will also be useful. Assuming that τ is a word of the form $a_{i_1} a_{i_2} \cdots a_{i_k}$, where for all j, $a_{i_j} \in \{*, 1, a_2, \ldots\}$, then ρ_τ is

$$M_4 \underbrace{S \cdots S}_{i_1+1} M_4 \underbrace{S \cdots S}_{i_2+1} M_4 \cdots M_4 \underbrace{S \cdots S}_{i_k+1} M_5$$

Notice that although τ and ρ_τ are both words, they are not equal. (Warning: We will speak of τ, ρ_τ, or ρ_T as being written on the tape and may give the picture

$\cdots$	ρ_τ	$\cdots$

But we do not mean that the symbols τ, ρ_τ, or ρ_T are themselves on the tape, but rather that the words they denote are. In fact, sometimes we will go so far as to say things like

"the tape is

$\cdots$	$*$	ρ_T	M_3	ρ_{q_i}	M_4	ρ_{a_j}	M_5	$*$	$\cdots$"

where M_3, M_4, and M_5 are the actual symbols on the tape and ρ_T, ρ_{q_i}, ρ_{a_j} denote the words on the tape.)

1.4 THE UNIVERSAL TURING MACHINE

Although we have considered only a very few problems for Tm's to do, we have already covered the landscape with Tm's, so to speak. Indeed, to take care of the set of functions U_i^n alone, we have infinitely many Tm's. It would certainly be much tidier if we could find one Tm that would do the work of all the others. That does not mean that the tapes fed to the super (or universal) machine would be just those fed to all the others. Rather, we will design a universal Tm T_U such that for all T and all τ, if $T(\tau) \equiv \tau'$, then T_U computes

$\rho_{\tau'}$ from an initial tape inscription containing ρ_T and ρ_τ. We give only a sketch of how this may be done. As we know from exercise 1.22, without loss of generality we may assume for any Tm that it computes from a half-infinite tape with the initial tape inscription τ written at the left end. Suppose that the Tm T which we wish T_U to imitate (or simulate) has alphabet $\{*, 1, a_2, \ldots, a_{n-1}\}$ and internal states $\{q_H, q_0, \ldots, q_{m-1}\}$ and that T computes from a half-infinite tape. The initial tape inscription for T_U is to be like

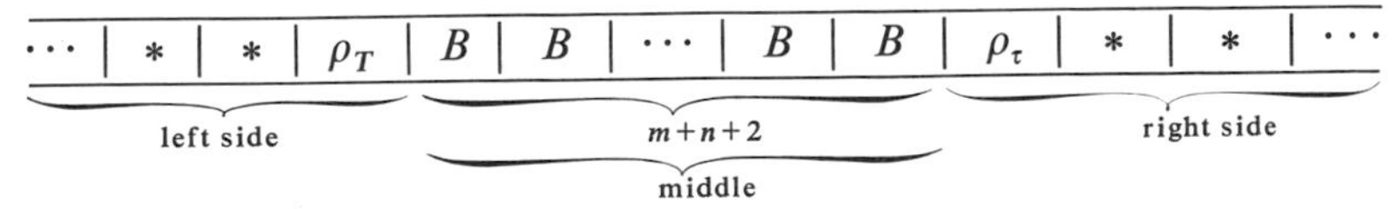

The Tm T_U is to do just what we do when we follow a computation through from a given word, an initial state, and a set of quintuples. That is:

1. We make a note of the initial state and the leftmost symbol of τ while leaving some indication as to which symbol we are working on in τ.
2. We look through the set of quintuples until we find one beginning with that same pair, as noted in step 1, of internal-state–tape-symbol.
3. Then we must do four things: (a) change the initial tape inscription according to the instruction at the place we had indicated in step 1, (b) note the new internal state, (c) move right or left, and (d) note the tape symbol now being read. Then, if not halt, move on to step 4.
4. We are in a situation similar to that of step 1 and can continue until a quintuple tells us to halt, if ever.

Since we already have copying and comparing machines, to design a Tm to work from the indicated tape along lines similar to those we ourselves follow is more tedious than difficult. We mention some of the difficulties that must be worked out and leave the actual writing down of quintuples to those unfortunate enough to be lost on a desert island.

1. Analogous to "making note of the initial state and leftmost tape symbol of τ" of step 1 above, T_U writes ρ_{q_0} on the middle section of its tape, and, supposing that a_i is the leftmost symbol of τ, T_U copies ρ_{ai} of ρ_τ onto the middle section. The tape then looks like this:

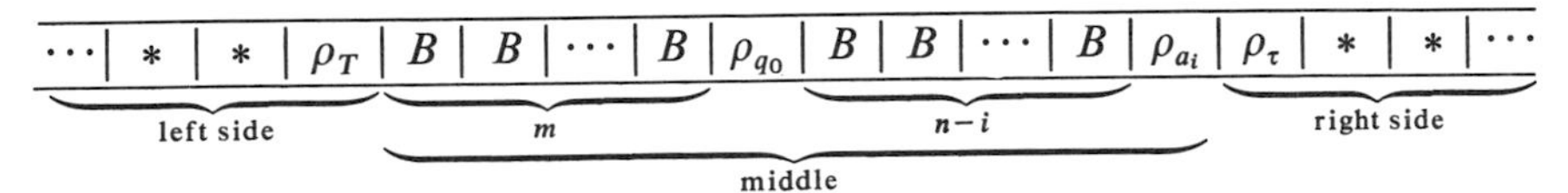

As also mentioned in step 1, we must leave some indication on the right side of the tape as to what symbol of τ we are working on. But since the ρ_{a_j}'s are separated by the marker M_4, we can change the M_4 preceding the ρ_{a_j} in question to M_6 and restore it later when we have finished with ρ_{a_j}.

2. Recall that the word ρ_τ begins and ends with markers and has markers between each string of $i+1$ S's for $\rho_{a_i} = \underbrace{SSS \cdots SS}_{i+1}$. Since T changes the initial tape inscription during the computation, say the tape has inscriptions $\tau, \tau_1, \ldots, \tau_k$ during the computation, then the right side of the tape of T_U will have successive inscriptions $\rho_\tau, \rho_{\tau_1}, \ldots, \rho_{\tau_k}$. If τ_i changes to τ_{i+1} by replacing a_j by a_k for $j < k$, then in going from ρ_{τ_i} to $\rho_{\tau_{i+1}}$, more space, so to speak, is needed. (For $k < j$, less space is needed.) There are at least two ways to handle this. One is to copy ρ_{τ_i} further right on the blank part of the tape and then copy it back into its original position adding (or deleting) the necessary S's at the same time so as to end with $\rho_{\tau_{i+1}}$ in the proper place on the right side of the tape. Another would be to have defined ρ_τ a little differently so that if n is the number of tape symbols for T, then between each pair of markers there would be space for $n+1$ S's. We choose not to do the latter because it seems nicer for the standard tape representation of τ to be independent of any particular machine.

3. Similar problems arise when we want to keep note of the current tape symbol in the middle part of the tape. What to do with either of the above schemes should be clear. We have left sufficient space in the middle of the tape to record any pair, internal state and tape symbol.

Exercise 1.23 There are many ways of setting up the universal Turing machine and the one given above is just one of the possibilities. (a) Suppose the standard tape representation of a Tm were not as defined. For example, it could be rather like the one given but with no B's. So, for $q_i s_j q_k s_t R$, we would have

$$\underbrace{QQQ \cdots Q}_{i+1}\underbrace{SSS \cdots S}_{j+1}\underbrace{QQQ \cdots Q}_{k+1}\underbrace{SSS \cdots S}_{t+1}R$$

How would the universal Turing machine be set up in terms of such a standard tape representation? (b) Consider standard tape representations for T and for τ that are altogether different from the one given and set up the universal Tm accordingly.

1.5 THE HALTING PROBLEM

Toward the end of section 1.1, two large questions about Tm's were raised. In the intervening sections, we began to investigate questions related to the first question. Now we shall consider the second: Given a Tm, how can I find out what mathematical problems it can answer? Suppose one is given a Tm, i.e., a finite set of quintuples, and wants to know what it will compute. Given a tape with $\bar{n}$, does it compute $\overline{n^2}$ or $\overline{n+2}$ or what? From $X\bar{n}$, $\overline{n+n}$, or what? But a more basic question is, Given a specific tape, does the Tm compute anything at all, i.e., does it ever halt? Putting a tape in, turning the machine on, and waiting, is a rather dubious approach; for, how long should one wait? Thus, we see that a very practical question is: Can we tell of any arbitrary Tm T and initial tape inscription τ, whether or not $T(\tau)$ is defined? However, there is a difficulty in the question since we may not agree as to what "we can tell" should mean. We would probably agree that "we can tell" if we had a special Tm which somehow could be given information about T and τ and then would print on its tape "yes" or "no" according to whether T would or would not halt on τ. The problem of determining whether or not $T(\tau)$ is defined for an arbitrary Tm T and initial tape inscription τ is known as *the halting problem for Turing machines*. That there can be no Tm to determine this is one of the basic results concerning Tm's and the theory of computability. The rest of this section will be devoted to proving this result.

Exercise 1.24 In exercise 1.14, the reader designed a Tm to tell whether or not an arbitrary word is an element of a given finite set of words. In that case, the finite set was written on the tape. However, we shall also be interested in whether or not certain things are elements of certain infinite sets. Since an infinite set cannot be written on a tape†, a new approach is needed. The set of positive integers is infinite, and the set of even positive integers is one of its infinite subsets. Can you design a Tm which will decide whether or not an arbitrary integer is even? Try more than one tape representation for the positive integers.

Exercise 1.25 Suppose that A is the set of all words over some alphabet Σ. Let B and C be infinite subsets of A such that $B \cap C$ is infinite. (See figure 1.1.)

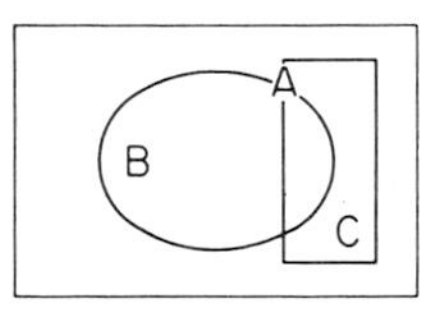

FIGURE 1.1

Imagine that you are trying to design a Tm to determine whether or not an arbitrary member of A is a member of B. A friend comes in to say that he has a Tm which can determine whether or not an arbitrary element of A

† Recall that only a finite number of squares may be nonblank.

belongs to C, *but* also that there is no Tm to tell whether or not an arbitrary member of C is in $B \cap C$. Do you keep working?

Theorem 1.1 (Unsolvability of the Halting Problem) There can be no Tm which can decide of an arbitrary Tm T and tape inscription τ whether or not $T(\tau)$ is defined.

PROOF (By assuming the theorem false, a contradiction will be derived.) Suppose there is such a machine T^* which for each T and τ can read a tape with ρ_T and τ on it and compute "yes" if $T(\tau)$ is defined and "no" if it is not. We may think of T^* as being a function of two variables ρ_T and τ:

$$T^*(\rho_T, \tau) = \begin{cases} \text{yes} & \text{if } \; T(\tau) \; \text{ defined} \\ \text{no} & \text{if } \; T(\tau) \; \text{ not defined} \end{cases}$$

But τ may be any possible tape inscription, and thus ρ_T itself. So, from the infinite set of pairs (ρ_T, τ) we may consider just the set (ρ_T, ρ_T) for all Tm's T and design a new Tm T^{**} based on T^*. The Tm T^{**} is to read all tapes (i.e., one for each Tm T) with only ρ_T as initial tape inscription, to write another copy of ρ_T on its tape, and then behave like T^*. Thus,

$$T^{**}(\rho_T) = \begin{cases} \text{yes} & \text{if } \; T(\rho_T) \; \text{ defined} \\ \text{no} & \text{if } \; T(\rho_T) \; \text{ not defined} \end{cases}$$

From T^{**}, we build another Tm T^{***} which acts like T^{**} but instead of writing "yes," it computes forever, and instead of writing "no," it writes "yes." Thus,

$$T^{***}(\rho_T) = \begin{cases} \text{never stops} & \\ \text{i.e., not defined} & \text{if } \; T(\rho_T) \; \text{ defined} \\ \text{yes,} & \\ \text{i.e., is defined,} & \text{if } \; T(\rho_T) \; \text{ not defined} \end{cases}$$

Since for each Tm T, T^{***} reads the tape with initial inscription ρ_T, certainly it will read the tape with initial inscription $\rho_{T^{***}}$. So in, in particular, consider $T^{***}(\rho_{T^{***}})$:

$$T^{***}(\rho_{T^{***}}) = \begin{cases} \text{undefined} & \text{if } \; T^{***}(\rho_{T^{***}}) \; \text{ defined} \\ \text{yes,} & \\ \text{i.e., defined,} & \text{if } \; T^{***}(\rho_{T^{***}}) \; \text{ not defined} \end{cases}$$

But this is a glorious contradiction and so we must conclude that our assumption, that the theorem is false, is false. Hence, the theorem is true. ∎

Exercise 1.26 The reader should be sure that if T^* were a Tm he could design T^{**} and T^{***}.

Exercise 1.27 To better appreciate Theorem 1.1, the reader would do well to pretend he does not know about this theorem and try to design a Tm to settle the halting problem.

Exercise 1.28 (The Printing Problem) Show that there can be no Tm which can decide of an arbitrary Tm T with alphabet Σ_1 and tape symbol $s_k \in \Sigma_1$, whether or not T computing from initial tape inscription τ will ever print s_k. (Hint: There are two possible approaches. (a) Imitate the proof of Theorem 1.1, arranging to get a contradiction about whether or not an arbitrary machine prints s_k; or (b) show that if the printing problem were solvable, then the halting problem would be solvable.)

Exercise 1.29 Prove that, for every *finite* set of words S on some alphabet Σ, there is a Tm T_S such that, for all words τ on Σ,

$$\begin{aligned} T_S(\tau) &= \text{yes} \quad &\text{if} \quad \tau \in S \\ T_S(\tau) &= \text{no} \quad &\text{if} \quad \tau \notin S \end{aligned}$$

1.6 A FEW REMARKS ABOUT DECISION PROBLEMS

We return to the halting problem so as to understand better the structure of the problem itself, that is, how it is presented. In order to abstract the framework, we consider the following facts:

1. There are infinitely many Tm's T. For any alphabet, there are infinitely many words τ on it. Hence, the set of ordered pairs $\{(T, \tau)\}$, as T varies over all Tm's and τ over all words on some given alphabet, is infinite.
2. There are infinitely many pairs (T, τ) for which $T(\tau)$ is defined.
3. So we have an infinite set A:

$$A = \{(T, \tau) \mid \text{all Tm's } T; \text{ all words } \tau\}$$

with an infinite subset B:

$$B = \{(T, \tau) \mid T(\tau) \text{ defined}\}$$

4. The problem, then, is to determine, for an arbitrary element $(T, \tau) \in A$, whether or not $(T, \tau) \in B$.

When a problem can be formulated along these lines—we wish to know of each element in a countably infinite set A, whether or not that element belongs to a certain set $B \subset A$—it is called a *decision problem*. A decision problem is said to be *solvable* (*decidable; Turing decidable*) if there is a Tm which computes yes or no for each element of A according to whether it does or does not belong to B. In this case, we also say there is a *decision procedure* or that there is a Tm which *decides* B. (If B is known to be finite, then the decision problem is solvable, as we have seen in exercise 1.29.) A decision problem is said to be *unsolvable* (*undecidable*, etc.) if there can be no such Tm. We also speak of B as being a *decidable* or *undecidable* set (in A), whichever is appropriate.

Theorem 1.2 If a set is Turing decidable, then it is Turing acceptable.

Exercise 1.30 Prove Theorem 1.2.

Exercise 1.31 (a) Which of the following are decision problems:

1. Given the set of all $n \times n$ matrices over the rational numbers, we wish to know which are nonsingular (i.e., have inverses).
2. We wish to know if there are 7 consecutive 7's in the decimal expansion of π.
3. We wish to know if there are positive integers $x, y, z, w, w > 2$, such that $x^w + y^w = z^w$.
4. Given the set of all quadruples of positive integers $\{(x, y, z, w)\}$ we wish to know for which quadruples it is true that $x^w + y^w = z^w$.

(b) Which of the above that are decision problems are solvable?

Exercise 1.32 Let $\Sigma = \{R, L, H, S, Q, B, M_1, \ldots, M_6\}$. (a) Can you design a Tm which can decide of an arbitrary word on Σ as initial tape inscription whether or not it is a ρ_T for some T? (b) Can you design a Tm which can decide whether or not an arbitrary word on Σ is in the form $\rho_T B^{m+n+2} \rho_\tau$, where m is the number of internal states of T, n is the number of tape symbols of T, and the tape symbols occurring in τ constitute a subset of the tape symbols of T?

Exercise 1.33 Let S be the set of all words on some alphabet Σ and let S_1 and S_2 be two subsets of S. Prove the following: (a) If there is a Tm T which decides $S_1 \cup S_2$ (in S) and there is also a Tm T' such that, if $\tau \in S_1$, then $T'(\tau)$ is defined and if $\tau \in S_2$, then $T'(\tau)$ is not defined, then there exists a Tm T'' such that $T''(\tau)$ is defined if and only if $\tau \in S_1$, i.e., T'' accepts S_1. (b) The set $\{\rho_T B^{m+n+2} \rho_\tau \mid T(\tau) \text{ is defined}\}$ is a Turing-acceptable set.

Theorem 1.3 There exists a set which is Turing acceptable but not Turing decidable.

Exercise 1.34 Prove Theorem 1.3.

1.7 WELL-KNOWN DECISION PROBLEMS

In this section, we discuss four of the more well-known and important examples of decision problems. The reader should read them thoughtfully. However, it is not too likely that he will understand them completely the first time through and so he is urged to reread them from time to time as he learns more. By the end of the book, if not before, he should have a clear idea of what they are about.

Hilbert's 10th Problem

This problem was posed by David Hilbert in 1900 and concerns what are known as Diophantine equations. That is, polynomial equations in any number of unknowns with coefficients from the integers, e.g.,

$$3x^2 - 2xy^3 + 7z^6 - 6 = 0.$$

In terms of the four steps above, the set A is the set of all Diophantine equations. The subset B is the set of all Diophantine equations with integral solutions. (That is, there are integers which, when substituted for the unknowns, make the polynomial identically zero.) The problem is to determine of an arbitrary member of A whether or not it belongs to B. Or, equivalently, to determine of an arbitrary Diophantine equation whether or not it has a solution in the integers. Although unsolvable problems were as yet unknown in 1900 and so Hilbert proposed the problem as one to be solved, i.e., an effective procedure given, even then he had a vague notion that there might be a concept like unsolvability. Only in 1970 was it finally shown to be an unsolvable problem. The basic theorems which together constitute the proof are to be found in J. Robinson (1952), Davis (1953), Davis *et al.* (1960), and Matijasevic (1970). The reader can consult Davis (1958, chapter 7) for a basic explanation of the problem and some of the early results.

The Truth Problem for First-Order Arithmetic and Hilbert's 10th Problem

In the discussion of Hilbert's 10th problem, we used our ordinary mathematical-English language, just as we do in most mathematical discussions. However, once in awhile it is useful to make everything extremely precise and in order to do so we introduce a "formal language" in place of our usual one. As a means of illustrating another decision problem and at the same time presenting Hilbert's 10th problem a little differently, we now give an example of a formal language which we call L_A. If the student keeps in mind certain comparisons with ordinary language, say English, he will understand better the description of the formal language. In English, we start with an alphabet; in a formal language, we start with a set of symbols. In English, before we can say what a sentence is, we must specify what nouns, verbs, etc. are, and then describe ways of putting these together to make sentences. As compared with English, the formal language L_A is quite simple since only terms and equations are defined as a preliminary to describing the manner in which these are to be put together to make formulas. *The symbols of* L_A are as follows:

For the integers: $\ldots, -3, -2, -1, 0, 1, 2, 3. \ldots$.
For variables over the integers: $x, y, z, x_1, y_1, z_1, \ldots$.
For operations: $+, \times$.
For connectives: $=, \wedge, \vee, \sim, (,)$.
Quantifiers: $\forall, \exists$.

The rules for forming the formulas of L_A are as follows:

Terms: An integer or a variable for an integer is a term. If t_1 and t_2 are terms, $(t_1 + t_2)$ and $(t_1 \times t_2)$ are terms. If t is a term and n is a positive integer, then t^n is a term.

Equations: If t_1 and t_2 are terms, then $(t_1 = t_2)$ is an equation.

Formulas: An equation is a formula. If s_1 and s_2 are formulas, then $(s_1 \wedge s_2)$, $(s_1 \vee s_2)$, and $(\sim s_1)$ are formulas.† If x is a variable for an integer and s is a formula, then $(\forall x)s$ and $(\exists x)s$ are formulas.‡

Examples §

Terms: -25, y, $(x + 5)$, $(x^2 + (7xy))$, $(4x(x + y))$.
Formulas:

1. $(\forall x)((x + 1) = 0)$, read "for all x, x plus one equals zero."
2. $(\exists x)((x + 1) = 0)$, read "there exists an x such that x plus one equals zero."
3. $(\exists x)((x + 7) = 12)$.
4. $(\forall x)(\exists y)((x + 1) = y)$.
5. $(\exists y)(\forall x)((x + 1) = y)$.
6. $((35 \times z^{21} \times x^{101}) - (73 \times x^3 \times z^2 \times y^{19}) + (16 \times y)) = 0$.
7. $(\exists x)((x^3 + (2 \times x^2) - x - 2) = 0)$.

† To be read respectively as "s_1 and s_2," "s_1 or s_2," and "not s_1."

‡ To be read respectively as "for all x, s" and "there exists an x such that s."

§ Some of the more obvious pairs of parentheses have been omitted. In place of "+ (− integer)," the usual "− integer" has been written.

8. $(\exists x)((x^4 + (3 \times x^2) + 2) = 0)$.
9. $(\forall x)((\sim((x - 1) = 0) \vee ((x \times x) = x)) \wedge (x = x))$.
10. $(\exists x)(\forall y)(((x^3 \times y^2 \times z) + 75) = 0)$.

By giving its symbols and rules for forming formulas, we have described L_A completely. We may carry the analogy between English and L_A further by considering when it makes sense to say "____ is true" or "____ is false." We cannot say of a noun, say "bird," that it is true (false). Similarly, we cannot say of a term, say $(x^3 - 5)$, that it is true (false). It does make sense to say of some (but not all, e.g., imperatives, questions, prayers, etc.) sentences that they are true (false). Similarly, it does make sense to say of some [but not all, e.g., $((x + 5) = y)$, $(\forall x)((x + y) = z))$] formulas that they are true, e.g., $(\forall x)(\exists y)((x \times y) = 1)$, and $((7 + 5) = 12)$. From all the formulas in L_A, we shall take only those for which "truth" makes sense. That means that for every variable in the equations of a formula there is a quantifier on that variable. We call such a formula a *sentence* in L_A. The truth or falsity of sentences of L_A is determined as follows: A sentence of the form $s_1 \wedge s_2$ is true if and only if both s_1 and s_2 are true; a sentence of the form $s_1 \vee s_2$ is false if and only if both s_1 and s_2 are false; a sentence of the form $\sim s$ is true (false) if and only if s is false (true); a sentence of the form $(\exists x)s$ is true if and only if there is an integer which, when substituted for x in s, makes it true; and a sentence of the form $(\forall x)s$ is true if and only if, no matter which integer is substituted for x in s, s is true. For example, consider the formulas (6), (10), (7), and (8) of the above examples. It does not mean anything to say that (6) and (10) are true, whereas it does make sense to say that (7) and (8) are true (false). Indeed, there is an integral solution (several: $x = 1$, $x = -1$, $x = 2$) of the equation in (7), so (7) is true, and there is no integral solution of the equation in (8), hence (8) is false.† The set of true sentences of L_A is usually referred to as first-order arithmetic. Recall that Hilbert's 10th problem concerns Diophantine equations and notice that the set of formulas of L_A that are equations is exactly the set of Diophantine equations. Further, corresponding to each Diophantine equation there is a sentence in L_A obtained from the equation by quantifying all the variables in the equation with existential quantifiers. So, for example, corresponding to the equation $(((x \times z) + (5 \times y) + z^3 - (2 \times x^2)) = 0)$ is the sentence

$$(\exists x)(\exists y)(\exists z)(((x \times z) + (5 \times y) + z^3 - (2 \times x^2)) = 0).$$

Call the set of sentences D. [In the examples, (2), (3), (7), and (8) belong to D.] Clearly, a sentence in D is true if and only if the equation to which it corresponds has an integral solution.

† *Exercise* 1.35: In the list of examples of formulas, determine which are sentences and, of these, which are true.

Take the set of sentences of L_A to be the set A of the outline. Then $D \subset A$ (Figure 1.2).† For the set B of the outline, we take the set of all true sentences in A. Of course, B automatically includes the members of D that are true. The decision problem for first-order arithmetic is to tell of an arbitrary member of A whether or not it is a member of B. *If* this decision problem were solvable, then the solvability of Hilbert's 10th problem would follow as a corollary.‡ *However*, in the 1930's it was shown that this decision problem is unsolvable and so one could not get at Hilbert's 10th problem by this approach. Of course, the unsolvability result does not imply that Hilbert's

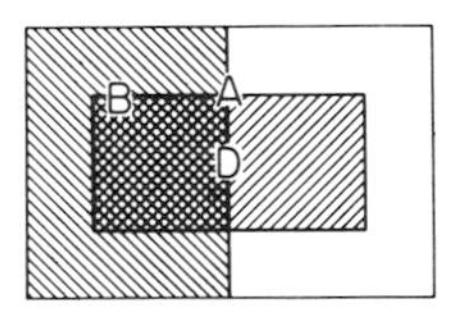

FIGURE 1.2

10th problem is unsolvable. The unsolvability of the decision problem for first-order arithmetic follows without too much difficulty from the undecidability of first-order number theory, which is like first-order arithmetic without symbols for the negative integers. The undecidability of first-order number theory was first shown by Church (1936). A different proof of the same result will be given in chapter XIV.

Exercise 1.38 Imagine a situation similar to that of exercise 1.25. A' is the set of all words on some alphabet Σ_1; B' and C' are infinite subsets with an infinite intersection. You are trying to design a Tm to decide of an arbitrary member of C' whether or not it is in $B' \cap C'$. This time, your friend dashes in to announce that he has just proved that there can be no Tm to decide of an arbitrary member of A' whether or not it belongs to B'. What do you do?

The Derivability Problem for Axiom Systems

The discussion here will be rather brief and informal, as the subject will be taken up with care later. Suppose we have a formal language, L, say L_A or something like it, and a fixed set of formulas in L taken as axioms. If, further, we have a specified set of rules (an effective procedure) by means of which we can operate on the axioms to derive other formulas in L and use these rules on the derived formulas, etc., then we have an axiom system T. The

† *Exercise* 1.36: Show, sketchily, that D is decidable (in A).

‡ *Exercise* 1.37: Using the previous exercise, prove the corollary: The solvability of first-order arithmetic implies the solvability of Hilbert's 10th problem.

formulas derived in this way are called the theorems of T. What we want to know is, Which formulas of L are theorems of T? In this case, the larger set A is the set of all formulas of L and the subset B is the set of just those formulas of A that are theorems of T. Thus, given an axiom system T, the problem is to determine of an arbitrary element of A whether or not it belongs to B. In this case, we will see that whether or not the derivability problem is solvable depends on the axiom system T.

The Word Problem for Semi-Thue and Thue Systems

Assume given once and for all a fixed, countable set of symbols $a_1, a_2, \ldots$. A finite set of these, $a_1, a_2, \ldots, a_n$, is an alphabet. Any finite concatenation of symbols from an alphabet is a word in the manner already defined. *A semi-Thue system* $\mathscr{S}$ is an alphabet Σ, a fixed *initial word* π_0 on Σ, and a finite set of ordered pairs (ρ_i, τ_i), $i = 1, 2, \ldots, m$, of words on Σ. The set of ordered pairs is called the *set of productions* of $\mathscr{S}$. We say a word π_k is derivable in $\mathscr{S}$ from π_0 (write: $\pi_0 \overset{\mathscr{S}}{\to} \pi_k$) if and only if (i) $\pi_0 \equiv \pi_k$, or (ii) π_0 has a subword ρ_i and π_k is identical to π_0 except that ρ_i has been replaced by τ_i providing (ρ_i, τ_i) is one of the productions, or (iii) there is a finite sequence of words $\pi_0, \pi_1, \ldots, \pi_k$, where for each i, $i = 0, 1, \ldots, k-1$, $\pi_i \overset{\mathscr{S}}{\to} \pi_{i+1}$. What we wish to determine of any semi-Thue system $\mathscr{S}$ and any word π_k on the alphabet of $\mathscr{S}$ is whether or not $\pi_0 \overset{\mathscr{S}}{\to} \pi_k$. For this problem, the larger set A is formed as follows: for each semi-Thue system $\mathscr{S}$, form the set of ordered pairs $(\mathscr{S}, \pi_k)$ for all words π_k on the alphabet of $\mathscr{S}$, then take the union of these sets over all semi-Thue systems. The subset B is just that subset of A whose members are $(\mathscr{S}, \pi_k)$, where $\pi_0 \to \pi_k$ in $\mathscr{S}$. So the decision problem, known as *the word problem for semi-Thue* systems, is to determine of an arbitrary member $(\mathscr{S}, \pi_k) \in A$ whether or not $(\mathscr{S}, \pi_k) \in B$.† In the next chapter, we will see that this is an unsolvable problem. As the proof is not too difficult, the student might try to do it now. (Hint: Think about how to make a semi-Thue system that looks like a Tm and then use the unsolvability of the halting problem.) Generalized Thue systems are like semi-Thue systems except that, for each production (ρ_i, τ_i), there is a corresponding production (τ_i, ρ_i) and there is no fixed initial word. A proof of the unsolvability of the word problem for generalized Thue systems will also be given in the next chapter. These two theorems were first proved by Post (1947). In chapter II, we will also consider some of the interesting variations on the themes of semi-Thue and generalized Thue systems.

† The student should reflect on the great similarities between this type of decision problem and that discussed in the previous paragraph under "The Derivability Problem for Axiom Systems." Indeed, one may give a unified discussion of both points of view as well as others. See Smullyan (1961).

1.8 ADDITIONAL EXERCISES†

Exercise 1.39 Let $\Sigma_n = \{*, 1, a_2, \ldots, a_n\}$ and let S_n be the set of all words on Σ_n. Prove that there exists a map $\varphi_n : S_n \to S_1$ so that for each Tm T with alphabet Σ_n, there exists a Tm T' on Σ_1 working from a half-infinite tape such that

$$\varphi_n(T(\tau)) \equiv T'(\varphi_n(\tau))$$

where $\tau \in S_n$ but contains no blanks.

Exercise 1.40 Prove the following: For every Tm T, there exists a Tm T' such that $T(\tau)$ is defined if and only if T' computing from a tape with τ as the initial tape inscription halts with the tape completely blank. Indicate that by $T'(\tau) \equiv *$.

Exercise 1.41 Prove that the problem of deciding of an arbitrary Tm T and initial tape inscription τ whether or not T halts with the tape completely blank, i.e., $T(\tau) \equiv *$, is unsolvable.

Exercise 1.42 (Unary–Decimal Converter) If the initial tape inscription is $X\underbrace{111 \cdots 1}_{n+1}Y$, design a machine to write n in decimal to the left of X. If the initial tape inscription is $Xd_1d_2 \cdots d_n Y$, where $d_i \in \{0, 1, \ldots, 9\}$, design a machine to write that number in unary to the left of X.

Exercise 1.43 Let S be the set of all words on the set of symbols $\{(\ , \ )\}$ and let R be the subset of S described as follows: (i) ( ) belongs to R, and (ii) if x and y belong to R, then (x) belongs to R and (xy) belongs to R. (a) Is R a Turing-decidable set? (b) In R, define R_2 to be the set of elements of R of length 2, i.e., $R_2 = \{(\ )\}$, R_4 to be the set of elements of R of length 4 together with the elements of R_2. Inductively, define R_{2n} to be the set of elements of R of length $2n$ together with the elements of $R_{2(n-1)}$. For each n, design a right-moving, "nonprinting" Tm (i.e., finite automata, see exercise 1.20) T_n which reads words $\tau \in S$ and halts to the right of τ if and only if $\tau \in R_n$; i.e., design T_n so that T_n accepts R_n. Do you think there is a finite automaton which accepts R?

Exercise 1.44 The customary definition of a finite automaton is that, instead of having a specific state q_H which causes the machine to halt for all symbols, any state causes the machine to halt if it is reading a blank square.

† Exercises 1.39, 1.40, and 1.41 will be needed in future chapters.

Thus, there are states $q_0, q_1, \ldots, q_{m-1}$, where q_0 is still the initial state, and certain of the q_i are designated as final states; it is assumed also that there are no blanks in the initial tape inscription. If the machine is in one of these designated states when it comes to a blank, then it is said to *accept* the word that was given as the initial tape inscription; otherwise, it rejects that word. In any case, the machine halts on the blank, so it is the state that it is in that determines accepting and rejecting.

(a) Prove that for any finite set of words on some alphabet, there is a finite automaton that accepts that set.

(b) Suppose that F is some finite set of natural numbers and that p and q are fixed natural numbers. Let $S = \{n \mid n \in F \text{ or } n = q + kp \text{ for any } k \in N\}$. Prove that for every such S, there is a finite automaton M_s whose only nonblank tape symbol is 1 such that M_s accepts the word 1^n if and only if $n \in S$.

(c) Suppose F is some finite set of natural numbers. Let p_i, q_i, $1 \leq i \leq t$ be fixed pairs of natural numbers. S is defined to be multiperiodic if $S = \{n \mid n \in F \text{ or } n = q_i + kp_i \text{ for some } i,\ 1 \leq i \leq t, \text{ and any } k \in N\}$. Show that for every multiperiodic set S, there is a finite automaton M_s such that M_s accepts the word 1^n if and only if $n \in S$.

(d) Show that if M is a finite automaton with 1 as its only nonblank tape symbol and $S_M = \{n \mid M \text{ accepts } 1^n\}$, then S_M is multiperiodic.

Chapter II

SEMI-THUE AND THUE SYSTEMS

2.1 INSTANTANEOUS DESCRIPTIONS OF A TURING MACHINE

Consider once more a Tm T operating at discrete intervals of time. If its alphabet is $\Sigma = \{*, 1, a_2, \ldots, a_{n-1}\}$ and its set of internal states is $Z = \{q_H, q_0, q_1, \ldots, q_{m-2}\}$, we may suppose that we give T a tape with initial tape inscription $\tau \equiv a_{i_1} a_{i_2} \cdots a_{i_p}$ and represent this, as in section 1.1, by

$$t = 0 \qquad \cdots \mid * \mid a_{i_1} \mid a_{i_2} \mid \cdots \mid a_{i_p} \mid * \mid \cdots$$
$$\uparrow q_0$$

But notice that this is easily represented by the word $q_0 a_{i_1} a_{i_2} \cdots a_{i_p}$ on $\Sigma \cup Z$. If T has the quintuple $q_0 a_{i_1} q_1 a_t R$, then at time $t = 1$, we have the word $a_t q_1 a_{i_2} \cdots a_{i_p}$. If we think of words on $\Sigma \cup Z$, then we can describe a certain subset of them, the *instantaneous descriptions of T*, as those words in which there is one and only one occurrence of an element of Z and immediately to the right of that occurrence there is at least one occurrence of an element of Σ. That is, an instantaneous description of T is a word of the form $\tau q a \tau'$ for $q \in Z$, $a \in \Sigma$, and τ, τ' words, possibly empty, on Σ.

We say that *instantaneous description δ_{i+1} follows according to Tm T from instantaneous description δ_i* (write $\delta_i \rightarrow \delta_{i+1}$) if:

1. For δ_i of the form $\tau q_j a_k \tau'$ (where τ and τ', possibly empty, are words on Σ), δ_{i+1} is either (a) $\tau a_k' q_j' \tau'$ if τ' is not empty and T has the quintuple $q_j a_k q_j' a_k' R$ or (b) $\tau a_k' q_j' *$ if τ' is empty and T has the quintuple $q_j a_k q_j' a_k' R$.
2. For δ_i of the form $\tau a_t q_j a_k \tau' (\tau, \tau'$ as above), δ_{i+1} is $\tau q_j' a_t a_k' \tau'$ if T has the quintuple $q_j a_k q_j' a_k' L$.
3. For δ_i of the form $q_j a_k \tau' (\tau'$ as above), δ_{i+1} is $q_j' * a_k' \tau'$ if T has the quintuple $q_j a_k q_j' a_k' L$.

Words of the form $q_0 \tau$ for τ on Σ are known as *initial instantaneous* *descriptions* and words of the form $\tau q_H *$ as *terminal instantaneous descriptions.* A finite sequence of instantaneous descriptions of T, $\delta_0, \delta_1, \ldots, \delta_t$, is a *computation of* T if $\delta_0 \to \delta_1$, $\delta_1 \to \delta_2, \ldots, \delta_{t-1} \to \delta_t$, where δ_0 is an initial instantaneous description. A finite sequence of instantaneous descriptions of T, $\delta_0, \delta_1, \ldots, \delta_t$, is a *terminal computation of* T if it is a computation of T and δ_t is a terminal instantaneous description. The *length of the computation* is the number of words occurring in it minus one; i.e., t.

We begin to see that by means of the definition of a computation in T we can look at Tm's not only from a machinelike point of view—tapes, tape symbols, internal states, etc.—but also from the point of view of sequences of words on an alphabet, one following from another according to some effective procedure.

Lemma 2.1 Let T be a Tm and suppose that at time 0, T is in state q_0 and τ_0 is the initial tape inscription. There is a time $t > 0$ such that T is in state q_i, τ_t is the word printed on the tape, and T is scanning the jth symbol of τ_t if and only if there is a computation of T of the form $\delta_0 \to \delta_1$, $\delta_1 \to \delta_2$, $\ldots, \delta_{t-1} \to \delta_t$, where $\delta_0 \equiv q_0 \tau_0$ and $\delta_t \equiv \tau_t' q_i \tau_t''$, where $\tau_t \equiv \tau_t' \tau_t''$ and $|\tau_t'| = j - 1$.

PROOF *Part I.* Show that if there is a $t > 0$ such that etc.$_1$, then there is a computation of T such as etc.$_2$. The proof is by induction on t. The cases for $t = 1$ and for $t - 1$ to t are similar; we consider the latter only. By the induction hypothesis, at time $t - 1$, T is in state $q_{i'}$, τ_{t-1} is written on its tape, and T is scanning the j'th symbol of τ_{t-1}. Thus, $\delta_{t-1} \equiv \tau_{t-1}' q_{i'} \tau_{t-1}''$ and $|\tau_{t-1}'| = j' - 1$. Suppose the j'th symbol is a_k and that T has the quintuple $q_{i'} a_k q_i a_k' R$. So, $\tau_{t-1} \equiv \tau_{t-1}' a_k \tau_{t-1}'''$. Then, at time t, T is in state q_i, has $\tau_t \equiv \tau_{t-1}' a_k' \tau_{t-1}'''$ on its tape, and is scanning the jth $(= j' + 1)$ symbol of τ_t. But then, that $\delta_t \equiv \tau_{t-1}' a_k' q_i \tau_{t-1}'''$ follows from δ_{t-1} by case 1 of the definition. If, instead of $q_i' a_k q_i a_k' R$, T has the quintuple $q_i' a_k q_i a_k' L$, then the argument follows in a similar fashion by case 2 (τ_{t-1}' not empty), or case 3 (τ_{t-1}' is empty).

Part II. To show the lemma in the other direction is also by induction on t, also tedious, also obvious, and will not be done here! ▮

Theorem 2.1 If T is a Tm and τ is a word on Σ of T, then $T(\tau) \equiv \tau'$ if and only if there is a terminal computation of T, $\delta_0, \delta_1, \ldots, \delta_t$, where $\delta_0 \equiv q_0 \tau$ and $\delta_t \equiv \tau' q_H *$.

Exercise 2.1 Prove the theorem.

We continue now to pursue the general idea of a sequence of words, one following from another according to some effective procedure.

2.2 SEMI-THUE SYSTEMS†

A generalized semi-Thue system is like the semi-Thue system defined in section 1.7, but without a fixed initial word. We define a *generalized semi-Thue system* S (abbreviate by gsTs) to be an alphabet and a finite set of ordered pairs (ρ_i, τ_i) of words on that alphabet. The set of ordered pairs is called the set of *productions of* S. A *derivation in* S is a finite sequence of words $\pi_0, \pi_1, \ldots, \pi_k$ such that for each j, $0 \leq j \leq k$, there are words ω and ω' on the alphabet of S such that $\pi_{j-1} \equiv \omega\rho_i\omega'$ and $\pi_j \equiv \omega\tau_i\omega'$, where (ρ_i, τ_i) is a production of S. We write $\pi \overset{S}{\rightarrow} \pi'$ if there is a derivation $\pi_0, \pi_1, \ldots, \pi_k$ where $\pi_0 \equiv \pi$ and $\pi_k \equiv \pi'$, and say π' is derivable from π_0. The *length of a derivation* is the number of words occurring in it minus one, i.e., k. If π is a word on the alphabet of a gsTs S with productions (ρ_i, τ_i), we say that (ρ_i, τ_i) *is applicable to* π if $\pi \equiv \omega\rho_i\omega'$. A semi-Thue system S *is monogenic for a set of words* W on its alphabet if for every $\pi \in W$, there is at most one production of S which is applicable to it.

Lemma 2.2 Let T be a Tm with alphabet Σ and internal states Z. There exists a gsTs S_T with alphabet $\Sigma \cup Z \cup \{h\}$, ($\Sigma \cap Z = \varnothing$ and $h \notin \Sigma \cup Z$), such that if $\delta_0, \delta_1, \ldots, \delta_t$ is a computation in T, then there is a derivation in S_T, $\pi_0, \pi_1, \ldots, \pi_t$, such that $\pi_i \equiv h\delta_i h$, for each i, $0 \leq i \leq t$.

PROOF We define the productions of S_T as follows: For every quintuple of T of the form $q_i a_j q_k a_t R$, let S_T have the production $(q_i a_j h, a_t q_k * h)$ and the set of productions $\{(q_i a_j x, a_t q_k x) \mid \text{as } x \text{ varies over } \Sigma\}$. Similarly, if T has the quintuple $q_i' a_j' a_k' a_t' L$, S_T is to have the production $(hq_i'a_j', hq_k'*a_t')$, and the set of productions $\{(xq_i'a_h', q_k'xa_t') \mid \text{as } x \text{ varies over } \Sigma\}$.

Exercise 2.2 Finish the proof of the lemma. (Hint: By induction on t, the length of the computation.) ▮

If T is a Tm and S_T is a gsTs that satisfies the lemma, then we call S_T the "Turing-gsTs associated with T" (!).

Lemma 2.3 If T is a Turing machine and S_T is its associated Turing-gsTs, then S_T is monogenic for the set of words of the form $h\tau_1 q_i \tau_2 h$, where τ_1 (possibly empty) and τ_2 are words on Σ of T and $q_i \in Z$ of T.

† These systems were defined and investigated in a series of papers starting about 1904 by Axel Thue. Subsequently, these and similar systems have been given his name.

PROOF Given a word of the form $h\tau_1 q_i \tau_2 h$, since τ_2 is not empty, it has a leftmost symbol, suppose it is a_j. By the definition of Tm, there is only one quintuple beginning with $q_i a_j$. Suppose it is $q_i a_j q_k q_t R$. Then, by the construction of S_T, there is a particular set of productions in S_T (see proof of Lemma 2.2) that have $q_i a_j$ in the first member. If the given word is of the form $h\tau_1 q_i a_j h$, then the only production applicable is $(q_i a_j h,\ a_t q_k *h)$. If the given word is such that the next-to-leftmost symbol of τ_2 is a_p, then the only production applicable is $(q_i a_j a_p,\ a_t q_k a_p)$. ▮

Lemma 2.4 Let T be a Tm and S_T be the Turing-gsTs associated with T. For all words π such that $hq_0\,\tau h \to \pi$ in S_T, π is of the form $h\tau_1 q_i \tau_2 h$ for τ_1 (possibly empty) and τ_2 on Σ of T.

PROOF The proof is by induction on the length of the derivation of π from $hq_0\,\tau h$. If $n = 0$, we just have $hq_0\,\tau h$ itself, which is in the proper form. Suppose that the lemma holds for derivations of length $n = k$; then we show that it holds for derivations of length $k + 1$. By the induction hypothesis, each of $hq_0\,\tau h, \pi_1, \ldots, \pi_k$, is of the proper form, so we may assume that π_k is of the form $h\tau_1 q_i a_j \tau_2' h$ (where τ_1 or τ_2' may be empty). By the construction and monogenicity of S_T, the one production applicable to π_k is in the set of productions $\{(q_i a_j h,\ a_t q_k *h)\} \cup \{(q_i a_j x,\ a_t q_k x) \mid \text{as } x \text{ varies over } \Sigma\}$, assuming that T has the quintuple $q_i a_j q_k a_t R$. Whichever it is, the result π_{k+1} of applying it is in the form $h\tau_1 a_t q_k x \tau_2' h$ for some x in Σ. ▮

Theorem 2.2 For every Tm T with alphabet Σ and set of internal states Z, there exists a gsTs S_T on $\Sigma \cup Z \cup \{h\}$ such that $T(\tau) \equiv \tau'$ if and only if $hq_0\,\tau h \to h\tau' q_H *h$ in S_T.

PROOF One direction follows from Lemma 2.2 and Theorem 2.1. For the S_T defined there, we must show that $hq_0\,\tau h \to h\tau' q_H *h$ in S_T implies that $T(\tau) \equiv \tau'$. Since there is a derivation and since S_T is monogenic for any word in a derivation from $hq_0\,\tau h$, there is one and only one production applicable at each step. But then by the construction of S_T in terms of T, there is a corresponding computation in T. The reader may complete the formalities of the proof by doing an induction on the number of steps in the derivation of $hq_0\,\tau h \to h\tau' q_H *h$ in S_T. ▮

The "special word problem for Turing-gsTs" is to decide whether or not $hq_0\,\tau h \to hq_H *h$ in S_T for an arbitrary Turing-gsTs S_T and arbitrary τ on Σ of T (associated with S_T).

Lemma 2.5 The special word problem for Turing-gsTs is unsolvable.

PROOF Recall exercise 1.41. Suppose that the special word problem for Turing-gsTs is solvable. We claim that if this is so, then there is a decision procedure to tell of an arbitrary Tm T and tape inscription τ on the alphabet of T whether or not T computing from τ halts on a completely blank tape. Suppose T and τ given; then, consider S_T and $hq_0\tau h$. By Theorem 2.2, $hq_0\tau h \rightarrow hq_H*h$ in S_T if and only if T halts on τ with the tape completely blank. Hence, if there is a decision procedure to tell whether or not $hq_0\tau h \rightarrow hq_H*h$ in S_T, it tells at the same time whether or not T halts on a completely blank tape if given τ. But, as we know from exercise 1.41, the latter problem is unsolvable, hence the special word problem for Turing–semi-Thue systems must also be unsolvable. ∎

The *word problem for generalized semi-Thue systems* is the problem of deciding whether or not $\pi \overset{S}{\rightarrow} \pi'$ for an arbitrary gsTs S and pair of words π and π' on the alphabet of S.

Theorem 2.3 The word problem for generalized semi-Thue systems is unsolvable.

Exercise 2.3 Prove Theorem 2.3. (Hint: Refer to exercise 1.25. For the theorem, what are the sets A, B, C, and $B \cap C$ of the exercise?)

Let us consider for a moment the form of the proof of Lemma 2.5. Let A be the set $\{(T, \tau) \mid \text{for all Tm's } T \text{ and words } \tau \text{ on } \Sigma \text{ of } T\}$; then, as a subset of A, let $B = \{(T, \tau) \mid T$ computing from τ as initial tape inscription halts with the tape completely blank$\}$. Let $A^* = \{(S_T, \tau) \mid$ for all Turing-gsTs and words τ on Σ of $T\}$ and let $B^* = \{(S_T, \tau) \mid hq_0\tau h \rightarrow hq_H*h \text{ in } S_T\}$. The essence of Theorem 2.2 is that $(T, \tau) \in B$ if and only if, for the associated Turing-gsTs S_T, $(S_T, \tau) \in B^*$. Further, given a Tm T, there is an effective procedure for constructing S_T. In general, we say that the decision problem for a set B (in A) is (one–one) reducible to the decision problem for a set B^* (in A^*) if there is an effective procedure which for each element $x \in A$ gives an element $x^* \in A^*$ such that $x \in B$ if and only if $x^* \in B^*$. If it is known that B is undecidable (in A), it then follows that B^* is undecidable (in A^*), as we saw in the proof of Lemma 2.5. In most cases, the method of proving that a decision problem is unsolvable is to show that the halting problem for Turing machines is reducible to it. There will be other examples in this chapter. More general questions about one decision problem being reducible to another will be considered further in chapter VII.

Recall the definition of a semi-Thue system in section 1.7, that is, it is a gsTs with a fixed initial word π_0. To say that π is derivable in a semi-Thue system S means that there is a derivation of π from the initial word of S.

A semi-Thue system S is *monogenic* if at most one production of S is applicable to the initial word of S, and if π is derivable in S, at most one production of S is applicable to π. The *word problem for semi-Thue systems* is the problem of deciding, for an arbitrary pair S and τ, where S is a semi-Thue system and τ a word, whether or not τ is derivable in S.

Theorem 2.4 The word problem for semi-Thue systems is unsolvable.

Exercise 2.4 Prove Theorem 2.4.

Theorem 2.5 (a) For every generalized semi-Thue system S, there exists a Tm T_S such that $T_S(\alpha, \beta)$ is defined if and only if $\alpha \overset{S}{\rightarrow} \beta$. (b) For every semi-Thue system S, there exists a Tm T_S such that $T_S(\beta)$ is defined if and only if β is derivable in S.

Exercise 2.5 Prove Theorem 2.5. (Hint for part a: Design T_S so that it carries out all possible one-step derivations from α, checks to see if β is among these results, and if not, carries out all one-step derivations from these results, checks against β, etc.).

Exercise 2.6 Let S be a gsTs with just one production. Show that the set $\{(\alpha, \beta) \mid \alpha \overset{S}{\rightarrow} \beta\}$ is a Turing-decidable set. (Hint: Think about lengths of words.)

2.3 THUE SYSTEMS

A *generalized Thue system* is a generalized semi-Thue system in which for every production (ρ_i, τ_i), there is also the production (τ_i, ρ_i), called the inverse of (ρ_i, τ_i). Thus, in a Thue system, $\pi \rightarrow \pi'$ if and only if $\pi' \rightarrow \pi$, and so we write $\pi \leftrightarrow \pi'$. The *word problem for generalized Thue systems* is the problem of deciding whether or not $\pi \leftrightarrow \pi'$ in S for an arbitrary generalized Thue system S and pair of words π, π' on the alphabet of S.

If S is a gsTs, let S^* be the generalized Thue system obtained from S by adding to the productions of S the inverses of all those productions. If S is a gsTs, define S^{-1} to be the gsTs with the same alphabet as S and with productions which are just the inverses of the productions of S. Then it makes sense to speak of S^* as $S \cup S^{-1}$.

Lemma 2.6 Let S_T be a Turing-gsTs. If π is a word of the form $h\tau_1 q_i \tau_2 h$ for τ_1 (possibly empty) and τ_2 on Σ of T and $\pi \leftrightarrow \pi'$ in the generalized Thue system S_T^*, then π' is of the form $h\tau_1' q_j \tau_2' h$ for τ_1' (possibly empty) and τ_2' on Σ of T.

Exercise 2.7 Prove the lemma.

Lemma 2.7 Let S_T be a Turing-gsTs and τ be a word on Σ of (the associated) T. Then $hq_0 \tau h \to hq_H * h$ in S_T if and only if $hq_H * h \to hq_0 \tau h$ in S_T^{-1}.

Exercise 2.8 Prove the lemma.

Exercise 2.9 If a gsTs S is monogenic for a set of words, is the gsTs S^{-1} also monogenic for that set?

Lemma 2.8 Let S_T be a Turing-gsTs and let $S_T{}^* = S_T \cup S_T^{-1}$. Then $hq_H * h \leftrightarrow hq_0 \tau h$ in $S_T{}^*$ if and only if $hq_H * h \to hq_0 \tau h$ in S_T^{-1}.

PROOF If $hq_H * h \to hq_0 \tau h$ in S_T^{-1}, then by the definition of $S_T{}^*$, $hq_H * h \leftrightarrow hq_0 \tau h$ also in $S_T{}^*$. On the other hand, let $\pi_0 \equiv hq_H * h, \pi_1, \ldots, \pi_k \equiv hq_0 \tau h$ be a derivation in $S_T{}^*$. Notice first that $\pi_0 \to \pi_1$ in S_T^{-1} because there is no production of S_T which is applicable to $hq_H * h$. Therefore, we may assume that either the entire derivation is in S_T^{-1}, in which case the lemma is true, or that $\pi_0 \to \pi_1, \ldots, \pi_{n-1} \to \pi_n$ in S_T^{-1} but that $\pi_n \to \pi_{n+1}$ in S_T; i.e., we consider the first use of a production from S_T in the derivation. Consider $\pi_{n-1}, \pi_n, \pi_{n+1}$. Since $\pi_{n-1} \to \pi_n$ by a production of S_T^{-1}, $\pi_n \to \pi_{n-1}$ by a production of S_T. But we are also assuming that $\pi_n \to \pi_{n+1}$ by a production of S_T. However, by Lemma 2.5, π_n is of the form $h\tau_1 q_j \tau_2 h$, and by Lemma 2.2, S_T is monogenic for words of that form. So the only production applicable is the inverse of the one used in obtaining π_n from π_{n-1}. Therefore, $\pi_{n-1} \equiv \pi_{n+1}$ and so $hq_H * h \to \pi_n$ in S_T^{-1} for any n, $0 \le n \le k$. ∎

Theorem 2.6 The word problem for generalized Thue systems is unsolvable.

Exercise 2.10 Prove this theorem.

Notice that it was easy to show (exercise 2.6) that the word problem for a gsTs with one production is solvable. It seems to be not so easy to show that the word problem for generalized Thue systems with one production is solvable; at least no one has yet been able to prove it solvable or unsolvable.

A *Thue system* is the same as a generalized Thue system, but it has a fixed initial word.

Theorem 2.7 The word problem for Thue systems is unsolvable.

Exercise 2.11 Define the word problem for Thue systems and prove Theorem 2.7.

2.4 THE POST CORRESPONDENCE PROBLEM†

A *Post correspondence system P* is given by a finite alphabet Σ and a finite set of pairs of words on Σ, (ρ_i, τ_i), $1 \le i \le m$. The system has a *solution* if there is a finite sequence of indices of pairs $i_1, i_2, \ldots, i_k$, $1 \le i_j \le m$, such that

$$\rho_{i_1}\rho_{i_2}\cdots\rho_{i_k} \equiv \tau_{i_1}\tau_{i_2}\cdots\tau_{i_k}$$

We will call such a word $\rho_{i_1}\rho_{i_2}\cdots\rho_{i_k}$ $(\tau_{i_1}\tau_{i_2}\cdots\tau_{i_k})$ a "solution word" for P. The *decision problem for Post correspondence systems* is the problem of deciding of an arbitrary P whether or not it has a solution. (The decision problem for Post correspondence systems is usually called the *Post correspondence problem.*)

Lemma 2.9 For every generalized semi-Thue system S and pair of words α and β on the alphabet of S, there exists a Post correspondence system P_S such that $\alpha \overset{S}{\rightarrow} \beta$ if and only if P_S has a solution.

PROOF *Part I.* Suppose S has alphabet $\Sigma = \{a_1, \ldots, a_n\}$ and productions (ρ_i, τ_i), $1 \le i \le m$. Let $\bar{\Sigma}$ denote $\{\bar{a}_1, \ldots, \bar{a}_n\}$, where $\Sigma \cap \bar{\Sigma} = \varnothing$. The system P_S is to have the alphabet $\Sigma \cup \Sigma \cup \{*, \bar{*}, [\ , \ ]\}$, where $\{*, *, [\ , \ ]\}$ is disjoint from Σ and from $\bar{\Sigma}$. The pairs of P_S are as follows:

1. $([\alpha*,\ [)$.
2. $(],\ \bar{*}\beta])$.
3. $(*,\ \bar{*})$.
4. $(\bar{*},\ *)$.

5. $(a_1, \bar{a}_1)$.	6. $(a_2, \bar{a}_2), \ldots,$	$4 + n$.	$(a_n, \bar{a}_n)$	("Set 5.")
$4 + n + 1$.	$(\bar{a}_1, a_1), \ldots,$	$4 + 2n$.	$(\bar{a}_n, a_n)$.	("Set 6.")
$4 + 2n + 1$.	$(\tau_1, \bar{\rho}_1), \ldots,$	$4 + 2n + m$.	$(\tau_m, \bar{\rho}_m)$.	("Set 7.")
$4 + 2n + m + 1$.	$(\bar{\tau}_1, \rho_1), \ldots,$	$4 + 2n + 2m$.	$(\bar{\tau}_m, \rho_m)$.	("Set 8.")

We may also refer to the pairs in 3, 4, set 5, and set 6 as the "copying pairs."

Part II. Prove that $\alpha \overset{S}{\rightarrow} \beta$ implies that P_S has a solution. We will build pairs of words L and R, where L will be a concatenation of left sides of pairs of P_S and R the concatenation of corresponding right sides of those pairs. If $\alpha \overset{S}{\rightarrow} \beta$, then there is a deduction $\alpha \rightarrow \pi_1$, $\pi_1 \rightarrow \pi_2, \ldots, \pi_{n-1} \rightarrow \beta$, so we can assume that $\alpha \equiv \omega\rho_i\omega'$ and $\beta \equiv \omega''\tau_j\omega'''$ for some ω, ω', ω'', ω''', i, j. By pair 1, we have

† Although this problem was first proposed and shown to be unsolvable by Post (1946), the presentation here follows that by Floyd (1964).

$L_1 \equiv [\omega\rho_i\omega'*$
$R_1 \equiv [$ Continuing with pairs in set 6,
$L_2 \equiv [\omega\rho_i\omega'*\bar{\omega}$
$R_2 \equiv [\omega$ and by one pair in set 8
$L_3 \equiv [\omega\rho_i\omega'*\bar{\omega}\bar{\tau}_i$
$R_3 \equiv [\omega\rho_i$ Then by pairs in set 6 and pair 3,
$L_4 \equiv [\omega\rho_i\omega'*\bar{\omega}\bar{\tau}_i\bar{\omega}'\bar{*}$
$R_4 \equiv [\omega\rho_i\omega'*$

Since $\pi_1 \to \pi_2$, there is a k and there are words μ, μ' such that $\bar{\omega}\bar{\tau}_i\bar{\omega}' \equiv \bar{\mu}\bar{\rho}_k\bar{\mu}'$. So, using the appropriate pairs from sets 5, 7, and 5 and pair 4, we have

$$L_5 \equiv [\omega\rho_i\omega'*\overline{\mu\rho_k\mu'*}\mu\tau_k\mu'*$$
$$R_5 \equiv [\omega\rho_i\omega'*\overline{\mu\rho_k\mu'*}$$

We can assume, inductively, that

$$L_6 \equiv [\alpha*\bar{\pi}_1*\cdots*\bar{\pi}_{n-1}\bar{*}$$
$$R_6 \equiv [\alpha*\bar{\pi}_1*\cdots*$$

and that $\bar{\pi}_{n-1} \equiv \overline{\omega''\rho_j\omega'''}$. So, using pairs from 5, 7, and 5, we have

$$L_7 \equiv [\alpha*\bar{\pi}_1*\cdots*\overline{\omega''\rho_j\omega'''*}\omega''\tau_j\omega'''$$
$$R_7 \equiv [\alpha*\bar{\pi}_1*\cdots*\overline{\omega''\rho_j\omega'''}$$

where $\beta \equiv \omega''\tau_j\omega'''$. Finally, by pair 2, we have a solution:

$$L_8 \equiv [\alpha*\bar{\pi}_1\bar{*}\cdots*\bar{\pi}_{n-1}\bar{*}\beta]$$
$$R_8 \equiv [\alpha*\bar{\pi}_1\bar{*}\cdots*\bar{\pi}_{n-1}\bar{*}\beta]$$

(Notice that if n is even and hence, in this development, we get π_{n-1} instead of $\bar{\pi}_{n-1}$, we can simply repeat it using pairs from set 5—e.g., the solution word would be $[\alpha*\bar{\pi}_1*\cdots\bar{*}\pi_{n-1}*\bar{\pi}_{n-1}\bar{*}\beta]$.)

Part III. Prove that if P_S has a solution, then $\alpha \overset{S}{\to} \beta$. Since pair 1 is the only pair in which both words have the same leftmost symbol, and pair 2 the only pair in which both have the same rightmost symbol, it is evident that if there is a solution, the leftmost pair used must be 1 and the rightmost 2. We suppose that

$$L_1 \equiv [\alpha* \quad \text{and} \quad R_1 \equiv [$$

Case 1. If there are no words ω, ω', and ρ_i such that $\alpha \equiv \omega\rho_i\omega'$, then we claim that for there to be a solution it must be that $\alpha \equiv \beta$. Notice that, in this case, sets 7 and 8 cannot be used. Until the rightmost pair, only copy pairs can be used. Thus,

$$L_2 = [\alpha*\overline{\alpha*}\alpha* \quad \text{and} \quad R_2 \equiv [\alpha*\overline{\alpha*}$$

This cycle ("a copy cycle") can be repeated any number of times. But the fact that there is a solution for P_S implies that pair 2 must finally be used, and so that $\alpha \equiv \beta$. In this case, of course, $\alpha \overset{S}{\rightarrow} \beta$.

Case 2. If there are words ω, ρ_i, and ω' such that $\alpha \equiv \omega\rho_i\omega'$, then there can be any number of "copy cycles" as above, or there can be uses of pairs from sets 7 and 8 in the manner described in Part II. Since it is impossible to have a solution word that does not end with $\bar{*}\beta]$, it is easy to see that the existence of a solution word implies the existence of a derivation of β from α in S. ▮

Theorem 2.8 The problem of deciding of an arbitrary Post correspondence system whether or not it has a solution is unsolvable. (Equivalently, the Post correspondence problem is unsolvable.)

2.5 SOME FORMAL GRAMMAR RAMIFICATIONS OF THE SEMI-THUE SYSTEM

In this section, we will be thinking only about semi-Thue systems. We have already considered a certain special kind of semi-Thue system, the Turing–semi-Thue system which "mimics" a Turing machine with a particular initial tape inscription. Among other interesting special kinds of semi-Thue systems are those that come under the general designation of formal grammars. We give the definitions as originally given by Chomsky (1959).

Type 0 Grammars The alphabet $\Sigma = \Sigma_1 \cup \Sigma_2$, where $\Sigma_1 \cap \Sigma_2 = \varnothing$, the initial word is a single symbol in Σ_1 (called the *start symbol*), and in the set of productions (ρ_i, τ_i), ρ_i is any nonempty word on Σ and τ_i is any word, possibly empty, on Σ. We call Σ_1 the *nonterminal alphabet* and Σ_2 the *terminal alphabet*. Type 0 grammars are often called *phrase-structure grammars.*

Type 1 Grammars Type 1 grammars are like type 0 grammars with the further restriction that each production (ρ_i, τ_i) has the form $(\pi_i a \pi_i', \pi_i \omega_i \pi_i')$ for $a \in \Sigma_1$, ω_i a nonempty word on Σ, and π_i, π_i' possibly empty words on Σ. A semi-Thue system of type 1 has come to be known as a *context-sensitive grammar*.

Type 2 Grammars Type 2 grammars are like type 1 grammars with the further restriction that each ρ_i is a single symbol in Σ_1 and τ_i is any nonempty word on Σ. A semi-Thue system of type 2 is also called a *context-free grammar*.

Type 3 Grammars Type 3 grammars have the same form as type 2 grammars except that τ_i must be either a single symbol in Σ_2 or of the form ax, where $a \in \Sigma_1$ and $x \in \Sigma_2$. The type 3 grammars are also known as the *regular grammars*.

In each of the four cases, a word on Σ_2 is called a *terminal word.* For a fixed i, $i = 0, 1, 2$, or 3, the set of terminal words derivable in a system of type i is known as a *type i language*. (Or by the corresponding name given above, such as a context-free language when $i = 2$.)

Exercise 2.12 Show that if L is a type 1 language, then L is a Turing-decidable set. [Hint: Notice that the form of the productions is such that for each (ρ_i, τ_i), $|\rho_i| \leq |\tau_i|$.]

Exercise 2.13 Define a context-free grammar whose language is just the set R of exercise 1.43.

Several natural problems arise in connection with these types of semi-Thue systems. One of particular interest is the following: For each i, define a coresponding type of machine (automaton) and concept of "accept" so that the set of words accepted by a machine of that type is a type i language, and conversely, each type i language is accepted by a machine of that type. We are in a position to show that the machines appropriate for type 0 languages are the Turing machines.†

Theorem 2.9 For every type 0 language L, there exists a Tm T_L such that $T_L(\tau)$ is defined if and only if $\tau \in L$.

Exercise 2.14 Prove the theorem. (Hint: Let G_L be the grammar that generates L. Design T_L so that with τ as its initial tape inscription it first puts markers on either side of τ, then prints the start symbol of G_L, and then carries out deductions in the manner described in exercise 2.5. If τ is the empty word, of course, T_L starts on a blank tape and should print markers on either side of a blank square.)

Theorem 2.10 For every Turing machine T, there exists a type 0 language L_T such that $T(\tau)$ is defined if and only if $\tau \in L_T$.

PROOF We already know that: (1) For every Tm T, there exists a Tm T' such that $T(\tau)$ is defined if and only if $T'(\tau) \equiv *$. (2) For every Tm T, there exists a Turing–semi-Thue system S_T such that $T(\tau) \equiv *$ if and only if

$$hq_0 \tau h \xrightarrow{S_T} hq_H *h$$

† It is the finite automata that "accept" the type 3 languages. Consult Rabin and Scott (1959) for the proof, or any of the books mentioned at the end of the section.

(3) For every Turing–semi-Thue system S_T, there exists a semi-Thue system S_T^{-1} such that

$$hq_0\tau h \xrightarrow{S_T} hq_H*h$$

if and only if

$$hq_H*h \xrightarrow{S_T^{-1}} hq_0\,\tau h$$

So what we will show is that for every such S_T^{-1}, there is a type 0 grammar G_T such that

$$hq_H*h \xrightarrow{S_T^{-1}} hq_0\,\tau h$$

if and only if τ belongs to the language generated by G_T. We will assume that either τ is empty (i.e., T starts on a blank tape and so τ is $*$) or τ contains no occurrence of $*$.

Part I. If T is a Tm and S_T^{-1} is the corresponding semi-Thue system as defined in section 2.3, then the alphabet of S_T^{-1} is $\Sigma \cup \{q_H, q_0, q_1, \ldots, q_{m-1}, h\}$, where Σ is the set of tape symbols of T (thus including $*$), $\{q_H, q_0, q_1, \ldots, q_{m-1}\}$ is the set of internal states of T (disjoint from Σ), and h is disjoint from both of these. Define G_T as follows: the terminal alphabet is Σ, the nonterminal alphabet is $\{q_H, q_0, \ldots, q_{m-1}, q_m, S, h\}$, where q_m and S are distinct from all other symbols. The starting symbol for G_T is S. The productions of G_T are:

1. (S, hq_H*h).
2. $\{(hq_0\,a_i, a_iq_m) \mid \text{for all } a_i \in \Sigma\}$.
3. $\{(q_m a_i, a_iq_m) \mid \text{for all } a_i \in \Sigma\}$.
4. $\{(a_iq_m h, a_i) \mid \text{for all } a_i \in \Sigma \text{ except } *\}$.
5. $(*q_m h, \text{——})$.
6. $\{(\rho_i, \tau_i) \mid (\rho_i, \tau_i) \text{ are the productions of } S_T^{-1}\}$.

Part II. Prove that

$$\text{if } \; hq_H*h \xrightarrow{S_T^{-1}} hq_0\,\tau h, \qquad \text{then } \; S \xrightarrow{G_T} \tau$$

Since $S \to hq_H*h$ in G_T, if the hypothesis holds, then $S \to hq_0\,\tau h$ in G_T.

Case 1. If τ is not simply $*$, then $\tau \equiv a_i\tau'$ for some $a_i \in \Sigma$ other than $*$ and τ' is possibly empty. Then, $hq_0\,a_i\tau'h \to a_iq_m\tau'h$ in G_T by a production in set 2. If τ' is not empty, then by the appropriate productions in set 3, $a_iq_m\tau'h \to a_i\tau'q_mh$ and so by the appropriate production of set 4, $a_i\tau'q_mh \to a_i\tau'$. If τ' is empty, then by set 4, $a_iq_mh \to a_i$. In any case, $S \to \tau$ in G_T.

Case 2. If τ is $*$, then $hq_0*h \to h*q_mh$ by the appropriate production of set 2. Then by 5, $h*q_mh \to$ ——.

Part III. Prove that

$$S \xrightarrow{G_T} \tau \qquad \text{implies} \qquad hq_H*h \xrightarrow{S_T^{-1}} hq_0\,\tau h$$

Since S is the initial symbol of G_T and (S, hq_H*h) is the only production in which S occurs at all, we can conclude that the first step of any deduction in G_T is $S \to hq_H*h$ and that production is not used in any other step of a deduction. The only way to get from hq_H*h to a word on Σ is via productions in sets 2, 3, and 4. This, in turn, requires that the deduction of τ be of the form

$$S \to hq_H*h \to \cdots \to hq_0\,a_i\rho \to \cdots \to \tau$$

for some $a_i \in \Sigma$ and word ρ. Suppose the $hq_0\,a_i\rho$ displayed is the first word of this form in the deduction. Then the productions used in going from hq_H*h to $hq_0\,a_i\rho$ must all be from set 6. Thus, $hq_0\,a_i\rho \equiv hq_0\,a_i\omega h$ and

$$hq_H*h \xrightarrow{S_T^{-1}} hq_0\,a_i\omega h$$

There are only two possibilities for the next step in the derivation of τ.

Case 1. If it is obtained by use of a production of set 2, then the only productions applicable thereafter are those in sets 2, 3, and 4. Thus, $\tau \equiv a_i\omega$, so we have that

$$hq_H*h \xrightarrow{S_T^{-1}} hq_0\,\tau h$$

Case 2. If it is obtained by use of a production of set 6, then, as before, the deductions in G_T and S_T^{-1} are the same until there is a word of the form $hq_0\,a_{i'}\rho'$. In this case, the argument is that just given. (That is, we are doing an induction on the number of steps in the deduction $S \to \tau$ that have the form $hq_0\,a_i\rho$.) ▮

A semi-Thue system (type i grammar) is *ambiguous* if there can be two different derivations of the same (terminal) word. It is well known that there are ambiguous context-free grammars. For example, let G have nonterminal alphabet $\{S, A\}$, terminal alphabet $\{[,], *\}$, start symbol S, and productions $(S, S])$, $(S, [A)$, $(S, [A])$, $(A, *)$. There are two different derivations of $[*]$ in G:

$$S \to S] \to [A] \to [*]$$
$$S \to [A] \to [*]$$

For fixed i, the *ambiguity problem for Type i grammars* is the problem of deciding of an arbitrary type i grammar whether or not it is ambiguous.

***Theorem 2.11*†** The ambiguity problem for type 2 grammars is unsolvable.

PROOF The theorem will be proved by showing that for every Post correspondence system P, there is a type 2 grammar G_P with the property that P has a solution if and only if G_P is ambiguous.

Part I. Let P be a Post correspondence system with alphabet Σ and pairs (ρ_i, τ_i), $1 \leq i \leq m$. Define G_P to have terminal alphabet $\Sigma \cup \{[, *, :,]\} \cup \{e_i | 1 \leq i \leq m\}$, where the three sets mentioned are all disjoint from one another. G_P has nonterminal alphabet $\{S\}$, so S is also its start symbol. G_P has productions:

1*L*. $(S, [S])$.
1*R*. $(S, [S*])$.
2*Li*. $\{(S, \rho_i S e_i *) | 1 \leq i \leq m\}$.
2*Ri*. $\{(S, \tau_i S * e_i) | 1 \leq i \leq m\}$.
3*Li*. $\{(S, \rho_i : e_i *) | 1 \leq i \leq m\}$.
3*Ri*. $\{(S, \tau_i : e_i) | 1 \leq i \leq m\}$.

Part II. Prove that if P has a solution, then G_P is ambiguous. Suppose a solution for P is given by

$$\rho_{i_1} \rho_{i_2} \cdots \rho_{i_{k-1}} \rho_{i_k} \equiv \tau_{i_1} \tau_{i_2} \cdots \tau_{i_{k-1}} \tau_{i_k}$$

Then we have the two following derivations in G_P of the same terminal word:

1. Using productions 1*L*, 2*L*, and 3*L*,

$$\begin{aligned} S &\to [S] \to [\rho_{i_1} S e_{i_1} *] \to [\rho_{i_1} \rho_{i_2} S e_{i_2} * e_{i_1} *] \\ &\to \cdots \to [\rho_{i_1} \rho_{i_2} \cdots \rho_{i_{k-1}} S e_{i_{k-1}} * \cdots e_{i_2} * e_{i_1} *] \\ &\to [\rho_{i_1} \rho_{i_2} \cdots \rho_{i_{k-1}} \rho_{i_k} : e_{i_k} * e_{i_{k-1}} * \cdots e_{i_2} * e_{i_1} *] \end{aligned}$$

2. Using the corresponding productions of 1*R*, 2*R*, and 3*R*,

$$\begin{aligned} S &\to [S*] \to [\tau_{i_1} S * e_{i_1} *] \to \cdots \to [\tau_{i_1} \tau_{i_2} \cdots \tau_{i_{k-1}} S * e_{i_{k-1}} * \cdots e_{i_2} * e_{i_1} *] \\ &\to [\tau_{i_1} \tau_{i_2} \cdots \tau_{i_{k-1}} \tau_{i_k} : e_{i_k} * e_{i_{k-1}} * \cdots e_{i_2} * e_{i_1} *] \end{aligned}$$

Part III. Prove that if G_P is ambiguous, then P has a solution. Suppose that there are two different derivations of the terminal word τ in G_P. Thus,

$$S \to \pi_1 \to \pi_2 \to \cdots \to \pi_k \to \tau$$

and

$$S \to \pi_1' \to \pi_2' \to \cdots \to \pi_{k'}' \to \tau$$

† There are proofs of this theorem by Chomsky and Schutzenberger (1963), by Cantor (1962), and by Floyd. The proof given here is based directly on Floyd (1964).

Claim first that if there is a derivation in G_P of a word of the form $\alpha S\beta$, then that derivation is unique. By induction on the number of steps of the derivation and by inspection of the productions, it is clear that the claim is justified. Therefore, if $\pi_1 \equiv \pi_1', \ldots, \pi_m \equiv \pi_m'$ but $\pi_{m+1} \not\equiv \pi_{m+1}'$, then for all i and j, $m < i \leq k$ and $m < j \leq k'$, $\pi_i \not\equiv \pi_j'$. In particular, $\pi_k \not\equiv \pi_{k'}'$. Since τ is a terminal word, it must be of the form $\tau \equiv \omega_1 : \omega_2$, where the only : occurring in τ is the one displayed. Notice that just to the right of : must be e_i for some i. We can thus conclude that the production used in $\pi_k \to \tau$ is $3Li$ and the production used in $\pi_{k'}' \to \tau$ is $3Ri$. (Note that they cannot both be $3Li$ or $3Ri$, for that would imply $\pi_k \equiv \pi_{k'}'$.) Thus, we have

$$\tau \equiv \alpha\rho_i{:}e_i{*}\beta \equiv \alpha'\tau_i{:}e_i\beta'$$

and $\alpha\rho_i \equiv \alpha'\tau_i$ and $*\beta \equiv \beta'$. We may assume that for some $j \geq 0$ the previous j steps of both derivations, i.e.,

$$\pi_{k-j} \to \cdots \to \pi_{k-1} \to \pi_k$$

$$\pi_{k'-j}' \to \cdots \to \pi_{k'-1}' \to \pi_{k'}'$$

are by means of corresponding productions of $2L$ and $2R$. Thus,

$$\tau \equiv \alpha_1\rho_{i_0}\rho_{i_1} \cdots \rho_{i_j}{:}\, e_{i_j}{*} \cdots e_{i_1}{*}e_{i_0}{*}\beta_1$$

and

$$\tau \equiv \alpha_1'\tau_{i_0}\tau_{i_1} \cdots \tau_{i_j}{:}\, e_{i_j}{*} \cdots e_{i_1}{*}e_{i_0}\beta_1'$$

This implies that $\beta_1' \equiv {*}\beta_1$. Assuming that the production used in $\pi_{k-j-1} \to \pi_{k-j}$ is not from $2L$, what can it be? Because of the arrangement of $*$'s, it cannot be from $1R$ or from $2R$. That leaves only $1L$, in which case it must be that $\pi_{k'-j-1}' \to \pi_{k'-j}'$ by $1R$. Thus,

$$\tau \equiv \alpha_2[\rho_{i_0}\rho_{i_1} \cdots \rho_{i_j}{:}\, e_{i_j}{*} \cdots e_{i_1}{*}e_{i_0}{*}]\beta_2$$

and

$$\tau \equiv \alpha_2'[\tau_{i_0}\tau_{i_0} \cdots \tau_{i_j}{:}\, e_{i_j}{*} \cdots e_{i_1}{*}e_{i_0}{*}]\beta_2$$

This implies that $\rho_{i_0}\rho_{i_1} \cdots \rho_{i_j} \equiv \tau_{i_0}\tau_{i_1} \cdots \tau_{i_j}$, thus giving a solution for P. ∎

For fixed i, the *intersection problem for type i languages* is the problem of deciding of an arbitrary pair of type i languages L_1 and L_2, whether or not $L_1 \cap L_2 = \varnothing$.

Theorem 2.12 The intersection problem for type 2 languages is unsolvable.

Exercise 2.15 Prove the theorem. [Hint: Refer to the proof of Theorem 2.11 and define two grammars G_L and G_R to be the two halves, so to speak, of

G_P so that (the language of G_L) $\cap$ (the language of G_R) $= \varnothing$ if and only if G_P is not ambiguous.]

The study of these types of semi-Thue systems and the machines that accept their languages is what is usually meant by "the study of formal grammars and languages." In this section, we have only considered those questions which could be answered directly from the development of the earlier sections. The interested reader can consult Ginsburg (1966), Gross and Lentin (1970), or Hopcroft and Ullman (1969a).

The reader intrigued by the Tag systems mentioned in the Introduction should now be able to read the proof of the unsolvability of the halting problem for Tag systems. The proof can be found in Minsky (1967, starting on p. 267) or in Cocke and Minsky (1964).

2.6 SOME ALGEBRAIC RAMIFICATIONS OF THE THUE SYSTEM

The generalized Thue system and the unsolvability of the word problem for generalized Thue systems have led to many questions in algebra, and, indeed, topology. We investigate one briefly and indicate some others.

A *semigroup* is a set X with an associative binary operation $*$. (This definition includes the fact that for all pairs x and $y \in X$, $x * y$ is also in X.)

Examples

1. Set of $n \times n$ matrices (over some field), with matrix multiplication.
2. Set of continuous (real-valued) functions on the real numbers with composition as the operation.
3. The set of all words on a set of symbols (finite or infinite) with juxtaposition (i.e., writing next to each other) as the operation.
4. The set of integers with $+$ as the operation.
5. The set is the rotations of a square by 0°, 90°, 180°, 270° and the operation is composition of rotations: $\{{}^A\square, \square^A, \square_A, {}_A\square\}$.

Indeed, everywhere we turn there are examples of semigroups; it is almost harder to find an example of something that is not than something that is.

The particular example we wish to consider has to do with generalized Thue systems directly. For any generalized Thue system S, the relation $\leftrightarrow$ is obviously an equivalence relation. Thus, we have an equivalence relation over the set of all words on the alphabet of S and we may speak of the equivalence classes of words in S. If τ is a word on the alphabet of S, then write $[\tau]$ for the equivalence class to which it belongs. Thus, $\rho, \pi \in [\tau]$ if and only if $\rho \leftrightarrow \pi \leftrightarrow \tau$.

Lemma 2.10 If S is a generalized Thue system and X is the set of equivalence classes of words on the alphabet of S, then X with the operation of juxtaposition is a semigroup.

PROOF Three things must be shown: (1) If $[\tau]$ and $[\rho]$ are equivalence classes, then $[\tau][\rho]$ is an equivalence class. But, as yet we have no definition for $[\tau][\rho]$, so we now define it to be the equivalence class $[\tau\rho]$. (2) If $\tau' \in [\tau]$ and $\rho' \in [\rho]$, then $[\tau'][\rho']$ must be the same equivalence class as $[\tau][\rho]$. That is, in terms of elements of X, $[\tau'][\rho'] = [\tau][\rho]$. By the definition in part 1, this means we must show that $\tau\rho \leftrightarrow \tau'\rho'$. But this is true since there is a derivation $\tau\rho \leftrightarrow \tau'\rho \leftrightarrow \tau'\rho'$ using the derivations $\tau \leftrightarrow \tau'$ and $\rho \leftrightarrow \rho'$. (3) Since the operation of a semigroup is associative, we must prove that $([\rho][\tau])[\pi] = [\rho]([\tau][\pi])$. But this fact follows from the first two. ∎

A *semigroup presentation* is a set of symbols, the *generators*, and a set of ordered pairs of words on those symbols, the *defining relations*. If S is a semigroup presentation, let G_S be the semigroup formed from S as in the lemma. We say *G_S is presented by S*. (If S is a generalized Thue system, then we say that G_S is *finitely presented*.) As nothing about finiteness was used in proving the lemma, we can conclude that every semigroup presentation presents a semigroup. (That is, the definitions do what they say they do!) The converse to the lemma is also true. However, as it is harder to prove, we just state it.

Theorem For every semigroup G, there is a semigroup presentation S such that G is isomorphic to G_S.

With this in mind, we may think of any semigroup as a set of equivalence classes of words on its generators, and so the following definition makes sense: The *word problem for semigroups* is the problem of determining of any two words τ and π on the generators of a presentation of the semigroup, whether or not $\tau \leftrightarrow \pi$; or equivalently, whether or not $[\tau] = [\pi]$. From Theorem 2.6, we have immediately the following theorem.

Theorem 2.13 The word problem for semigroups is unsolvable.

A *group* is a semigroup with the additional properties that (a) there is a special element e in the set X such that for all $x \in X$, $x * e = x$, and (b) for all $x \in X$, there exists a $y \in X$ such that $x * y = e$. Examples of groups are:

1. Examples 4 and 5 of semigroups.
2. The set of all one-to-one functions (from reals to reals) with the operation of composition.

3. The set of nonsingular $n \times n$ matrices (over some field) with matrix multiplication.

4. For each n, the group presented by the generators $\{s, t\}$ and the relations $s^n = e$, $t^2 = e$, $tst = s^{-1}$.

As in the case for semigroups, it is also true that a generalized Thue system is a finite presentation of a group. There is also a theorem saying that for every group there is a group presentation describing it.† The word problem for groups can thus be defined as the problem of determining, for any two words on the generators of a presentation of the group G, whether or not these words represent the same group element. That the word problem for finitely presented groups is unsolvable has been shown by both Novikov (1958) and Boone (1959).‡ However, Magnus (1932) has shown that the word problem for groups presented by generalized Thue systems with just one production is solvable. There are various group properties, such as being cyclic, being finite, being nontrivial, being free, being simple, being solvable, etc. A very natural decision problem thus arises: Can we tell of an arbitrary group presentation whether or not the group presented by it has property P? For a certain class of properties (which includes all those mentioned), Rabin (1958)§, using the unsolvability of the word problem for groups, has shown this decision problem to be unsolvable. In Magnus, *et al.* (1966) groups are studied from the point of view of group presentations.

We mention one other path where the generalized Thue system leads. In topology, groups are associated with topological objects. Thus, decision problems about topological objects can be formulated and studied in terms of the groups associated with them. One of the early results of this nature is by Markov (1958), showing that the following decision problem (roughly stated) is undecidable: To decide of any given pair of n-dimensional ($n \geq 4$) differentiable manifolds whether or not they are homeomorphic. More recent results of this nature can be found in Boone *et al.* (1968), which includes clear introductory remarks and an extensive bibliography.

† These theorems are fairly difficult. The reader who knows some group theory or wants to learn some can consult Rotman (1965), pp. 235–241, Magnus *et al.* (1966, chapters 1 and 2), or Kurosh (1956, Vol. 1, pp. 124–129).

‡ Chapter 12 of Rotman (1965) is devoted to a proof of the unsolvability of the word problem for groups.

§ The reader with some knowledge of group theory is encouraged to read this very well written paper.

Chapter III

ENUMERATIONS AND GÖDEL NUMBERING

3.1 ENUMERATIONS AND COUNTABLE SETS

Most of the time in this book we deal with sets whose cardinality is less than or equal to $\aleph_0$; that is, they are finite or in one-to-one correspondence with the natural numbers, $N = \{0, 1, 2, \ldots\}$. Such a set is said to be *countable*. When we say that a set S is in one-to-one correspondence with N, we mean that there is a function $f: N \to S$ which is one-to-one and onto. When there is such a function, we also say it *enumerates the set* S. We write s_0 for $f(0)$, s_1 for $f(1)$, etc. The set may be a set of anything, for example, Tm's. Exercise 1.18 was to show that the set of nonsimilar Tm's is countably infinite. The reader is probably familiar with the standard proof for showing that the set of rationals is countably infinite. Both these examples will be considered in detail in the next few pages.

Suppose we have a finite alphabet Σ and take the set S to be the set of all words on Σ. Since Σ is finite, it can be ordered in any way—just as the alphabet we use for English is. The set of all words of any specific length is also finite, so it too can be ordered arbitrarily. But, we prefer a tidier method which, since the symbols are ordered, is similar to, but not exactly like, the way that we order words in a dictionary. For each k, let S_k be the set of all words of length k. Using the order of the given alphabet, order S_k lexicographically. For example, if $k = 3$ and the alphabet is that for English, the first element is *aaa*, the second is *aab*, ..., the 27th is *aba*, etc. Then, to order all S, line up the elements of Σ, followed by the elements of S_2, etc. In ordering S, we have also shown it to be in one-to-one correspondence with N. We will refer to this type of ordering by saying that "the elements of S are ordered by length and for each length by lexicographic order."

Exercise 3.1 Given $\Sigma = \{a, b\}$ and S the set of all words on Σ, give the first nine or so members of the ordering of S by length and lexicographic ordering.

Exercise 3.2 How does this method differ from the ordering in a real dictionary? Try to order all the words on some finite alphabet in *exactly* the same way that a real dictionary does. What happens?

We know what is meant by an *ordered pair* of natural numbers (n_0, n_1). That is, for $n_0, n_1, m_0, m_1 \in N$, $(n_0, n_1) = (m_0, m_1)$ if and only if $m_i = n_i$, $i = 0, 1$, and, in particular $(n_0, n_1) \neq (n_1, n_0)$ unless $n_0 = n_1$. This idea can be extended in the obvious way to (*ordered*) *k-tuples* $(n_0, n_1, \ldots, n_{k-1})$. The n_i in the k-tuple are the *terms* (or *components*) of the k-tuple.

Exercise 3.3 Is it true that if the set of all k-tuples on N, as $k = 1, 2, \ldots$, is countably infinite, then the set of all words on a countably infinite set of symbols is countably infinite?

Theorem 3.1 The set of all k-tuples on N, as $k = 1, 2, \ldots$, is countably infinite.

PROOF (1) Show that for each positive integer k, the set S_k of all k-tuples of N is countably infinite: By induction on k. For $k = 1$, $S_1 = N$, so obviously true. For $k + 1$, we assume that S_k is countable, i.e., in one-to-one correspondence with N, so the correspondence gives the enumeration $s_{k,0}, s_{k,1}, \ldots$. Any element $s_{k,i}$ is a k-tuple, so any pair $(s_{k,i}, j)$, for some $j \in N$, is a $(k + 1)$-tuple. Thus, $S_{k+1} = \{(s_{k,i}, j) \mid i, j \in N\}$. We may consider the elements of S_{k+1} in the following array:

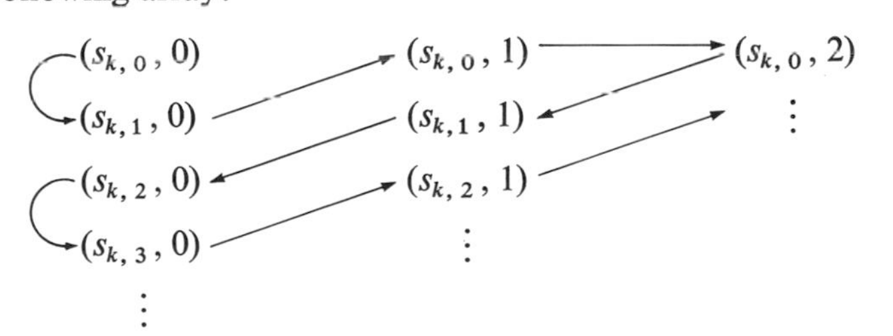

Associating the pair $(s_{k,0}, 0)$ to the natural number 0, $(s_{k,1}, 0)$ to 1, and so forth following the arrows, we pick up all pairs in S_{k+1} and for each pair there is a unique natural number. This process provides a one-to-one correspondence between N and S_{k+1}. The procedure of setting out a rectangular array and following along in the order indicated is a very common procedure; we call it a *zigzag method.*†

† Although a very standard procedure, it does not seem to have a very standard name. The method was first used by Cantor. The word "diagonal" is sometimes associated with this procedure. But since "diagonal" is used in connection with another procedure, it is better not to use it here.

(2) Show that $\bigcup_{k \in N - \{0\}} S_k$ is countable: We show this explicitly (though the reader should be able to do it as an exercise), again using a zigzag method:

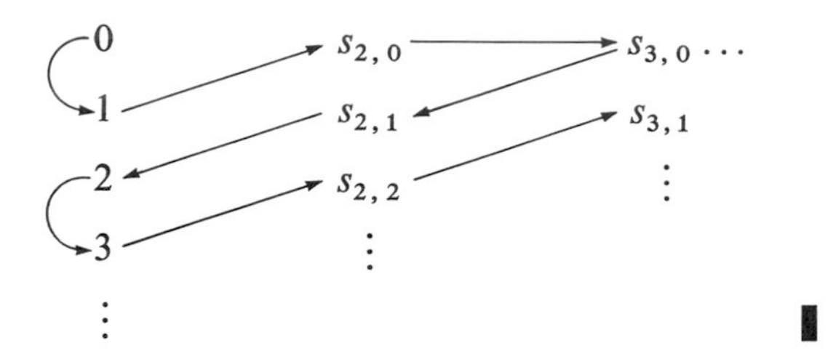

∎

Exercise 3.4 (a) Such elaborate methods are not needed to show that the finite union of countable sets is countable. Prove this. (b) Is the countable union of countable sets countable?

Exercise 3.5 Assume we have a countably infinite set of symbols $a_0, a_1, \ldots$. In the enumeration of S_k in the proof of Theorem 3.1, we can tell to what natural number each k-tuple corresponds. (a) What happens if we try to order each S_k lexicographically? (b) What happens if we try to order all of S in the way a dictionary would?

In case the reader was unable to do Exercise 1.18, it is proved as the next theorem. In fact, two proofs will be given because the procedures of both proofs are very useful.

Theorem 3.2 There are $\aleph_0$ equivalence classes of nonsimilar Turing machines.

FIRST PROOF In general, if a set is partitioned into equivalence classes $E_1, E_2, \ldots$ and if $x_1 \in E_1$, $x_2 \in E_2, \ldots$ (just one x_i from each E_i), then we call the set $\{x_i | x_i \in E_i\}$ a set of representatives. Naturally, any set of representatives is in one-to-one correspondence with any other set of representatives. Usually, when one wants to find the cardinality of a set of equivalence classes, one works from a set of representatives. Since "similarity" as defined for Tm's is an equivalence relation, in order to show that the set of equivalence classes is countably infinite, we will show that a certain set of representatives is. Of course, if two Tm's are similar, they have the same number of quintuples, so the following way of looking at sets of representatives makes sense. Let P_1 be any set of representatives of nonsimilar Tm's with one quintuple, P_2 any set of representatives of similar Tm's with two quintuples, etc. We need only show (1) that for each i, P_i is finite and then (2) that

$$\bigcup_{i \in N - \{0\}} P_i$$

is in one-to-one correspondence with N. It is easy to see that (2) follows from (1) because if all P_i are finite, say $P_i = \{T_{i,1}, T_{i,2}, \ldots, T_{i,n_i}\}$, we can string out their elements: $T_{1,1}, T_{1,2}, \ldots, T_{1,n_1}, T_{2,1}, \ldots$. So it remains to be shown that each P_i is finite. For each i, there are only finitely many pairs n, m such that $i = nm$. For each pair, let $S_{n,m}$ be the set of all words of length $5nm$ on the alphabet

$$\{q_H, q_0, q_1, \ldots, q_{m-1}\} \cup \{*, 1, a_2, \ldots, a_{n-1}\} \cup \{R, L\}$$

$S_{n,m}$ is obviously finite and so then is the set

$$U_i = \bigcup_{\substack{n,m \text{ such} \\ \text{that } nm=i}} S_{n,m}$$

We claim that U_i has a set of representatives of the i-quintuple Tm's as a subset. If $R_{n,m}$ is the set of all Tm's with $\Sigma = \{*, 1, \ldots, a_{n-1}\}$ and $Z = \{q_H, q_0, \ldots, q_{m-1}\}$, then $R_{n,m} \subseteq S_{n,m}$. But, as we noticed on section 1.3, for every Tm with n tape symbols and $m + 1$ internal states (the extra 1 for q_H), there is a Tm similar to it belonging to $R_{n,m}$. Thus, there is a set of representatives P_i which is a subset of

$$\bigcup_{\substack{n,m \text{ such} \\ \text{that } nm=i}} R_{n,m}$$

which is itself a subset of U_i.

Second Proof A set of representatives of the equivalence classes of non-similar Tm's may be thought of as a subset of the set of all words on a certain countably infinite set of symbols:

$$\{*, 1, a_2, \ldots\} \cup \{q_H, q_0, q_1, \ldots\} \cup \{R, L\}$$

Once we have an enumeration $\tau_0, \tau_1, \tau_2, \ldots$ of all the words on that set of symbols, we can get an enumeration $P = \{T_0, T_1, \ldots\}$ of a set of representatives of the equivalence classes by looking through the list of τ_i's as follows: Is τ_0 a Tm? If so, set $T_0 = \tau_0$; if not, look at τ_1. Is τ_1 a Tm? Etc. Thus, for the smallest i such that τ_i is a Tm, set $T_0 = \tau_i$. Then, T_1 is the τ_j for the next $j > i$ such that τ_j is a Tm and τ_j is not similar to T_0, etc. ▌

Exercise 3.6 Could you build a Tm to do this enumeration? That is, $T(\bar{n}) \equiv \rho_{T_n}$.

3.2 GÖDEL NUMBERINGS

Suppose we have a countable set of symbols (finite or infinite): $\Sigma = \{a_0, a_1, a_2, \ldots\}$. Let S be the set of all words on Σ; let S^* be the set of all k-tuples, for all $k > 0$, of elements of S.

Example 1 Let the set of symbols be the English alphabet $\Sigma = \{a, b, c, \ldots, x, y, z\}$. Then, S properly contains all words in English. S^* can be thought of as containing all sentences in English.

Example 2 For some Tm T, let Σ be the union of the set of all tape symbols of T and the set of q_i's which are the internal states of T. Then S includes the set of all instantaneous descriptions of T. S^* contains the set of all computations of T.

Example 3 Let $\Sigma = \{0, 1, 2, 3, 4, 5, 6, 7, 8, 9\}$. Then S includes among its members the usual decimal representation of each natural number. S^* is the set of all k tuples of S and thus includes the decimal representation of the set proved to be countable in Theorem 3.1.

As we know, for any finite Σ, we can show that the corresponding S^* is countable, by using a zigzag argument. Then we have a one-to-one onto map $f: N \to S^*$, so we may also consider f^{-1}, which is one-to-one from S^* onto N. Are these effectively computable functions, by whatever standards of effectiveness we may choose? Yes, for, given a member of S^*, we can find an $n \in N$ to correspond to it by starting the zigzag process and continuing until we hit that member of S^*, which we are assured of doing eventually. On the other hand, given an $n \in N$, we can find the member of S^* corresponding to n by doing the first n steps of the zigzag procedure. Thus, the procedures for computing both f^{-1} and f seem quite effective.

Exercise 3.7 Does a similar argument work equally well for the case when Σ is a countably infinite set of symbols?

Suppose $\Sigma = \{a_0, a_1, a_2, \ldots\}$ is any countable set of symbols and S and S^* are the associated set of words and k-tuples of words. In general, any effective one-to-one map from S to N (or from S^* to N) is called a *Gödel numbering of S* (or of S^*). We will consider a particular Gödel numbering which is very much like the one used by Gödel. It is based on the following theorem, known as the fundamental theorem of arithmetic (also as the unique factorization theorem).

Theorem Any positive integer $n > 1$ can be factored in a unique way as

$$n = p_{i_1}^{e_1} p_{i_2}^{e_2} \cdots p_{i_m}^{e_m}$$

where $p_{i_1} < p_{i_2} < \cdots < p_{i_m}$ are primes and each $e_j > 0$.

[For a proof, consult Herstein (1964, p. 19) or any elementary modern algebra or number theory book.] Define the mapping $\# : S^* \to N$ inductively from the members of Σ as follows: $\# a_i = p_i$, where p_i is the ith prime taking $p_0 = 2$. If $\tau \in S$, then τ is of the form $a_{i_0} a_{i_1} \cdots a_{i_k}$, so define

$$\# \tau = 2^{\# a_{i_0}} 3^{\# a_{i_1}} \cdots p_k^{\# a_{i_k}}$$

Notice that this mapping $\#$ is not onto N. Notice also that $a_i \in \Sigma$ and therefore has the Gödel number p_i and, as a word of length 1, $a_i \in S$ and so has Gödel number 2^{p_i}. Finally, $(a_i) \in S^*$ since there is a 1-tuple in S^* whose only term is a_i.

Exercise 3.8 Suppose $\Sigma = \{a, b, c\}$, and use the Gödel numbering $\#$ just defined. (a) Is $5 \in \#(\Sigma)$, $5 \in \#(S)$, $5 \in \#(S^*)$? Is $2^5 \in \#(S)$? $2^5 \in \#(S^*)$? $2^{2^5} \in \#(S^*)$? (b) To what elements, if any, of $S \cup S^*$ do the following numbers correspond under $\#$?

$$2^5 3^5 5^5; \qquad 2^{11} 3^7; \qquad 11^{11},$$

$$2^{2^5 3^2} 3^{2^2} 5^{2^3 3^3} 7^{2^5 3^5 5^5}; \qquad 2^5 3^{2^5 3^2}$$

We emphasize that the importance of Gödel numbering is that it is a procedure by which, given an element of S (or S^*, or $S \cup S^*$), one can effectively compute a number $n \in N$ and, conversely, given any $n \in N$, one can effectively decide if it corresponds to some element of S (or S^* or $S \cup S^*$), and if so, to which one. If we can Gödel-number sets of words that are of interest for some reason, then we can put our investigations of these sets into a framework of sets of natural numbers and number-theoretic functions. In particular, decision problems about sets of words become decision problems about sets of natural numbers.

Exercise 3.9 Let Σ be a countable set of symbols. Suppose that we wished to consider not only all words on Σ, that is S, and the set of all k-tuples of words S^*, but also all k-tuples of k-tuples. Call that set S^{**}. Naturally, we can then think of all k-tuples of elements of S^{**}, etc. That is, inductively, $x \in S^{(i+1)*}$ if x is a k-tuple whose terms are elements of S^{i*}. Let the union of all such sets be called $G(\Sigma)$. Give at least one Gödel numbering of $G(\Sigma)$.

Exercise 3.10 Refer to example 2 above and give a Gödel numbering of all computations for a fixed Tm T.

Exercise 3.11 Let $\mathscr{T}$ be the set of all Tm's operating on a half infinite tape with alphabet $\{*, 1\}$. (a) Give a Gödel numbering of $\mathscr{T}$ making sure that the set of Gödel numbers of $\mathscr{T}$ is decidable in N. (For the decidable part, the proof can be in terms of an effective procedure instead of the actual design of a Tm.) (b) Give a Gödel numbering of the set $\{\rho_T \mid T \in \mathscr{T}\}$.

Exercise 3.12 Suppose Σ is finite and we want a Gödel numbering of $S \cup S^*$ on Σ. Outline how to design a Tm T such that $T(x) \equiv \bar{n}$ for $x \in S \cup S^*$ and such that if $T(x) \equiv T(y)$ then $x = y$. (Warning: Complicated.)

Chapter IV

RECURSIVE FUNCTIONS

4.1 PRELIMINARIES

Our concern is still with the idea of effective procedure. The Tm certainly provides a reasonable formulation of this idea and one which appeals to our programming point of view. But it is rather foreign to our calculus, analysis point of view. In mathematics, we are accustomed to thinking of functions as being represented by something that looks like $f(x) = \cdots$. So we leave Tm's for awhile and try to develop a formulation of effectively computable function from this more familiar point of view.

We will be working with number-theoretic functions. Let N denote the set of natural numbers and N^k abbreviate $\underbrace{N \times N \times \cdots \times N}_{k}$. An $(n+1)$-*ary number-theoretic function* f is a rule which associates to each $(n+1)$-tuple $(x_0, x_1, \ldots, x_n)$ of some subset of N^{n+1} a unique element of N which is denoted by $f(x_0, x_1, \ldots, x_n)$. As we will be studying only number-theoretic functions, usually we will just say "function," except for some examples using real numbers. The *graph of an* $(n+1)$-*ary function* f is a set $\{((x_0, x_1, \ldots, x_n), f(x_0, x_1, \ldots, x_n))\} \subseteq N^{n+2}$, such that (a) $(x_0, x_1, \ldots, x_n)$ varies over all the domain of f, and (b) no two members of the set having the same entries for $x_0, x_1, \ldots, x_n$ have different entries for $f(x_0, x_1, \ldots, x_n)$. The definition of equal functions given in chapter I says that two functions are equal if their graphs are equal. Many people do not distinguish between a function and its graph and we too will sometimes ignore the distinction. But there are important distinctions, especially for the purposes of our studies in this book. As indicated in the Introduction, a rule is sometimes something concrete that we may be able to handle, whereas a graph, at least of the more common functions, is not, since it is infinite.

Suppose that we are considering some $(n+1)$-ary function f. The domain of f can be divided up into chunks so that we take fixed values for $x_1, x_2, \ldots, x_n$

and let x_0 vary over N. That is, by fixing particular values $n_1, n_2, \ldots, n_n$ (for $x_1, x_2, \ldots, x_n$), we obtain the unary function $f(x_0, n_1, n_2, \ldots, n_n)$. Thus, instead of working with the $(n+1)$-ary function $f(x_0, x_1, x_2, \ldots, x_n)$, we will work with infinitely many unary functions $f(x_0, n_1, n_2, \ldots, n_n)$, one for each n-tuple $(n_1, n_2, \ldots, n_n)$. When we do this, we are taking $x_1, x_2, \ldots, x_n$ as *parameters*. For a geometric interpretation of this idea, we consider for a moment real-valued functions of two or more real variables. We can think of the graph of such a function as representing a surface in 3- (or more) space. If we take all the variables but one as parameters, we are, geometrically, considering the curve formed by the intersection of a plane and the surface. Since it is easy to draw, we illustrate this with a quarter of a hemisphere. Let $f(x, y) = (3^2 - x^2 - y^2)^{1/2}$ for $x, y \geq 0$ and take x to be a parameter and for the drawing give it the value 2 (see figure 4.1).

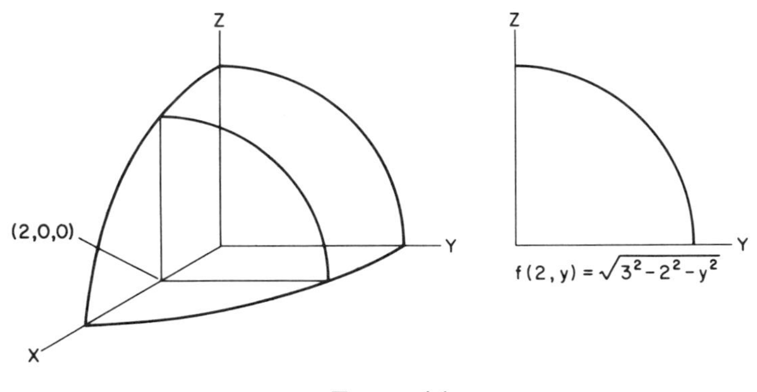

FIGURE 4.1

In the scheme we are about to develop, we will assume that $x_1, x_2, \ldots, x_n$ $(n > 0)$ are parameters. That is, before any computation begins, there are actual natural numbers supplied for $x_1, x_2, \ldots, x_n$ which will be fixed throughout the computation. For each set of fixed values for $x_1, x_2, \ldots, x_n$, x_0 will take the values of the natural numbers. For example, restricting our consideration once more to N, suppose $f(x_0, x_1, x_2) = x_0 + (x_1 \cdot x_2)$, where x_1 and x_2 are parameters. Then,

$$\begin{aligned} f(0, x_1, x_2) &= 0 + (x_1 \cdot x_2) \\ f(1, x_1, x_2) &= 1 + (x_1 \cdot x_2) = f(0, x_1, x_2) + 1 \\ &\vdots \\ f(n+1, x_1, x_2) &= (n+1) + (x_1 \cdot x_2) = ((x_1 \cdot x_2) + n) + 1 \\ &= f(n, x_1, x_2) + 1 \end{aligned}$$

Notice that, in this particular example, once $f(0, x_1, x_2)$ is computed, we merely need to add one to obtain each successive function value. We would surely agree that the function of the example is effectively computable.

Recall what is meant by the composition of functions. If f and g are functions of one variable, then we can define a new function h by $h = g \circ f$, where $g \circ f(x)$ is defined to be $g(f(x))$. For example, if $f(x) = x - 5$ and $g(x) = x^2$, then $h(x) = g \circ f(x) = g(f(x)) = g(x - 5) = (x - 5)^2$. We can extend this idea of composition to functions of several variables. Suppose we have $f_0, f_1, \ldots, f_m$, which are functions of $n + 1$ variables, and g, which is a function of $m + 1$ variables. Then we can define a new function h of $n + 1$ variables by

$$h(x_0, x_1, \ldots, x_n) = g(f_0(x_0, \ldots, x_n), f_1(x_0, \ldots, x_n), \ldots, f_m(x_0, \ldots, x_n))$$

If we agree that $g, f_0, \ldots, f_m$ are effectively computable, we must certainly agree that h is.

We will start with a very small number of functions and use the simple ideas introduced above in putting them together to form more and more complicated functions.

4.2 DEFINITION OF PRIMITIVE RECURSIVE FUNCTIONS

We define three simple kinds of functions as our initial functions:

1. *The zero function*: $Z(x_0) = 0$.
2. *The successor function*: $S(x_0) = x_0 + 1$.
3. *The projection (pick-out) functions*, for all i and n, $0 \le i \le n$:

$$U_i^{n+1}(x_0, x_1, \ldots, x_i, \ldots, x_n) = x_i.$$

An $(n + 1)$-ary function f *is defined primitive recursively* if one of the following is true: (a) it is the zero function, the successor function, or a projection function; or (b) it is defined by the *recursion scheme*

$$f(0, x_1, \ldots, x_n) = g(x_1, \ldots, x_n)$$
$$f(n + 1, x_1, \ldots, x_n) = h(f(n, x_1, \ldots, x_n), n, x_1, \ldots, x_n)$$

where g and h are primitive recursively defined; or (c) it is defined by *composition*

$$f(x_0, x_1, \ldots, x_n) = h(g_0(x_0, \ldots, x_n), \ldots, g_k(x_0, \ldots, x_n))$$

where $h, g_0, \ldots, g_k$ are all primitive-recursively defined functions. An $(n + 1)$-ary function f *is a primitive recursive* (pr) *function* if it is equal to a function that is defined primitive recursively.†

Now that we have the definition of a primitive recursive function (prf) before us, the natural questions are: What functions are primitive recursive (pr)? Are there any functions which are not pr?

4.3 SOME SIMPLE PRIMITIVE RECURSIVE FUNCTIONS

1. *Addition*: Suppose we wish to add any natural number n to a specific number x_1 (thus taking x_1 as a parameter). We can think of this as adding 1 to x a total of n times: $((((x_1 + 0) + 1) + 1) \cdots + 1)$. In this way, once we solve the $(x_1 + 0)$ problem (so to speak!) we can use that answer to get the $(x_1 + 1)$ answer and that to get the $((x_1 + 1) + 1)$ answer, etc. So we have something like

$$+(0, x_1) = x_1$$
$$+(n + 1, x_1) = (x_1 + n) + 1$$

This is not quite a prf according to the definition, but the following is:

$$+(0, x_1) = U_2{}^2(0, x_1)$$
$$+(n + 1, x_1) = S\big(U_1{}^3(+(n, x_1), n, x_1)\big)$$

It is a prf because $+$ is in the recursion scheme form, where the right side of the first equation is a prf and the right side of the second equation is a prf because it is defined by composing prf's.

2. *Multiplication*: Suppose we want to multiply any natural number $(n + 1)$ times a fixed natural number x_1. Since $(n + 1) \cdot x = n \cdot x_1 + x_1$, and since addition is pr, if we knew how to compute $n \cdot x_1$, we would be done. That is,

$x_1 \cdot 0 = 0,\ x_1 \cdot 1 = x_1,\ x_1 \cdot 2 = x_1 \cdot 1 + x_1, \ldots, x_1 \cdot (n + 1) = x_1 \cdot n + x_1, \ldots$

So,

$$\times\,(0, x_1) = Z(x_1)$$
$$\times\,(n + 1, x_1) = h\big(\times\,(n, x_1), n, x_1\big)$$

† In chapter I, a Tm was defined to be *given* by a certain finite set of words on a certain set of symbols. We wish to have available a similar method of *giving* (or presenting) a primitive recursive function; that is, we wish to be able to use a certain set of symbols (just as ink marks) which represent primitive recursive functions. We also wish to be able to maintain the distinction between a function and its graph when it seems necessary. It is for these reasons that the definition of a primitive recursive function given here appears in a slightly different form from the definition generally given in the literature. The primitive recursively defined functions are to be thought of as the symbols which represent the primitive recursive functions. This idea is in accordance with the discussion of symbols and meaning in the Introduction.

where the definition of h is that

$$h(x_0, x_1, x_2) = +\big(U_1^{\,3}(x_0, x_1, x_2), U_3^{\,3}(x_0, x_1, x_2)\big)$$

and it is therefore a prf.

3. *Factorial*—**Exercise 4.1:** Show that the function $f(n) = n!$ is pr.
4. *Zero test*:

$$\overline{\text{sg}}(0) = S(0)$$
$$\overline{\text{sg}}(n+1) = Z\big(U_2^{\,2}(\overline{\text{sg}}(n), n)\big)$$

Check to see that $\overline{\text{sg}}(0) = 1$ and for $n > 0$, $\overline{\text{sg}}(n) = 0$.

5. *Nonzero test*—**Exercise 4.2:** Show that the following function is pr:

$$\text{sg}(n) = \begin{cases} 0 & n = 0 \\ 1 & n \neq 0 \end{cases}$$

6. *Exponentiation*: If we want to take some number x to a power which is a natural number, we notice that

$$x^0 = \begin{cases} 1 & \text{if} \quad x > 0 \\ 0 & \text{if} \quad x = 0 \end{cases}$$

and $x^{n+1} = x^n \cdot x$. Formally, we have

$$\text{Exp}(0, x_1) = \text{sg}(x_1)$$
$$\text{Exp}(n+1, x_1) = h\big(\text{Exp}(n, x_1), n, x_1\big)$$

where h is defined by

$$h(x_0, x_1, x_2) = \times\big(U_1^{\,3}(x_0, x_1, x_2), U_3^{\,3}(x_0, x_1, x_2)\big)$$

and is certainly a prf.

7. *Parity*:

$$\text{P}(0) = 0$$
$$\text{P}(n+1) = \overline{\text{sg}}\big(U_1^{\,2}(\text{P}(n), n)\big)$$

Check to see that

$$\text{P}(n) = \begin{cases} 0 & \text{for} \quad n \quad \text{even} \\ 1 & \text{for} \quad n \quad \text{odd} \end{cases}$$

8. *Half*: We want

$$[n/2] = \begin{cases} n/2 & \text{if} \quad n \quad \text{is even} \\ (n-1)/2 & \text{if} \quad n \quad \text{is odd} \end{cases}$$

We see that this function is pr as follows:

$$[0/2] = 0$$
$$[(n+1)/2] = h([n/2], n)$$

where, by definition, $h(x_0, x_1) = +(U_1^2(x_0, x_1), P(U_2^2(x_0, x_1)))$.

9. *The predecessor function*—**Exercise 4.3:** Show that the following function is pr:

$$\mathrm{pd}(x_0) = \begin{cases} 0 & \text{if } x_0 = 0 \\ x_0 - 1 & \text{if } x_0 > 0 \end{cases}$$

10. *Limited subtraction*: We will show that the following function is pr:

$$\dot{-}(x_0, x_1) = \begin{cases} x_1 - x_0 & \text{if } x_1 \geq x_0 \\ 0 & \text{if } x_1 < x_0 \end{cases}$$

To do so, we look at the first $x_1 + 2$ values:

$$\begin{aligned} \dot{-}(0, x_1) &= x_1 \\ \dot{-}(1, x_1) &= x_1 - 1 \\ &\vdots \\ \dot{-}(x_1 - 1, x_1) &= 1 \\ \dot{-}(x_1, x_1) &= 0 \\ \dot{-}(x_1 + 1, x_1) &= 0 \\ &\vdots \end{aligned}$$

As x_0 increases, each function value is one less than the preceding until the value of zero is reached. Thus, using pd, we have

$$\dot{-}(0, x_1) = U_2^2(0, x_1)$$
$$\dot{-}(n+1, x_1) = \mathrm{pd}(U_1^3(\dot{-}(n, x_1), n, x_1))$$

11. "*Other limited subtraction*"—**Exercise 4.4:** Show that the following function is pr:

$$\dot{\div}(x_0, x_1) = \begin{cases} x_0 - x_1 & \text{if } x_0 \geq x_1 \\ 0 & \text{if } x_0 < x_1 \end{cases}$$

Henceforth, we will write $(x_1 \dot{-} x_0)$ for $\dot{-}(x_0, x_1)$ and $(x_0 \dot{-} x_1)$ for $\dot{\div}(x_0, x_1)$.

12. "*Equals function*"—**Exercise 4.5:** Show that the following function is pr:

$$=(x_0, x_1) = \begin{cases} 1 & \text{if } x_0 = x_1 \\ 0 & \text{if } x_0 \neq x_1 \end{cases}$$

The examples of this section have been quite simple and presented in full detail. Of course, there are many more examples and the student may consult various books, especially Kleene (1952, chapter IX) or Péter (1967). Actually, once we saw basically how a function worked, e.g., $x(n+1) = xn + x$, it was pretty certain that the function could be shown to be pr. In the next section, we will investigate some functions which are not so easily seen to be pr.

Exercise 4.6 Prove that if an $(n+1)$-ary function f' is pr, then it is defined for all $(n+1)$-tuples of N^{n+1}. (Hint: A careful proof is by induction on the form of the primitive-recursively defined function f, which must exist and equal f' since f' is given as being pr. Thus, one must show that (1) each of the three basic functions is defined everywhere, i.e., Z and S are defined for each element of N and U^{n+1} is defined for each element of N^{n+1}, (2) assuming that g is defined for each element of N^{n+1} and h for each element of N^{n+2}, then any function defined using the recursion scheme is defined for all elements in N^{n+1}, and (3) assuming that h is defined for each element of N^{k+1} and $g_0, \ldots, g_k$ are all defined for each element of N^{n+1}, then any function defined from them by composition is defined for each element of N^{n+1}. Warning: If you understand the induction, the exercise is trivial.)

Exercise 4.7 In chapter I, once Turing machines were defined, we adopted the notion of "similar" meaning "same" except for the use of different symbols and then settled on a standardized notation for Turing-machine quintuples. From there, we proved that there are just $\aleph_0$ many nonsimilar Tm's. Try to do the analogous steps for primitive-recursively defined functions. (An exercise similar to this is worked in the last section of this chapter.)

4.4 SOME USEFUL PRIMITIVE RECURSIVE FUNCTIONS

Notice that in all the prf's considered so far, we think of finding the function value $f(n+1)$ in terms of the value $f(n)$. Consequently, we always start at 0, find $f(0)$, go to 1, find $f(1)$, and so on. We may think of the argument values as stages in our knowledge of the function. At n, we are finding out a little more about f, knowing, so far, only $f(0), \ldots, f(n-1)$. In this view, we are purposely confusing the function with its graph. That is, we may have written before us some very elaborate primitive-recursively defined function, but have no idea off-hand as to what the graph looks like. Hence, we compute function values, one step at a time, $f(0), f(1), \ldots$, learning, step by step, a little more. Much of the discussion in this section will be presented from this point of view.

Let S_{n+1} be any subset of N^{n+1}; we say that S_{n+1} is an $(n+1)$-*ary number-theoretic relation*. If f is an $(n+1)$-ary function such that

$$f(x_0, x_1, \ldots, x_n) = \begin{cases} 1 & \text{if} \quad (x_0, x_1, \ldots, x_n) \in S_{n+1} \\ 0 & \text{if} \quad (x_0, x_1, \ldots, x_n) \notin S_{n+1} \end{cases}$$

then f is defined to be *the characteristic function of* S_{n+1}. If, in addition, f is a prf, then S_{n+1} is defined to be *a primitive recursive relation* (*predicate*).

Exercise 4.8 Is the set $\{(x, x) | \text{all } x \text{ on } N\}$ a primitive recursive relation?

Exercise 4.9 Prove that for any fixed x_1, the set $\{(x_0, x_1) | \text{all } x_0 > x_1\}$ is a primitive recursive relation.

Exercise 4.10 Prove that the set $\{x | x = 2n \text{ for some } n \in N\}$ is a primitive recursive relation.

The functions discussed in the following will be shown roughly to be primitive recursive. That is, we will not go into all the detail of using the projection functions and the recursion scheme in its precise form. The manner we will adopt is that used most of the time by most of the people working on prf's.

1. *Functions defined according to conditions:* A familiar example of such a function is any step function, such as

$$f(x) = \begin{cases} -1 & \text{for} \quad x < 0 \\ 0 & \text{for} \quad x = 0 \\ 1 & \text{for} \quad x > 0 \end{cases}$$

Notice that the conditions cover all possibilities since for any number x, $x < 0$, $x = 0$, or $x > 0$, and that for a given x, only one condition is true. That is, the conditions are exhaustive and mutually exclusive. Another example is

$$f(n) = \begin{cases} 2n & \text{if} \quad n \quad \text{is even} \\ 3n & \text{if} \quad n \quad \text{is odd} \end{cases}$$

From exercise 4.10, we have the prf k_1 such that

$$k_1(n) = \begin{cases} 1 & \text{if} \quad n \text{ is even} \\ 0 & \text{if} \quad n \text{ is odd} \end{cases}$$

and so $k_2(n) = \overline{\text{sg}}\,(k_1(n))$ is also pr. Using k_1 and k_2, we can define this function f equally well by

$$f(n) = \begin{cases} 2n & \text{if} \quad k_1(n) = 1 \\ 3n & \text{if} \quad k_2(n) = 1 \end{cases}$$

Note further that for all n, $k_1(n) + k_2(n) = 1$.

Suppose that we have prf's $g, h_1, \ldots, h_m, k_1, \ldots, k_m$, where for all (x_0, x_1) $\sum_{i=1}^{m} k_i(x_0, x_1) = 1$, and we define f by

$$f(0, x_1) = g(x_1)$$
$$f(n+1, x_1) = \begin{cases} h_1(f(n, x_1), n, x_1) & \text{if} \quad k_1(n+1, x_1) = 1 \\ \vdots & \\ h_m(f(n, x_1), n, x_1) & \text{if} \quad k_m(n+1, x_1) = 1 \end{cases}$$

We see that f is pr because we can define f' by

$$f'(0, x_1) = g(x_1)$$
$$f'(n+1, x_1) = \sum_{i=1}^{m} h_i(f'(n, x_1), n, x_1) \cdot k_i(n+1, x_1)$$

which is pr, and check to see that $f = f'$.

Exercise 4.11 Prove that the second example given above is a prf.

Exercise 4.12 Show that the following function is pr:

$$f(n, x_1, x_2) = \begin{cases} (x_1 \dot{-} n) \cdot x_2 & \text{if} \quad x_1 > n \\ (x_1 \cdot n) + x_2 & \text{if} \quad x_1 \leq n \end{cases}$$

Exercise 4.13 Suppose that the functions $g_0, g_1, \ldots, g_k, h$ are primitive recursive functions. Show that f is primitive recursive, where f is defined by

$$\begin{aligned} f(0, x) &= g_0(x) \\ f(1, x) &= g_1(x) \\ &\vdots \\ f(k, x) &= g_k(x) \\ f(n+1, x) &= h(f(n, x), n, x) \qquad \text{for} \quad n \geq k \end{aligned}$$

2. *Functions with specific values at a finite number of places:* Suppose, for example, we want a function such that $f(2) = 7$, $f(4) = 5$, $f(6) = 13$, and $f(9) = 10$, and that it does not matter what the value of f is elsewhere—suppose constant elsewhere. Is f a prf? In order to see that it is, we will define nine other functions, which are obviously pr, by looking at the values of f through $f(10)$:

$f(0) = k$	$g_0(0) = k$	and	$g_0(n+1) = g_1(n)$
$f(1) = k$	$g_1(0) = 7$	and	$g_1(n+1) = g_2(n)$
$f(2) = 7$	$g_2(0) = k$	and	$g_2(n+1) = g_3(n)$
$f(3) = k$	$g_3(0) = 5$	and	$g_3(n+1) = g_4(n)$
$f(4) = 5$	$g_4(0) = k$	and	$g_4(n+1) = g_5(n)$
$f(5) = k$	$g_5(0) = 13$	and	$g_5(n+1) = g_6(n)$
$f(6) = 13$	$g_6(0) = k$	and	$g_6(n+1) = g_7(n)$
$f(7) = k$	$g_7(0) = k$	and	$g_7(n+1) = g_8(n)$
$f(8) = k$	$g_8(0) = 10$	and	$g_8(n+1) = k$
$f(9) = 10$			
$f(10) = k$			

Since all g_i are pr, we see that the function f' defined as follows is a prf:

$$f'(0) = k, \qquad f'(n+1) = g_0(n)$$

Exercise 4.14 Check to see that the function f' defined just above is equal to the f originally given.

Exercise 4.15 Think about how to generalize this procedure to functions defined to have specific values at any finite number of places.

3. *Functions defined by bounded minimization:* We introduce a new idea which corresponds roughly to the idea of "according to what we know so far" about some function. Write $\mu_{j \le n}[h(j) = 0]$ to mean "the least j less than or equal to n such that $h(j) = 0$." Given a function h, this idea can be used to define a new function f as follows:

$$f(n) = \begin{cases} \mu_{j \le n}[h(j) = 0] & \text{if there is such a } j \\ 0 & \text{if there is no such } j \end{cases}$$

Suppose we have a specific n and function h and are computing the function values $h(0)$, $h(1)$, ... Then, the first time, *if ever*, there is a $j \le n$ such that $h(j) = 0$, we make note of that j. It is that value that is denoted by $\mu_{j \le n}[h(j) = 0]$. If there is such a j, then $f(n) = j$; but if after computing $h(0), \ldots, h(n)$ there was no j such that $h(j) = 0$, then $f(n) = 0$. [Notice that, given h and n, the maximum number of steps needed to find $f(n)$ is known before the computation of $f(n)$ begins].

For a familiar example, we take a function from the nonnegative reals to the reals. Consider the parabola given by $h(x) = (8/9)(x - 3)^2 - 2$. We can define a new function $f\colon N \to$ positive reals in terms of h as follows:

$$f(n) = \mu_{x \le n}[h(x) = 0], \qquad \text{otherwise} \quad 0$$

In this case, we are looking for the smallest positive real root of h. Notice that h has roots at $x = 3/2$ and $x = 9/2$. At $n = 0$, we look only at $x = 0$; but $h(0) > 0$, so $f(0) = 0$. At $n = 1$, we look at all $x \le 1$ and find that h is positive in that interval, so $f(1) = 0$. But at $n = 2$, we look at all $x \le 2$ and find that at $x = 3/2$, $h(3/2) = 0$, and therefore $f(2) = 3/2$. For $n = 3$ and 4, since $3/2 < 4$ and $h(3/2) = 0$, $f(n) = 3/2$. In fact, this is true for all $n \ge 2$ because, although there is another root greater than 3/2, once the first one is encountered, we stick with it because it is the smallest.

Exercise 4.16 Draw the graphs of the functions h and f of the example just given.

We turn our attention once more to number-theoretic functions. If h is pr, then we will show that the following function is pr:

$$f(n, x_1) = \begin{cases} \mu_{y \le n}[h(y, x_1) = 0] & \text{if } y \text{ exists} \\ 0 & \text{if no such } y \text{ exists} \end{cases}$$

Exercise 4.17 Suppose $h(y, x_1) = (x_1 \dot{-} y)$ and f is defined as above. Compute $f(0, x_1)$ and $f(n, x_1)$ for $n < x_1$, and compute $f(x_1, x_1)$ and $f(n, x_1)$ for $n > x_1$.

In order to show that f is pr, we will see what f really says by considering how the value of f is computed at successive values. Thus, at $n + 1$, we assume $f(n, x_1)$ already computed. For $f(n + 1, x_1)$ to have the value $(n + 1)$, three things must be true: (1) $h(n + 1, x_1) = 0$, (2) $f(n, x_1) = 0$, and (3) $h(0, x_1) \neq 0$. That (1) must be true is obvious. But the fact that $h(n + 1, x_1) = 0$ is not sufficient to conclude that $f(n + 1, x_1) = n + 1$, because there may have been an earlier value $y < n + 1$ such that $h(y, x_1) = 0$. Had that been the case, then $f(y, x_1) = y = f(y + 1, x_1) = \cdots = f(n, x_1)$. So we must check to see if $f(n, x_1) = 0$; that is, (2) must be true. However, even that is not sufficient, for $f(n, x_1)$ could be zero for two reasons: because no value has been found so far *or* because $h(0, x_1) = 0$ so 0 *is* the value. And so (3) must be true also. In all other cases, $f(n + 1, x_1)$ will be the same as $f(n, x_1)$, i.e., either no value has been found, so $f(n, x_1) = 0$, or one has been found and $f(n, x_1)$ is that number. Define the function f' by

$$f'(n + 1, x_1) = \begin{cases} n + 1 & \text{if } h(n + 1, x_1) = 0 \text{ and } f'(n, x_1) = 0 \\ & \text{and } h(0, x_1) \neq 0 \\ f'(n, x) & \text{otherwise, i.e., if} \\ & h(n + 1, x_1) \neq 0 \text{ or } f'(n, x_1) \neq 0 \\ & \text{or } h(0, x_1) = 0 \end{cases}$$

Exercise 4.18 Finish the proof that f is a prf.

Exercise 4.19 Prove that f is primitive recursive if g and h are primitive recursive and f is defined by

$$f(n, x) = \begin{cases} \mu_{j \le g(n,x)}[h(j, x) = 0] & \text{if there is such a } j \\ 0 & \text{otherwise} \end{cases}$$

Exercise 4.20 Let the function $[\sqrt{x}]$ be defined to have as its value the largest natural number y such that $y^2 \le x$. Show that $[\sqrt{x}]$ is a primitive recursive function.

Exercise 4.21 Let $E(x)$ be defined by saying $E(x) = y$ if there is a z such that $z^2 + y = x$ and $(z + 1)^2 > x$. We call E "the excess over a square function." Show that E is a primitive recursive function by verifying that $E(x) = x \dot{-} [x^{1/2}]^2$.

4. *Pairing and projection functions:* Very often, it is very convenient to be able to keep track of a pair (or more) of numbers by means of one number and to be able to retrieve the original pair from that one number. An additional useful property is that the function which pairs the two into one be onto N; naturally, it must be one-to-one.† Also naturally, for our purposes, such functions should be primitive recursive. There are various possibilities. The pairing and projection functions presented here, traditionally called J, K, L, are defined so that $K(J(x, y)) = x$ and $L(J(x, y)) = y$.‡ (They are sometimes referred to as Cantor's pairing and projection functions.)

Let

$$J(x, y) = \tfrac{1}{2}\big((x + y)^2 + 3x + y\big)$$

Notice that $((x + y)^2 + 3x + y)$ is always even and therefore $J(x, y) = [\frac{1}{2}((x + y)^2 + 3x + y)]$ and so is a primitive recursive function. We will show that J is one-to-one from N^2 onto N. We first show that J is onto N. Recall Gauss's formula $\sum_{i=1}^{k} i = k(k + 1)/2$. For *any* z, let r be the largest number such that $\sum_{i=1}^{r} i \leq z$ and let x be the difference; that is, $z = x + \sum_{i=1}^{r} i$. Of course, $x \leq r$, otherwise r would not be the largest. Let y be the difference, i.e., $x + y = r$. Thus, $z = x + \sum_{i=1}^{x+y} i = x + \frac{1}{2}(x + y + 1)(x + y)$. But then,

$$\begin{aligned}\tfrac{1}{2}\big(2x + (x + y + 1)(x + y)\big) &= \tfrac{1}{2}(2x + x^2 + 2xy + y^2 + y + x)\\ &= \big[\tfrac{1}{2}\big((x + y)^2 + 3x + y\big)\big] = J(x, y)\end{aligned}$$

We now show that J is one-to-one from N^2. Suppose $z = J(x, y)$. Then, from $2z = (x + y)^2 + 3x + y$, it follows that $8z + 1 = (2x + 2y + 1)^2 + 8x$ and so $(2x + 2y + 1)^2 \leq 8z + 1$. On the other hand, one can check that $8z + 1 < (2x + 2y + 3)^2$. Therefore, $[(8z + 1)^{1/2}]$ is either $2x + 2y + 1$ or $2x + 2y + 2$. Since $[\frac{1}{2}(2x + 2y + 2)] = [\frac{1}{2}(2x + 2y + 3)] = x + y + 1$, we have

$$\tfrac{1}{2}([(8z + 1)^{1/2}] + 1) = x + y + 1$$

or

$$x + y = \big[\tfrac{1}{2}[(8z+1)^{1/2}] + 1\big) - 1\big] \qquad (*)$$

† At the end of paragraph 5, "Some functions concerning primes," the reader can easily give a pairing function with the appropriate corresponding projection functions; however, that pairing function will not be onto N.

‡ Other possible pairing functions are, for example: (i) $C(n, m) = 2^n + 2^{n+m+1}$ [Minsky (1967, p. 180)]. This function has the pleasant property that when represented in binary there are just n zeros to the left of the first 1 and m zeros between the first 1 and the second 1. (ii) $V(n, m) = 2^n(2m + 1) - 1$ [Grzegorczk (1961, p. 34)]. The presentation here follows Davis (1958, p. 43).

Since $3x + y = ((x + y)^2 + 3x + y) - (x + y)^2$, we also have

$$3x + y = 2z - ([\tfrac{1}{2}[(8z + 1)^{1/2}] + 1] - 1)^2 \qquad (**)$$

Suppose x, y and x', y' are numbers such that $J(x, y) = J(x', y')$. Then both pairs satisfy (*) and (**), so

$$x + y = x' + y'$$

and

$$3x + y = 3x' + y'$$

Subtracting one from the other, we find that $x = x'$, so $y = y'$. We will use (*) and (**) also to show that there are functions K and L such that $K(J(x, y)) = x$ and $L(J(x, y)) = y$. Denote the right side of (*) by $f(z)$ and the right side of (**) by $g(z)$. Then, solving these equations simultaneously, we have

$$x = \tfrac{1}{2}(g(z) - f(z))$$
$$y = f(z) - \tfrac{1}{2}(g(z) - f(z))$$

Noticing that $g(z)$ is just $2z - (f(z))^2$ and that for any m, $m^2 + m$ is even, we have

$$x = [\tfrac{1}{2}(2z - (f(z))^2 - f(z))] \qquad (***)$$

and

$$y = [\tfrac{1}{2}((f(z))^2 + 3f(z) - 2z)] \qquad (****)$$

Let $K(z)$ be the right side of (***) and $L(z)$ be the right side of (****). Both are obviously primitive recursive.

Exercise 4.22 Extend this idea to define functions J^m and K_i, $1 \le i \le m$, where J^m is from N^m onto N and each K_i is unary and satisfies $K_i(J^m(x_1, \ldots, x_m)) = x_i$.

Exercise 4.23 Show that an n-ary function defined to have specific values at a finite number of places is primitive recursive.

5. *Some functions concerning primes*: In the section on Gödel numbering in Chapter III, it was essential, for the Gödel numbering #, to be able to refer to "the ith prime" and "the exponent of the ith prime in the prime factorization of n." We would like to see that the functions $p(n)$, the nth prime (we have usually just written p_n) and $P(n, x_1)$, the exponent of the nth prime in the prime decomposition of x_1, are pr. We consider them one at a time.

For $p(n)$: The definition of a *prime number* p is that there are no natural numbers other than 1 and p that divide it. A natural number m *divides* a natural number n, if there is a natural number k such that $mk = n$.Obviously, $k \leq n$. So, given any number n, to determine if it is a prime we need, at most, to try dividing n by all the numbers m, $1 < m < n$. Once we can determine for some number n that it is the ith prime, we can find the $(i + 1)$th prime from the ith prime, since for every prime p, there exists at least one prime between p and $p! + 1$, and hence the $(i + 1)$th prime is the smallest p', such that $p < p' < p! + 1$. We see that these three concepts can be formalized and shown to be pr as follows.

(a) Informally,

$$d(m, n) = \begin{cases} 1 & \text{if } m \text{ divides } n \\ 0 & \text{if } m \text{ does not divide } n \end{cases}$$

Let $=(x_0, x_1)$ be the function of exercise 4.5 such that

$$=(x_0, x_1) = \begin{cases} 1 & \text{if } x_0 = x_1 \\ 0 & \text{if } x_0 \neq x_1 \end{cases}$$

We will see that d is a prf by verifying that

$$d(m, n) = \sum_{1 < k < n} (=(km, n))$$

Define

$$\begin{aligned} F(0, m, n) &= 0 \\ F(1, m, n) &= 0 \\ F(2, m, n) &= F(1, m, n) + (=(2m, n)) \\ &\vdots \\ F(j + 1, m, n) &= F(j, m, n) + (=((j + 1)m, n)) \end{aligned}$$

Notice that if $m > 1$ once $j \geq n$, $(=(jm, n)) = 0$, so F is a constant from then on (or before). Finally,

$$d(m, n) = F(U_2{}^2(m, n) \dot{-} 1, U_1{}^2(m, n), U_2{}^2(m, n))$$

(b) Informally,

$$\Pr(n) = \begin{cases} 1 & \text{if } n \text{ is a prime} \\ 0 & \text{if } n \text{ is not a prime} \end{cases}$$

Exercise 4.24 Prove that Pr is a prf; i.e. $\{n \mid n \text{ is a prime}\}$ is a primitive recursive relation.

(c) Finally, we use the recursion scheme and functions already proved to be pr to show that the function whose value at n is the nth prime is pr:

$$p(0) = 2$$
$$p(n + 1) = \mu m_{p(n) < m < p(n)! + 1}[1 \dot{-} \Pr(m) = 0]$$

For $P(n, x)$: For a given natural number x, $P(n, x)$ is to have as its value the exponent of the nth prime in the prime decomposition of x. Obviously, this exponent will be less than x itself. To say that n is not a factor of x is the same as saying that the exponent of n in the prime decomposition of x is 0. Further, $p(n)^{P(n, x)}$ divides x, but $p(n)^{P(n, x)+1}$ does not divide x. So,

$$P(n, x) = \big(\mu m_{0 < m < x}[d((p(n)^m, x) \cdot d((p(n))^{m+1}, x) = 0]\big) \cdot d((p(n)), x)$$

6. *The history (course-of-values) function:* In the prf's so far, the value of the function at $n + 1$ depended directly on the value at n, but not directly on the value at any $m < n$. Consider the following example:

$$f(0) = 5$$
$$f(n + 1) = (f(n) \cdot f(n \dot{-} 2)) + f(n \dot{-} 5)$$

Since $+$, $\cdot$ and $\dot{-}$ are pr and since f is defined using only these three functions at its own function values at earlier stages, one *feels* that f should be a prf. That it is a prf will be a consequence of the following discussion.

Let h be any $(k + 1)$-ary number-theoretic function; there is no requirement that it be pr. Define the function $\tilde{h}$ by

$$\tilde{h}(0, x_1, \ldots, x_k) = 1$$
$$\tilde{h}(n + 1, x_1, \ldots, x_k) = \prod_{0 \le i \le n} p_i^{(h(i, x_1, \ldots, x_k)) + 1}$$

For a function h, the function $\tilde{h}$ so defined is known as the *history* (or *course-of-values*) *function* of h for the obvious reason that at any n, the history of h at all stages less than n can be read in the function value of $\tilde{h}$ at n. Since h might have the value zero at m and since ${p_m}^0 = 1$, we add the one to the exponent so all the primes up to the nth will occur as factors. Thus, for any m, $0 \le m < n$,

$$h(m, x_1, \ldots, x_k) = P(m, \tilde{h}(n, x_1, \ldots, x_k)) \dot{-} 1$$

If g is a prf and f is defined by

$$f(n, x_1, \ldots, x_k) = g(\tilde{f}(n, x_1, \ldots, x_k), n, x_1, \ldots, x_k)$$

where $\tilde{f}$ is the history function of f, then f is a prf. The proof, in three parts, follows. For the sake of convenience in reading and writing, we take k, the number of parameters, equal to zero.

(a). For the same prf g above, define the function f' as follows:

$$f'(0) = 1$$
$$f'(n+1) = \prod_{0 \le i \le n} p_i^{g(f'(i))+1}$$

In order to verify that f' is pr, we look at some values:

$$\begin{aligned}
f'(0) &= 1 \\
f'(1) &= p_0^{g(f'(0))+1} = 2^{g(1)+1} \\
f'(2) &= p_0^{g(f'(0))+1} \cdot p^{g(f'(1))+1} = 2^{g(1)+1} \cdot 3^{g(2^{g(1)}+1)+1} \\
&\vdots \\
f'(n+1) &= \underbrace{p_0^{g(f'(0))+1} \cdot p_1^{g(f'(1))+1} \cdot \ldots}_{f'(n)} \cdot p_n^{g(f'(n))+1} \\
&= f'(n) \cdot p_n^{g(f'(n))+1}
\end{aligned}$$

So,

$$f'(0) = 1$$
$$f'(n+1) = f'(n) \cdot p_n{}^{g(f'(n))+1}$$

Since g, multiplication, powers of primes, and successor are all primitive recursive, f' is also primitive recursive.

(b) Define f'' as follows:

$$f''(0) = g(1), \qquad f''(n+1) = P(n+1, f'(n+2)) \dot{-} 1$$

Since g, P, f', and $\dot{-}$ are pr, f'' is pr.

(c) To finish the proof that f is pr, we show that $\tilde{f} = f'$ and $f = f''$. The proof is by induction on n, the variable, and both equalities will be shown simultaneously. At $n = 0$: $f'(0) = 1 = \tilde{f}(0)$ and $f''(0) = g(1)$ and $f(0) = g(\tilde{f}(0)) = g(1)$. We assume that the equalities hold at $n = k$, that is $f'(k) = \tilde{f}(k)$ and $f(k) = f''(k)$, and show that they hold at $n = k + 1$:

$$f'(k+1) = p_0^{g(f'(0))+1} \cdots p_k^{g(f'(k))+1}$$

from which, by the induction hypothesis,

$$f'(k+1) = p_0^{g(\tilde{f}(0))+1} \cdots p_k^{g(\tilde{f}(k))+1}$$

but, by the definition of $\tilde{f}$,

$$f'(k+1) = p_0^{f(0)+1} \cdots p_k^{f(k)+1}$$

so, by the definition of $\tilde{f}$,

$$f'(k+1) = \tilde{f}(k+1)$$

Since

$$f''(k+1) = P(k+1, f'(k+2)) \dot{-} 1$$

by the definition of f',

$$\begin{aligned} f''(k+1) &= P(k+1, p_0^{g(f'(0))+1} \cdots p_k^{g(f'(k))+1} p_{k+1}^{g(f'(k+1))+1}) \dot{-} 1 \\ &= g(f'(k+1)) \end{aligned}$$

Since

$$f(k+1) = g(\tilde{f}(k+1))$$

and we have already shown that $f'(k+1) = \tilde{f}(k+1)$, we have that $f(k+1) = f''(k+1)$

Exercise 4.25 Let f be defined by

$$f(0) = 5, \qquad f(n+1) = f(n) \cdot f(n \dot{-} 2) + f(n \dot{-} 5)$$

(a) Compute the values of f at $n = 0, 1, 2, 3, 4, 5, 6, 7$. (b) Prove that f is pr.

Exercise 4.26 For any function f, is $\tilde{f}$, its history function, monotonic increasing?

Exercise 4.27 Show that the function f defined as follows is pr:

$$f(0) = 5, \qquad f(n+1) = (n+1)^{f(n)} + n \cdot f(n \dot{-} 3)$$

7. *Simultaneously defined functions*: The situation now is that two functions f_0 and f_1 are being defined so that at stage $n+1$ the values of both have been computed at stage n, and that that information is needed for the computation of each at $n+1$. However, neither depends on the other at $n+1$. Perhaps if one thinks of the New York and Toronto Stock Exchanges with each minute a unit, then, since they open at the same time, what happens at each at the seventh minute depends on what happened at both during the first six minutes. The example may be a little farfetched, but it is just such functions which are used to describe the behavior of Turing machines; we will see an example in chapter V. These functions are also used in developing very powerful methods to prove important theorems in recursive function theory; one of these will be given in section 8.5.

If g_0, h_0, g_1, and h_1 are prf's, and f_0 and f_1 are defined by

$$\begin{aligned} f_0(0, x) &= g_0(x), \qquad & f_0(n+1, x) &= h_0(f_0(n, x), f_1(n, x), n) \\ f_1(0, x) &= g_1(x), \qquad & f_1(n+1, x) &= h_1(f_1(n, x), f_0(n, x), n) \end{aligned}$$

then f_0 and f_1 are pr. We shall see that this is true by defining a new function

$$j(n, x) = \begin{cases} f_0(n/2, x) & \text{if } n \text{ is even} \\ f_1(\frac{1}{2}(n-1), x) & \text{if } n \text{ is odd} \end{cases}$$

and showing that j is pr. In order to do this, define the function j' as follows:

$$j'(0, x) = g_0(x), \qquad j'(1, x) = g_1(x)$$

$$j'(n+2, x) = \begin{cases} h_0(j'(n, x), j'(n+1, x), n/2) & \text{if } n \text{ is even} \\ h_1(j'(n, x), j'(n \dot{-} 1, x), (n \dot{-} 1)/2) & \text{if } n \text{ is odd} \end{cases}$$

Exercise 4.28 To finish the proof that f_0 and f_1 are pr, (a) prove that j' is pr, (b) prove that $j = j'$, and (c) prove that $f_0(n, x) = j'(2n, x)$ and $f_1(n, x) = j'(2n+1, x)$.

Exercise 4.29 For any $m > 2$, given that the functions $g_0, g_1, \ldots, g_{m-1}$ and $h_0, h_1, \ldots, h_{m-1}$ are pr, show how to define $f_0, f_1, \ldots, f_{m-1}$ simultaneously so that they are pr, using the case $m = 2$ as a guide.

8. *Iterative functions*: When we have a function h whose domain includes its image, it makes sense to write $h^n(x)$, meaning

$$\underbrace{h(h \cdots h}_{n}(x))$$

For an example, consider $h(x) = 3x^2 - 5x$. Then,

$$h(1) = -2 \quad h^2(1) = h(-2) = 22, \quad h^3(1) = h(h^2(1)) = h(22) = 1342, \quad \text{etc.}$$

With this example in mind, suppose that g and h are any prf's and

$$f(0, x) = g(x), \qquad f(n+1, x) = h^{n+1}(f(0, x))$$

then, as we shall see, f is pr. To arrive at this conclusion, we compute f for a few values:

$$\begin{aligned} f(0, x) &= g(x) \\ f(1, x) &= h(f(0, x)) \\ f(2, x) &= h^2(f(0, x)) = h(h(f(0, x))) = h(f(1, x)) \\ f(3, x) &= h^3(f(0, x)) = h(h(h(f(0, x)))) = h(f(2, x)) \\ &\vdots \end{aligned}$$

If we define the prf f' by

$$f'(0, x) = g(x), \qquad f'(n+1, x) = h(U_1{}^3(f'(n, x), n, x))$$

it is easy to check that $f = f'$ and is therefore pr.

Exercise 4.30 Suppose g and h are pr and $f(0, x) = g(x)$, $f(n+1, x) = f(n, h(x))$. Show that f is pr.

Consider the similar but more complicated case of the function f given by

$$f(0, x_1, x_2) = g(x_1, x_2)$$
$$f(n+1, x_1, x_2) = f(n, h_1(x_1, x_2), h_2(x_1, x_2))$$

where g, h_1, and h_2 are assumed to be pr. We will see that f is also pr, by relying heavily on the discussion just above and the functions J, K, and L. By computing f at the first few steps,

$$\begin{aligned} f(0, x_1, x_2) &= g(x_1, x_2) \\ f(1, x_1, x_2) &= f(0, h_1(x_1, x_2), h_2(x_1, x_2)) = g(h_1(x_1, x_2), h_2(x_1, x_2)) \\ f(2, x_1, x_2) &= f(1, h_1(x_1, x_2), h_2(x_1, x_2)) \\ &= f(0, h_1(h_1(x_1, x_2), h_2(x_1, x_2)), h_2(h_1(x_1, x_2), h_2(x_1, x_2))) \\ &= g(h_1(h_1(x_1, x_2), h_2(x_1, x_2)), h_2(h_1(x_1, x_2), h_2(x_1, x_2))) \end{aligned}$$

we see that it has the kind of repetitiveness of exercise 4.30, but that it is working on two variables, not just one. Since we know from exercise 4.30 that if k_1 and k_2 are prf's, then k_0 given by

$$k_0(0, y) = k_1(y), \qquad k_0(n+1, y) = k_0(n, k_2(y))$$

is pr, perhaps we can put together all the information about x_1 and x_2 into the single integer y. But this is just what the pairing function can do by letting

$$y = J(x_1, x_2)$$

Since J, K, L and h_1, h_2, g are pr,

$$k_2(y) = J(h_1(K(y), L(y)), h_2(K(y)), L(y)))$$

is pr. Thus, the function f' defined by

$$f'(0, y) = g(K(y), L(y)), \qquad f'(n+1, y) = f'(n, k_2(y))$$

is pr.

Exercise 4.31 Finish the proof that f is pr.

Exercise 4.32 Outline how to carry out the proof of primitive recursiveness for

$$f(0, x_1, \ldots, x_n) = g(x_1, \ldots, x_n)$$
$$f(n+1, x_1, \ldots, x_n) = f(n, h_1(x_1, \ldots, x_n), \ldots, h_n(x_1, \ldots, x_n))$$

for g, h_1, $\ldots$, h_n all pr.

Exercise 4.33 Suppose that g is a primitive recursive function and that f is defined by $f(0, x) = g(x)$ and $f(n + 1, x) = f(n, f(n, x))$. Prove that f is a prf.

While few would disagree that every primitive recursive function is a total effectively computable function, the converse is open to question.

Theorem 4.1 Not all total effectively computable functions are primitive recursive.

Exercise 4.34 Prove the theorem. (Hint: Assume that all total effectively computable functions are pr and derive a contradiction. The proof is analogous to Cantor's proof of the uncountability of the reals; indeed, rather like the proof of the unsolvability of the halting problem.The method common to these proofs, and many more to come, is known as "diagonalization.")

Since the formalism of primitive recursive functions has not gotten us to our goal of representing all effectively computable functions by means of simple building blocks and simple building methods, we continue our search.

4.5 TOTAL, REGULAR, AND PARTIAL FUNCTIONS, AND UNBOUNDED MINIMIZATION

An $(n + 1)$-ary function f which is defined for all $(x_0, \ldots, x_n) \in N^{n+1}$ is called a *total function*. If there may be a $(x_0, \ldots, x_n) \in N^{n+1}$ for which f is not defined, then f is a *partial function*. If f is an $(n + 1)$-ary function such that for every n-tuple $(x_1, \ldots, x_n) \in N^n$, there exists at least one $n \in N$ such that $f(n, x_1, \ldots, x_n) = 0$, then f is said to be a *regular function*. We will illustrate these three ideas with geometric examples.

Example 1 A function which is total and regular: Let g be defined on ordered pairs of nonnegative reals with function values in the reals as follows:

$$g(x, y) = \frac{1}{(1 + x)}(y - 1)^2 - 2$$

As a surface in that part of 3-space for which $x \geq 0$, $y \geq 0$, and z has any real value, it is a trough, the bottom of which lies on the line parallel to the x axis through the point $(0, 1, -2)$. If we slice the surface with planes perpendicular to that line, we get parabolas which are opening wider and wider as slices are taken at larger and larger x. Since each parabola has its vertex at $(x, 1, -2)$ and opens upward, there is at least one point where the parabola pierces the positive part of the x, y plane, i.e., where $g(x, y) = 0$ for $x, y \geq 0$. First, we notice that g is certainly defined for all pairs (x, y) of positive real numbers. Second, we see that, taking x as a parameter and y as a variable, for every $x \geq 0$, there is a $y \geq 0$ such that $g(x, y) = 0$, so g is a regular function.

Example 2 A function which is total but not regular: Again we have a parabolic trough (this time it does not get wider) which is tilted so that for $x > 5$, the trough lies above the x, y plane. Let

$$h(x, y) = \tfrac{5}{4}(y - 1)^2 - 5 + x$$

Certainly h is total, but it is not regular, since the function values for $x > 5$ are all positive.

Before giving an example of a partial function, we ask the reader to recall the bounded minimization operation given in section 4.4. We shall consider what it means to remove the restriction of being "bounded." We say that f *is defined using the minimization operation on* k if

$$f(x_0, \ldots, x_n) = \begin{cases} \mu_y[k(y, x_0, \ldots, x_n) = 0] & \text{if there is such a } y \\ \text{undefined} & \text{otherwise} \end{cases}$$

where k is a total function.† Thus, $f(x_0, \ldots, x_n)$ has as its function value the smallest y for which $k(y, x_0, \ldots, x_n) = 0$ if there is such a y; otherwise, $f(x_0, \ldots, x_n)$ is not defined. In order to understand better how this works, we investigate f in case the function k is the function g of the first example.

Example 3 Another total function: Let

$$f(x) = \begin{cases} \mu_y[g(x, y) = 0] & \text{if there is such a } y \\ \text{undefined} & \text{otherwise} \end{cases}$$

for the regular function $g(x, y) = (4/(1 + x))(y - 1)^2 - 2$. We will take $x \in N$ and y in the nonnegative reals. Thus, $f\colon N \to$ nonnegative reals. (Notice that x is a variable for the function f but a parameter for g.) Computing a few values, we find that: At $x = 0$, $g(0, y) = 4(y - 1)^2 - 2$, so $g(0, y) = 0$ for $y = (2 \pm \sqrt{2})/2$. As both are positive and μ says to take the smaller, $f(0) = (2 - \sqrt{2})/2$. At $x = 1$, $g(1, y) = (4/2)(y - 1)^2 - 2$, so $g(1, y) = 0$ for $y = 0$ and $y = 2$, so $f(1) = 0$. For all $x > 1$, there is one $y < 0$ and one $y > 0$ such that $g(x, y) = 0$, therefore f has that $y > 0$ as its function value. It is important to notice that since g is a regular function, f is a total function.

† Notice that $x_1, \ldots, x_n$ are parameters for both f and k, x_0 is a variable for f but a parameter for k, and, of course, y is a variable for k.

Example 4 A partial function: Now let

$$f(x) = \begin{cases} \mu_y[h(x, y) = 0] & \text{if there is such a } y \\ \text{undefined} & \text{otherwise} \end{cases}$$

for the total, but not regular, function $h(x, y) = \frac{5}{4}(y - 1)^2 - 5 + x$. As above, we take $x \in N$, while y is in the nonnegative reals. At $x = 0$, $h(0, y) = 0$ for $y = -1$ and $y = 3$, so $f(0) = 3$. At $x = 1$, $h(1, y) = 0$ for $y = (5 \pm 4\sqrt{5})/5$, so $f(1) = (5 + 4\sqrt{5})/5$, the only positive root. For $x = 2$ and 3, there is one positive root and one negative root, so the function value of f is the positive root. At $x = 4$, $h(4, y) = 0$ for $y = (5 \pm \sqrt{10})/5$, so $f(4) = (5 - \sqrt{10})/5$. At $x = 5$, the roots of $h(5, y)$ coincide at $y = 1$, so $f(5) = 1$. For $x > 5$, there are no real roots, so f is not defined. Thus, f is a partial function.

We emphasize that there is a great difference between functions defined by bounded minimization and those defined by unbounded minimization. In the case of bounded minimization, there is but one graph needed in order to determine *all* function values of f and there is a preestablished bound on how far out on the graph we will have to look; to compute the value of f at each n, we need consider at most the first n values of that graph. In the case of unbounded minimization, for each step n, there is a different graph (associated with n) to which we must refer in order to determine the function value $f(n)$. If f is defined using minimization on a function known to be regular, at each n we need only look at a finite piece of the associated graph since the regularity of the function assures us that sooner or later the function value will be zero, but there is no preestablished bound on how far we must look as in the bounded minimization case. On the other hand, if f is defined using minimization on a function only known to be total, at some steps n we may be looking forever.

Exercise 4.35 Give an example of a binary function which is regular but not total.

4.6 RECURSIVE AND PARTIAL RECURSIVE FUNCTIONS

A function g is defined partial-recursively if (i) it is the zero function, the successor function, or a projection function; (ii) it is defined by composing functions which are defined partial-recursively; (iii) it is defined by the recursion scheme from functions which are defined partial-recursively; or (iv) it is defined using the minimization operation on a function which is defined partial-recursively (and is total).

An $(n + 1)$-ary number-theoretic function f is a *partial recursive function* if it equals a function which is defined partial-recursively.

A function g is defined recursively if (i) it is the zero function, the successor function, or a projection function; (ii) it is defined by composing functions which are defined recursively; (iii) it is defined by the recursion scheme from functions which are defined recursively; or (iv) it is defined using the minimization operation on a function which is defined recursively and which is regular (and is total).

An $(n + 1)$-ary number-theoretic function f is a *recursive function* if it equals a function which is defined recursively.

Exercise 4.36 Prove that if f is recursive, then f is total.

Exercise 4.37 Prove that the set of partial recursive functions includes the set of recursive functions.

Exercise 4.38 Give an example of a partial recursive function which is not total.

Exercise 4.39 (a) Let $f(n, x)$ be defined to be the smallest y such that $n^y \geq x$. Show that f is a partial recursive function. For $x > 0$, are $f(0, x)$ and $f(1, x)$ defined? (b) Let $g(n, x)$ be defined to be the smallest y such that $(n + 2)^y \geq x$. Is g a recursive function? Is g a primitive recursive function?

Exercise 4.40 Suppose h is a partial recursive function which is not total and g is a recursive function and $f = h \circ g$; can f be total?

Exercise 4.41 Let k be some fixed natural number. If g is a binary function such that for all $x \in N$, there exists a $y \in N$ such that $g(y, x) = k$, then we say that g is a *k-regular function*. There is a similar definition for any n-ary function for $n \geq 2$. (a) Show that for any k-regular recursive function g, there is a recursive function h such that $g(y, x) = k$ if and only if $h(y, x) = 0$. Then, of course, h is regular. (b) Show that if f is defined by use of minimization on a k-regular recursive function, then f is a recursive function. (c) If f is defined by using minimization in the form $\mu_y[g(y, x_0, \ldots, x_n) = k]$, where g is a total partial recursive function, show that f is partial recursive.

Exercise 4.42 Show that for every partial recursive function there exist $\aleph_0$-many partial-recursively defined functions equal to it.

Exercise 4.43 As in exercise 4.7, devise a standard set of symbols and conventions for representing the class of all functions defined partial-recursively. Call the class so represented $\mathscr{K}$. (This exercise will be done in the next section.)

Exercise 4.44 Show that the cardinality of the class $\mathscr{K}$ is $\aleph_0$.

Exercise 4.45 Give a Gödel numbering of $\mathscr{K}$.

Exercise 4.46 Show that the class of all number-theoretic functions has cardinality strictly greater than $\aleph_0$.

We end the section by noting that we have the following string of proper inclusions of classes of functions:

$$\text{primitive recursive} \underset{(1)}{\subset} \text{recursive} \underset{(2)}{\subset} \text{partial recursive} \underset{(3)}{\subset} \text{number theoretic}$$

That the inclusions hold follows from the definitions. That the inclusions are proper can be seen as follows: (1) by Theorem 4.1; (2) the existence of partial recursive functions which are not total; (3) different cardinalities—exercises 4.44 and 4.46.

4.7 REMARKS ON EXERCISES 4.7 AND 4.43

In those two exercises, the reader was asked to establish a notational convention for representing the prf's and the partial-recursively defined functions, respectively. That is, we think of them simply as sets of ink marks on the paper. This discussion will be in terms of partial recursively defined functions. (Refer to the footnote to the definition of primitive recursive functions, p. 70.) We need a set of symbols, an effective way of distinguishing words we want as opposed to those we do not want, and an effective way of distinguishing finite sequences of those words that are to represent members of $\mathscr{K}$ from those that do not. What follows is a rough description of just one of the ways in which this can be done.

The symbols are:

$\mathbf{S}, \mathbf{Z}, \mathbf{U}_1{}^1, \mathbf{U}_1{}^2, \mathbf{U}_2{}^2, \mathbf{U}_1{}^3, \mathbf{U}_2{}^3, \mathbf{U}_3{}^3, \ldots, \mathbf{U}_1{}^n, \ldots, \mathbf{U}_{n-1}{}^n, \mathbf{U}_n{}^n, \ldots,$
$\mathbf{0}, \mathbf{n}, \mathbf{y}, \boldsymbol{\mu}$
$(\,), [\,]\ =$

$\mathbf{x}_0, \mathbf{x}_1, \mathbf{x}_2, \ldots$ (numerical variable symbols)

$\left.\begin{array}{l} \mathbf{f}_0{}^1, \mathbf{f}_1{}^1, \mathbf{f}_2{}^1, \ldots \\ \mathbf{f}_0{}^2, \mathbf{f}_1{}^2, \ldots \\ \vdots \\ \mathbf{f}_0{}^k, \mathbf{f}_1{}^k, \ldots \\ \vdots \end{array}\right\}$ (function symbols)

The words we want are:

Terms: (1) For any i, $\mathbf{x_i}$ is a term; $\mathbf{0}$ and $\mathbf{n}$ and $\mathbf{y}$ are terms. (2) If $\tau_1, \ldots, \tau_n$ are terms, then $\mathbf{S}(\tau_1)$, $\mathbf{Z}(\tau_1)$, and $\mathbf{U_i^n}(\tau_1, \cdots, \tau_n)$ are terms (for any i and n such that $0 \leq i \leq n$). Also, $\mathbf{f_i^n}(\tau_1, \cdots, \tau_n)$ (for any i and n) are terms. (3) If $\tau_1, \ldots, \tau_n$ are terms, then $\boldsymbol{\mu}_\mathbf{y}[\mathbf{f_i^{n+1}}(\mathbf{y}, \tau_1, \ldots, \tau_n) = \mathbf{0}]$ is a term.

Equations: If τ_1 and τ_2 are terms, $\tau_1 = \tau_2$ is an equation.

The sequences we want: The set of sequences which is to repesent the members of $\mathscr{K}$ is to be a certain set of sequences of equations as just defined. For these sequences, we adopt the convention that whenever we use numerical variable symbols, if we need n of them, we take the first n in the list, and whenever we use function symbols, we also take initial segments of the appropriate list; e.g., if we need two one-place function symbols, three two-place function symbols, and one five-place function symbol, they will be $\mathbf{f_0^1}$, $\mathbf{f_1^1}$, $\mathbf{f_0^2}$, $\mathbf{f_1^2}$, $\mathbf{f_2^2}$, and $\mathbf{f_0^5}$, in order of occurrence. This convention is to hold throughout the rest of the definition.

A partial-recursively defined function is a finite sequence of equations $E_0, E_1, \ldots, E_p$, where $E_0, E_1, \ldots, E_p$ satisfy the following properties:

(I) E_0 is of the form $\mathbf{f_0^1}(\mathbf{x_0}) = \mathbf{Z}(\mathbf{x_0})$ or $\mathbf{f_0^1}(\mathbf{x_0}) = \mathbf{S}(\mathbf{x_0})$ or

$$\mathbf{f_0^n}(\mathbf{x_0}, \cdots, \mathbf{x_{n-1}}) = \mathbf{U_i^n}(\mathbf{x_0}, \cdots, \mathbf{x_{n-1}}).$$

(II) Inductively, E_i is in one of the following forms:

1. $\mathbf{f_j^1}(\mathbf{x_0}) = \mathbf{Z}(\mathbf{x_0})$ or $\mathbf{f_j^1}(\mathbf{x_0}) = \mathbf{S}(\mathbf{x_0})$ or

$$\mathbf{f_j^n}(\mathbf{x_0}, \cdots, \mathbf{x_{n-1}}) = \mathbf{U_i^n}(\mathbf{x_0}, \cdots, \mathbf{x_{n-1}}).$$

2. $\mathbf{f_j^{k+1}}(\mathbf{0}, \mathbf{x_1}, \ldots, \mathbf{x_k}) = \mathbf{f_m^k}(\mathbf{x_1}, \ldots, \mathbf{x_k})$, where E_{i+1} is

$$\mathbf{f_j^{k+1}}(\mathbf{S}(\mathbf{n}), \mathbf{x_1}, \ldots, \mathbf{x_k}) = \mathbf{f_{m'}^{k+2}}(\mathbf{f_j^{k+1}}(\mathbf{n}, \mathbf{x_1}, \ldots, \mathbf{x_k}), \mathbf{n}, \mathbf{x_1}, \ldots, \mathbf{x_k})$$

or E_i is

$$\mathbf{f_j^{k+1}}(\mathbf{S}(\mathbf{n}), \mathbf{x_1}, \ldots, \mathbf{x_k}) = \mathbf{f_{m'}^{k+2}}(\mathbf{f_j^{k+1}}(\mathbf{n}, \mathbf{x_1}, \ldots, \mathbf{x_k}), \mathbf{n}, \mathbf{x_1}, \ldots, \mathbf{x_k})$$

where E_{i-1} is $\mathbf{f_j^{k+1}}(\mathbf{0}, \mathbf{x_1}, \ldots, \mathbf{x_k}) = \mathbf{f_m^k}(\mathbf{x_1}, \ldots, \mathbf{x_k})$ and in both cases $\mathbf{f_m^k}$ occurs on the left side in some $E_{i'}$ of the sequence for $i' < i$ and $\mathbf{f_{m'}^{k+2}}$ also occurs on the left side in some $E_{j'}$ of the sequence for $j' < i$.

3. $\mathbf{f_j^{k+1}}(\mathbf{x_0}, \ldots, \mathbf{x_k}) = \mathbf{f_q^m}(\mathbf{f_{j1}^{k+1}}(\mathbf{x_0}, \ldots, \mathbf{x_k}), \ldots, \mathbf{f_{jm}^{k+1}}(\mathbf{x_0}, \ldots, \mathbf{x_k}))$, where $\mathbf{f_q^m}$, $\mathbf{f_{j1}^{k+1}}, \ldots, \mathbf{f_{jm}^{k+1}}$ occur on the left sides of equations earlier in the sequence.

4. $\mathbf{f_j^{k+1}}(\mathbf{x_0}, \ldots, \mathbf{x_k}) = \boldsymbol{\mu}_\mathbf{y}[\mathbf{f_q^{k+2}}(\mathbf{y}, \mathbf{x_0}, \ldots, \mathbf{x_k}) = \mathbf{0}]$, where $\mathbf{f_q^{k+2}}$ occurs on the left side of some equation earlier in the sequence.

Chapter V

EQUIVALENCE OF RECURSIVE AND TURING-COMPUTABLE FUNCTIONS

5.1 TURING COMPUTABILITY

An $(n + 1)$-ary number-theoretic function f is *Turing computable in general* (Tcg) if there exists a Tm T_f that, when given a tape with $x_0, x_1, \ldots, x_n$ as initial tape inscription,† (1) will halt with the value $f(x_0, x_1, \ldots, x_n)$ as the answer computed if $f(x_0, x_1, \ldots, x_n)$ is defined, or (2) will not halt if $f(x_0, x_1, \ldots, x_n)$ is not defined. If T is a Tm whose existence proves that a function f is Tcg, we will write $f \sim T$, and we say "T computes f." If an $(n + 1)$-ary number-theoretic function f is total and is Tcg, then it is *Turing computable* (Tc). So if f is Tc and $f \sim T$, then T halts for all $x_0, x_1, \ldots, x_n$.

As we know, for every Tm T, there are infinitely many Tm's T' such that T is equivalent to T'. Hence, if a function f is Tcg (or Tc), there exist infinitely many Tm's T' such that $f \sim T'$. By exercise 1.39, we know that, without loss of generality, we can restrict our attention to Tm's on $\{*, 1\}$ working on a half-infinite tape. Call the set of all such Tm's $\mathscr{T}$. We can thus amend the definition of Tcg to say, f is Tcg if there exists a Tm $T_f \in \mathscr{T}$ such that $f \sim T_f$.

Recall that the set $\mathscr{K}$ is the set of all partial-recursively defined functions in some standardized notation. Given the formalism for Tm's and the formalism for functions defined partial-recursively, it is natural to ask if the following are true:

1. For every $g \in \mathscr{K}$, there is a Tm, $T_g \in \mathscr{T}$ such that $g \sim T_g$.
2. For every $T \in \mathscr{T}$, there is a $g_T \in \mathscr{K}$ such that $g_T \sim T$.

That the answers are yes is usually taken as evidence in favor of each formalism being an adequate formalization of effective computability. Once

† Throughout this paragraph, recall that the symbols actually written on the tape naturally depend on the choice of tape notation.

these statements are seen to be true, we will be able to conclude without much difficulty: (1) A function is Tc if and only if it is recursive, and (2) a function is Tcg if and only if it is partial recursive.

That Turing computability encompasses effective computability is also substantiated by the fact that Turing computability can be shown to imply each of the machinelike developments of computability which have been proposed so far. Among these, two that are often used are Shepherdson–Sturgis (1963) machines and Wang (1957) machines. Both formulations are much more like programs than Turing machines are, especially in terms of jump operations. Each is equivalent to Tm's; that is, anything that can be computed by a Shepherdson–Sturgis machine or a Wang machine can be computed by a Tm and conversely.

5.2 RECURSIVE IMPLIES TURING COMPUTABLE

Theorem 5.1 If $g \in \mathscr{K}$, then there exists a Tm $T \in \mathscr{T}$ such that $g \sim T$.

PROOF If a function is partial-recursively defined, then by definition it is one of the three initial functions or is formed from these by finitely many uses of the recursion scheme, composition, and the minimization operation provided minimization is used only on total functions. Thus, the form of the proof of this theorem is an induction on the formation of partial-recursively defined functions: (1) Show that for each of the three initial functions, there is a Tm that computes it and (2) show that if there are Tm's that compute the partial-recursively defined functions $h, g_1, \ldots, g_m$, respectively, then there is (a) a Tm to compute the function f defined by composition from $h, g_1, \ldots, g_m$, and (b) a Tm to compute the function f defined by the recursion scheme from g_1 and g_2, and (c) a Tm to compute the function f defined by using minimization of the function h when h is a total function.

Exercise 5.1 Prove Theorem 5.1. ∎

Theorem 5.2 If f is a recursive function, then f is Turing computable.

Theorem 5.3 If f is a partial recursive function, then f is Turing computable in general.

Exercise 5.2 Prove Theorems 5.2 and 5.3.

Exercise 5.3 If f is an $(n + 1)$-ary function, define the 1-ary function $\bar{f}$ by

$$\bar{f}(x) = f(K_0(x), \ldots, K_m(x))$$

where the K_i are the functions defined in exercise 4.22. Show that if f is Tc (or Tcg), then $\bar{f}$ is Tc (or Tcg).

5.3 TURING COMPUTABLE IMPLIES RECURSIVE

Theorem 5.4 If $T \in \mathscr{T}$, then there exists a $g \in \mathscr{K}$ such that $T \sim g$.

PROOF There are various proofs of this theorem, but the one given by Minsky (1967, pp. 170–184) is probably the most amusing. What follows, up to Theorem 5.5, is a presentation of a somewhat modified version of that proof.

If a function f is Tc, then there is a Tm T_f such that $T_f \sim f$. In order to show that such a function is also recursive, we must examine exactly how a Tm works, step by step, and try to construct a recursive function that follows the same procedure. (Since $T \in \mathscr{T}$, T has $\{*, 1\}$ for its alphabet and works on a half-infinite tape.)

We assume time $t = 0$ to be the moment the Tm is turned on with the following conditions holding: (1) the machine is in internal state q_0, (2) the tapehead is reading the leftmost square of tape, (3) the initial tape inscription n is printed in binary (with $*$ for 0) on the leftmost squares of tape with the order of least significance to most significance taken (for the following discussion) from left to right. At $t = 1$, the Tm will have performed the first print-move operation, etc. At any time t during a computation, we have a situation which we represent by

$$\underbrace{| \; b_s \; | \cdots | \; b_3 \; | \; b_2 \; | \; b_1 \; | \; b_0 \; |}_{m_t} \; \underset{\substack{\uparrow \\ q_t}}{s_t} \; \underbrace{| \; c_0 \; | \; c_1 \; | \; c_2 \; | \cdots | \; c_r \; |}_{k_t} \cdots$$

That is, b_i's, c_i's, and s_t are $*$ or 1, b_s is the symbol in the leftmost square of tape, and c_r the rightmost 1. The tapehead is reading s_t and the Tm is in internal state q_t. Since everything is binary, the string $b_s \cdots b_3 b_2 b_1 b_0$ is a binary number, call it m_t, as is the string $c_0 c_1 c_2 \cdots c_r$, called k_t. (Notice that the order of significance of the bits is taken to be in opposite order, so that the bit of least significance is next to s_t.) What the situation will be at time $t + 1$ depends, of course, on the rest of the quintuple that begins $q_t s_t$. Suppose it is $q_t s_t q_{t+1} s_{t'} R$. Then, we have

$$\underbrace{| \; b_s \; | \cdots | \; b_1 \; | \; b_0 \; | \; s_{t'} \; |}_{m_{t+1}} \; \underset{\substack{\uparrow \\ q_{t+1}}}{c_0} \; \underbrace{| \; c_1 \; | \cdots | \; c_r \; |}_{k_{t+1}} \cdots$$

and we see that

$$\begin{aligned} m_{t+1} &= 2m_t + s_{t'} \\ k_{t+1} &= [k_t/2] \\ s_{t+1} &= c_0 = \mathrm{P}(k_t) \end{aligned}$$

Recall that

$$[k/2] = \begin{cases} k/2 & \text{if } k \text{ even} \\ (k-1)/2 & \text{if } k \text{ odd} \end{cases}$$
$$\mathrm{P}(k) = \begin{cases} 0 & \text{if } k \text{ even} \\ 1 & \text{if } k \text{ odd} \end{cases} \tag{1}$$

Exercise 5.4 Give the situation at $t + 1$ had the quintuple been $q_t s_t q_{t+1} s_{t'} L$.

As we noted early in chapter I, a set of quintuples constituting a Tm is a function, since for each beginning pair $q_i s_j$ there is a unique triple $q_k s_t X$ for $X \in \{R, L\}$. Obviously, for each Tm T, this can be broken up into three functions Q_T, S_T, and D_T defined for all pairs (q_i, s_j) of internal states and tape symbols of T. If $q_i s_j q_k s_t R$ is a quintuple of T, then

$$\begin{aligned} Q_T(q_i, s_j) &= q_k \\ S_T(q_i, s_j) &= s_t \\ D_T(q_i, s_j) &= R \end{aligned} \tag{2}$$

We can assign natural numbers to these things in a simple-minded way by assigning 0 to the direction R, 1 to the direction L, 0 to $*$ and 1 to 1, and, if T has m internal states, assign i to q_i for $0 \leq i \leq m - 2$ and assign $m - 1$ to q_H, but let $m - 1$ be referred to as H for this discussion. Let $\tilde{s}_i$ denote 0 or 1 according to whether s_i is $*$ or 1. Using this assignment, there are three number-theoretic functions corresponding to the functions in (2):

$$\begin{aligned} Q_T{}^*(i, \tilde{s}_j) &= k \\ S_T{}^*(i, \tilde{s}_j) &= \tilde{s}_t \\ D_T{}^*(i, \tilde{s}_j) &= \begin{cases} 0 & \text{if } D_T(q_i, s_j) = R \\ 1 & \text{if } D_T(q_i, s_j) = L \end{cases} \end{aligned} \tag{3}$$

Exercise 5.5 Are these three functions pr? (Hint: Recall part 2 of section 4.4.)

Using the functions in (3), we shall see how to write the equations in (1) in the form of functions for which t is a variable and n, the initial tape inscription, is a parameter. There will be four functions: one for internal state; one for m, what is written on the tape to the left of the tapehead; one for k, what is written on the tape to the right of the tapehead; and one for s, the symbol being read by the tapehead.

$q_T(0, n) = 0$ since a Tm always starts in q_0

$m_T(0, n) = 0$ since there is no tape left of the tapehead at $t = 0$.

$k_T(0, n) = \left[\frac{n}{2}\right]$ since the tapehead is on the leftmost symbol of the initial tape inscription n at $t = 0$.

$s_T(0, n) = P(n)$ since the leftmost square is the least significant figure of n and is the square being read at $t = 0$.

$$
\begin{aligned}
q_T(t+1, n) &= Q_T^*(q_T(t, n), s_T(t, n)) \\
m_T(t+1, n) &= 2m_T(t, n) + S_T^*(q_T(t, n), s_T(t, n)) \cdot (1 \dot{-} D_T^*(q_T(t, n), s_T(t,n))) \\
&\quad + [m_T(t, n)/2] \cdot D_T^*(q_T(t, n), s_T(t, n)) \\
k_T(t+1, n) &= (2k_T(t, n) + S_T^*(q_T(t, n), s_T(t, n))) \cdot D_T^*(q_T(t, n), s_T(t, n)) \\
&\quad + [k_T(t, n)/2] \cdot (1 \dot{-} D_T^*(q_T(t, n), s_T(t, n))) \\
s_T(t+1, n) &= P(k_T(t, n)) \cdot (1 \dot{-} D_T^*(q_T(t, n), s_T(t, n))) \\
&\quad + P(m_T(t, n)) \cdot D_T^*(q_T(t, n), s_T(t, n))
\end{aligned}
$$

Thus, we have four functions defined simultaneously from pr functions and so they are pr. (Refer to part 7 of section 4.4.)

A Tm T finishes a computation from initial tape inscription n when it enters internal state q_H. Further, the answer lies on the tape just left of the tapehead. Thus, the answer is $m_T(t^*, n)$ for $q_T(t^*, n) = H$, where t^* is the smallest t such that $q_T(t, n) = H$. The functions q_T, m_T, k_T, and s_T are total because all prf's are total. All this can be represented by the function

$$f_T(n) = m_T(\mu_t[q_T(t, n) = H], n)$$

Notice that q_T is H-regular if and only if T halts on n. Hence, f_T is certainly a partial recursive function, and may be a recursive function. By the construction of f_T, it is clear that $f_T \sim T$. Since f_T is a partial recursive function, there exists a $g \in \mathscr{K}$ such that $f_T = g$ and so $g \sim T$. ∎

Theorem 5.5 If f is Turing computable in general, then f is a partial recursive function.

Exercise 5.6 Prove the theorem by proving the following lemma: If f is an $(n+1)$-ary Tcg function and $T_{\bar{f}} \in \mathscr{T}$ such that $T_{\bar{f}} \sim f$, and $f_{T\bar{f}}$ is the partial recursive function defined by

$$f_{T\bar{f}}(n) = m_{T\bar{f}}(\mu_t[q_{T\bar{f}}(t, n) = H], n)$$

then $f = f_{T\bar{f}}$ and, since $\bar{f} \circ J^{n+1} = f$, f is a partial recursive function. (Hint: It is just a matter of carefully following through the definitions.)

Theorem 5.6 If f is Turing computable, then f is a recursive function.

PROOF It is the same as the proof of Theorem 5.5 except to emphasize that the Tm T halts if and only if the function q_T of the construction is H-regular. ▮

5.4 SOME RESULTS OF THE EQUIVALENCE

In proving that to every partial-recursively defined function there corresponds a Tm that computes it, an effective construction was given so that an actual Tm to compute the function was produced. Also, an effective construction was given which, for any Tm, produces an actual partial-recursively defined function that the Tm computes. Henceforth, we assume that our standard Gödel numbering of Tm's $\mathcal{T}$ and of partial-recursively defined functions $\mathcal{K}$ is onto N. Henceforth, also let the elements of $\mathcal{T}$ be denoted by T_i, where i is the Gödel number of the Tm. Similarly, an element of $\mathcal{K}$ is denoted by g_j, where j is the Gödel number of that element of $\mathcal{K}$. For a Tm T_i, we denote by g_{T_i} the element of $\mathcal{K}$ with the smallest Gödel number such that $T_i \sim g_{T_i}$. Conversely, for a partial-recursively defined function g_i, we denote by T_{g_i} the element of $\mathcal{T}$ with the smallest Gödel number such that $T_{g_i} \sim g_i$; of course, T_{g_i} is T_k for some k.

Theorem 5.7 If f is a partial recursive function, then it is equal to a partial-recursively defined function in which the μ-operation is used just once.

Theorem 5.8 If a partial recursive function is total, then it is recursive.

Exercise 5.7 Prove the theorems.

Theorem 5.9 There is a partial recursive function that cannot be extended by any recursive function.

PROOF Suppose that for every $g_i \in \mathcal{K}$, there is a $g_{i'} \in \mathcal{K}$ such that $g_{i'}$ is recursive and extends g_i, i.e., for all n, if $g_i(n)$ is defined, $g_{i'}(n) = g_i(n)$. Define f by

$$f(i) = \begin{cases} g_i(i) + 1 & \text{if } g_i(i) \text{ is defined} \\ \text{undefined} & \text{otherwise} \end{cases}$$

f is certainly partial recursive and so there exists a p such that $f = g_p$. Then by the assumption, there exists a p' such that $g_{p'}$ is recursive and extends g_p. That is, for all n,

$$g_{p'}(n) = \begin{cases} g_p(n) & \text{if } g_p(n) \text{ is defined} \\ \text{something} & \text{otherwise} \end{cases}$$

By the definition of g_p,

$$g_p(p') = \begin{cases} g_{p'}(p') + 1 & \text{if } g_{p'}(p') \text{ is defined} \\ \text{undefined} & \text{otherwise} \end{cases}$$

Since $g_{p'}$ is total, it is defined and so we have $g_p(p') = g_{p'}(p') + 1$; but by the definition of $g_{p'}$, $g_p(p') = g_{p'}(p')$. Contradiction. ▌

Whatever class of functions may be encompassed by the idea of total effectively computable functions, surely the class of recursive functions must be included. The facts that the recursive functions and the Turing-computable functions are the same class of functions, and that no one has yet produced a function which all agree to as being a total effectively computable function but not a recursive function, are often taken as weighty evidence in support of the following conjecture, known as *Church's thesis*†: Any total effectively computable function is a recursive function. Of course, "total effectively computable function" is not defined precisely; thus, perhaps the thesis should not be called a conjecture, since it cannot be proved. It might better be referred to as a working hypothesis. The student is encouraged to use Church's thesis in proving theorems and solving problems. However, he should not use it without giving some thought to backing up his arguments with actual Tm's or recursive functions. Let us give the name *the extended Church thesis*‡ to that version of Church's thesis that says that the class of partial effectively computable functions coincides with the class of partial recursive functions. Of course, this cannot be proved either, but we have as evidence in favor of its truth that the class of partial recursive functions coincides with the class of functions that are Turing computable in general. The reader is also encouraged to use the extended Church thesis, but to do so with considerable caution, as there are serious pitfalls. One in particular is of the following nature: since we are dealing with procedures that may not terminate, we can safely say, "*if* the procedure terminates, do so-and-so"; but we *cannot* say, "if the procedure does not terminate, do so′-and-so′." For examples of the possible difficulties, do exercises 5.9 and 5.10 below.

† For Church's original discussion, see Church (1936), sections 1 and 7. Further interesting discussions of this topic can be found in Davis (1958, p. 10), Kleene (1952, pp. 300, 331, see his index for more), Rogers (1967, section 1.7), and Shoenfield (1967, section 6.5).

‡ This is not a standard name. "Church's thesis" is sometimes used to include what is here called "the extended Church thesis." Consult the references of the previous footnote, in particular Kleene (1952, p. 331).

The effective interchangeability of partial recursive functions with Tm's allows us to use whichever seems easier in our approach to solving problems. It also allows us to use what we already have proved about Tm's in proving things about partial recursive functions.

Theorem 5.10 There exists a $u \in N$ such that for all $n, m \in N$,

$$g_u(n, m) = g_n(m) \qquad \text{for} \quad g_u, g_n \in \mathscr{K}$$

Exercise 5.8 Prove the theorem.

Theorem 5.11 There is no recursive function f such that for all unary functions $g_i \in \mathscr{K}$,

$$f(i, n) = \begin{cases} 1 & \text{if} \quad g_i(n) \quad \text{is defined} \\ 0 & \text{if} \quad g_i(n) \quad \text{is not defined} \end{cases}$$

Exercise 5.9 Prove the theorem.

Exercise 5.10 Show carefully that the function f defined by

$$f(i) = \begin{cases} g_i(i) + 1 & \text{if} \quad g_i(i) \quad \text{is defined} \\ 0 & \text{otherwise} \end{cases}$$

is not a recursive function.

5.5 THE *s-m-n* THEOREM, THE RECURSION THEOREM, AND SELF-REPRODUCING MACHINES

It is natural to wonder if anything interesting happens when we consider recursive functions applied to the Gödel numbers of elements of $\mathscr{T}$ or $\mathscr{K}$. For example, if f is any recursive function, is there any particularly interesting relationship between T_i and $T_{f(i)}$? Or, from another point of view, is the set of Gödel numbers of Tm's which are related in some interesting way the image of some recursive function? Two examples of this kind of question are: (1) Is there a recursive function f such that for all $g_i \in \mathscr{K}$, the image of $g_{f(i)}$ is the set $\{j \mid g_j = g_i\}$? (2) Is there a recursive function f such that for all $g_i \in \mathscr{K}$,

$$g_{f(i)}(j) = \begin{cases} 1 & g_i = g_j \\ 0 & \text{otherwise?} \end{cases}$$

In both cases, the answer is no. The reader is, of course, invited to prove them. [For the first problem, consult Rogers (1967, p. 33).]

Many of the proofs used in future chapters of this book will be based on Turing machines or Church's thesis and so we will only rarely use the next two theorems. (They will be used in section 6.6.) However, the reader who expects to study recursive function theory beyond this book should be acquainted with them since they are widely used in formal recursive function theory proofs. An amusing immediate result will be given also. The first of these allows us to take a partial recursive function of $n + m$ variables and change to another partial recursive function which takes m of them as parameters and the remaining n as variables. This extremely useful theorem is known as *Kleene's s-m-n theorem*, also as the parametrization theorem or the iteration theorem.

Theorem 5.12 For every m, $n \geq 1$, there is an $(m + 1)$-ary recursive function s_n^m such that for all $x_1, \ldots, x_m, x_{m+1}, \ldots, x_{m+n}$, if $g_i \in \mathscr{K}$, then

$$g_{s_n^m(i, x_1, \ldots, x_m)}(x_{m+1}, \ldots, x_{m+n}) = g_i(x_1, \ldots, x_m, x_{m+1}, \ldots, x_{m+n})$$

PROOF Let $T_{g_i} \sim g_i$ and design the following Tm T. No matter what is written on its tape, when started, T reads the leftmost nonblank symbol, moves to the left, and prints out $x_1, \ldots, x_m$. Thus, the tape that has $x_{m+1}, \ldots, x_{m+n}$ as initial tape inscription will have at this stage $x_1, \ldots, x_m$, $x_{m+1}, \ldots, x_{m+n}$ printed on it. T then moves to the leftmost symbol of x_1 and thereafter behaves like T_{g_i}. Since $T \in \mathscr{T}$, there are functions $g_j \in \mathscr{K}$ such that $T \sim g_j$. Using the effective constructions of this chapter and Church's thesis, it is easy to see that s_n^m is just the recursive function such that for one of these j, $s_n^m(i, x_1, \ldots, x_m) = j$. ∎

The next theorem, also due to Kleene, is known as the *recursion theorem* and has several forms [Kleene (1952, section 66), Davis (1958, p. 147), Rogers (1967), chapter 11)]. We present the simplest form. Notice that it is in the form, often found in other parts of mathematics, of a fixed-point theorem.

Theorem 5.13 For every recursive function f, there is an $n \in N$ such that

$$g_n(x) = g_{f(n)}(x) \qquad \text{for all} \quad x$$

for $g_n, g_{f(n)} \in \mathscr{K}$.

PROOF Define the function H by

$$H(u, x) = \begin{cases} g_{g_u(u)}(x) & \text{if} \quad g_u(u) \quad \text{is defined} \\ \text{undefined} & \text{otherwise} \end{cases}$$

By the extended Church thesis, H is a partial recursive function, and so for some i, $H = g_i$, i.e.,

$$g_i(u, x) = \begin{cases} g_{g_u(u)}(x) & \text{if } g_u(u) \text{ is defined} \\ \text{undefined} & \text{otherwise} \end{cases}$$

By the *s-m-n* theorem,

$$g_{s_1{}^1(i,u)}(x) = g_i(u, x) \qquad (*)$$

But $s_1{}^1$ is a recursive function, so for some k, $s_1{}^1 = g_k$, and so by another use of the *s-m-n* theorem,

$$s_1{}^1(i, u) = g_k(i, u) = g_{s_1{}^1(k,i)}(u)$$

Call this last $G(u)$; so, from (*),

$$g_i(u, x) = g_{G(u)}(x)$$

For any recursive function f, since G is recursive, $f \circ G$ is recursive, and so, for some p, $f \circ G = g_p$. Thus, for all x and z,

$$g_{f(G(z))}(x) = g_{g_p(z)}(x)$$

In particular, consider what happens at $z = p$:

$$g_{f(G(p))}(x) = g_{g_p(p)}(x) = g_{s_1{}^1(i,p)}(x) = g_{G(p)}(x)$$

The middle equality holds by the definition of H because g_p is recursive and so $g_p(p)$ is defined. Therefore, the n in question is $G(p)$. ▮

The final theorem of the chapter is an interesting application of the recursion theorem to Turing machines to show the existence of a Tm T whose only function is to write out on its tape ρ_T its own standard tape description. This is sometimes interpreted as saying that T reproduces itself. The theorem and proof presented here are essentially the same as in Rogers (1967, p. 188). On p. 189, Rogers also presents another version of "self-reproducing" machines which the reader might like to read. Two lemmas are needed for the proof of the theorem.

Lemma 5.1 (a) There exists a recursive function h such that for all $g_i, g_j \in \mathscr{K}$, $g_{h(i,j)} = g_i \circ g_j$, where $g_{h(i,j)} \in \mathscr{K}$. (b) There exists a recursive function h' such that for all $T_i, T_j \in \mathscr{T}$, $T_{h'(i,j)} = T_i \circ T_j$, where $T_{h'(i,j)} \in \mathscr{T}$. (c) There exists a recursive function h'' such that for all $T_i \in \mathscr{T}$ and $g_j \in \mathscr{K}$, $T_{h''(i,j)} = T_i \circ T_{g_j}$. ▮

Lemma 5.2 There exists a Turing machine T such that for all n, $T(n) \equiv \rho_{T_n}$.

Exercise 5.11 Prove Lemmas 5.1 and 5.2.

Theorem 5.14 There exists a Turing machine with Gödel number p, T_p, such that for all x, $T_p(x) \equiv \rho_{Tp}$.

PROOF Let q be the Gödel number of the Turing machine of Lemma 5.2. Define the function H by $H(x, y) = h''(q, x)$ for h'' of Lemma 5.1. Certainly, H is a partial recursive function, so for some i, $g_i(x, y) = h''(q, x)$. By two uses of the s-m-n theorem, as in the proof of the previous theorem, there is a recursive function G such that $g_{G(x)}(y) = g_i(x, y) = h''(q, x)$. By the recursion theorem, since G is a recursive function, there is an n such that $g_n = g_{G(n)}$, so for all y,

$$g_n(y) = g_{G(n)}(y) = h''(q, n)$$

Then

$$\begin{aligned} T_{h''(q,n)}(y) &= T_q(T_{g_n}(y)) && \text{by the definition of } h'' \\ &= T_q(h''(q, n)) && \text{since } g_n(y) = h''(q, n) \\ &\equiv \rho_{T_{h''(q,n)}} && \text{by the definition of } T_q \end{aligned}$$

Thus, $h''(q, n)$ is the p in the statement of the theorem. ▮

Chapter VI

INSIDE RECURSIVE FUNCTIONS

6.1 REMARKS

Four special topics about recursive functions and one about partial recursive functions are presented in this chapter. Of these, only section 6.3 is required later in the book, and then only in section 14.4. Some knowledge of section 6.4 is needed for section 6.5. Sections 6.3–6.6 are taken directly from the original papers. The papers in question are nicely written and the student should read the originals. The papers are largely computational; the reader who wishes to learn the basic methods for investigating classes of recursive functions is advised to study them. In sections 6.4–6.6 the notion of "computational complexity" is introduced and a few specific examples of that notion are studied in detail.

6.2 ACKERMANN'S FUNCTION †

Although we have already learned that not all total effectively computable functions are primitive recursive, a second proof will be considered because the construction of the function and the investigation of its properties are interesting and instructive. Before taking up the second proof, we consider for a moment the idea of double induction. Suppose we have a function $f(n, m)$, where both n and m are variables. If we wish to prove that the function has some particular property, the most natural approach is to do an induction on n and m simultaneously. Consider the integer points on the n, m plane (figure 6.1). In an induction on one variable, we must get from j to $j + 1$. But, for an induction on two variables, we must get from (j, k) to $(j + 1, k + 1)$.

† The material in this section is based on the presentation of the Ackermann function given by Hermes (1965, pp. 84–88).

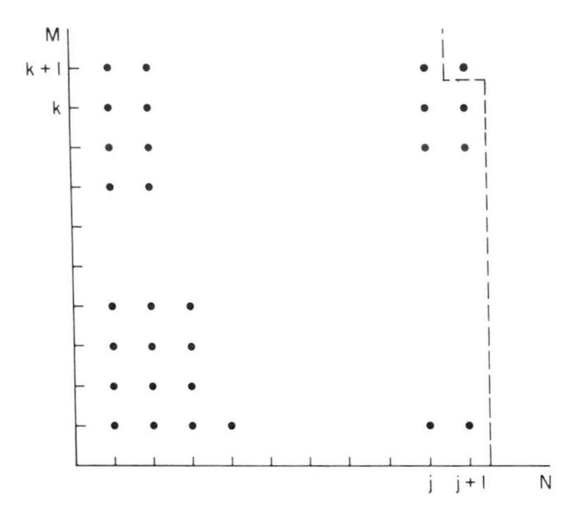

FIGURE 6.1

The usual way, which is the way in which we will be using double induction in the discussion of the function A below, is to do an "outer induction" on n and an "inner induction" on m: (1) Show that at $n = 0$, the property holds at $m = 0$ and if the property holds at $m = k$, i.e., for $f(0, k)$, then it holds at $m = k + 1$; (2) assume the property to be true for all $n \leq j$, then, to show that it is true at $(j + 1, k + 1)$ prove that it is true at $(j + 1, 0)$, and then, assuming that it is true at $(j + 1, k)$ (everything left of the line in the figure), show that it is true at $(j + 1, k + 1)$.

Recall how we started building prf's in section 4.2. From the successor function, we got addition; from addition, multiplication; from multiplication, exponentiation. In each case, the new function was constructed from the previous one in exactly the same way. Let $f_0(x_0, x_1) = S(x_0)$, $f_1(x_0, x_1) = +(x_0, x_1)$, $f_2(x_0, x_1) = \times(x_0, x_1)$, and $f_3(x_0, x_1) = \text{Exp}(x_0, x_1)$. Consider the following sequence of primitive recursive functions:

$$f_0(m, x) = S(m)$$

$$+\begin{cases} f_1(0, x) = x \\ f_1(m + 1, x) = f_0\,(f_1(m, x), x) \end{cases}$$

$$\times\begin{cases} f_2(0, x) = 0 \\ f_2(m + 1, x) = f_1(f_2(m, x), x) \end{cases}$$

$$\text{Exp}\begin{cases} f_3(0, x) = 1 \\ f_3(m + 1, x) = f_2(f_3(m, x), x) \end{cases}$$

$$\vdots$$

$$\begin{cases} f_{n+1}(0, x) = g_{n+1}(x) \qquad \text{for some prf} \quad g_{n+1} \\ f_{n+1}(m + 1, x) = f_n(f_{n+1}(m, x), x) \end{cases}$$

$$\vdots$$

It is interesting to consider the function F which accounts for all the f_i's, that is, $F(n, m, x) = f_n(m, x)$, and to ask if it is pr. Notice that by defining $F(n+1, m+1, x) = F(n, F(n+1, m, x), x)$, we have an inductive kind of definition of F which accounts for all the f_i's except where $n = 0$ and where $m = 0$, and which certainly satisfies our idea of effectively computable. Since it can be defined inductively, one might expect to be able to show that F is primitive recursive. On the other hand, since the inductive definition depends on two variables instead of one as in the prf's that we have considered so far, one might equally well suspect that F is not primitive recursive. The function F was originally constructed by Ackermann and is known as the *Ackermann function*. Rather than showing that the Ackermann function is not pr, we will investigate a slightly different function.

Define the function A of two variables as follows:

$$A(0, m) = m + 1$$
$$A(n+1, 0) = A(n, 1)$$
$$A(n+1, m+1) = A(n, A(n+1, m))$$

In order to show that A is not pr, we must prove the following:

Lemma 6.1 For every k-ary primitive recursive function f, there exists a C such that $f(n_1, \ldots, n_k) < A(C, \sum_1^k n_i)$.

But, in order to prove the lemma, a number of properties of A must be established. The abbreviation x' for $x + 1$ will be used throughout.

Property 0 For any natural numbers p, q, r, notice that if $p > q > r$, then $p > r'$.

Property 1 For all n, m, $m < A(n, m)$.

PROOF (Outer induction on n) (1) At $n = 0$, by the definition of A, $A(0, m) = m' > m$. Suppose true at $n = j$ and show at $n = j'$ by an inner induction on m. At $m = 0$,

$$\begin{aligned} A(j', 0) &= A(j, 1) && \text{by the definition of } A \\ &> 1 && \text{by the outer induction hypothesis} \end{aligned}$$

Therefore, $A(j', 0) > 0$. Assume true at $n = j'$ and $m = k$ and show for $m = k'$.

$$\begin{aligned} A(j', k') &= A(j, A(j', k)) && \text{by the definition of } A \\ &> A(j', k) && \text{by the outer induction hypothesis} \\ &> k && \text{by the inner induction hypothesis} \end{aligned}$$

Therefore, $A(j', k') > k'$, by property 0.

Property 2 For all n, m, $A(n, m) < A(n, m')$.

Exercise 6.1 Prove property 2.

Property 3 For all n, m, $A(n, m') \leq A(n', m)$.

PROOF (By induction on m) At $m = 0$, $A(n, 1) = A(n', 0)$, by the definition of A. At $m = k$, assume that $A(n, k') \leq A(n', k)$ and show that $A(n, k'') \leq A(n', k')$. By property 1, $k' < A(n, k')$, therefore $k'' \leq A(n, k')$, which, by induction hypothesis, $\leq A(n', k)$. Therefore, since $k'' \leq A(n', k)$, by property 2,

$$A(n, k'') \leq A(n, A(n', k)) = A(n', k')$$

the equality following by the definition of A.

Property 4 For all n, m, $A(n, m) < A(n', m)$.

Exercise 6.2 Prove property 4.

Property 5 For all m, $A(1, m) = m + 2$.

Exercise 6.3 Prove property 5.

Property 6 For all m, $A(2, m) = 2m + 3$.

PROOF At $m = 0$,

$$\begin{aligned} A(2, 0) &= A(1, 1) && \text{by the definition of } A \\ &= 1 + 2 && \text{by property 5} \\ &= 2 \cdot 0 + 3 \end{aligned}$$

Assume true at $m = k$ and show at k'.

$$\begin{aligned} A(2, k') &= A(1, A(2, k)) && \text{by the definition of } A \\ &= A(1, 2k + 3) && \text{by the induction hypothesis} \\ &= 2k + 3 + 2 && \text{by property 5} \\ &= 2k' + 3 \end{aligned}$$

Property 7 For any $n_1, n_2, \ldots, n_r$, there exists an n^* such that for all m,

$$\sum_{i=1}^{r} A(n_i, m) \leq A(n^*, m)$$

PROOF (By induction on r) At $r = 1$, n^* can be n_1. Suppose that the property holds for $r = t$, and show that it does at $r = t'$; i.e., show that for all $n_1, \ldots, n_{t'}$, there exists an n^{**} such that $\sum_{i=1}^{t'} A(n_i, m) \leq A(n^{**}, m)$.

$$
\begin{aligned}
\sum_{i=1}^{t} & A(n_i, m) + A(n_{t'}, m) \\
&\leq A(n^*, m) + A(n_{t'}, m) && \text{by the induction hypothesis} \\
&\leq A(N, m) + A(N, m) && \text{for } \quad N = \max(n^*, n_{t'}) \\
&< 2A(N, m) + 3 && \\
&= A\big(2, A(N, m)\big) && \text{by property 6} \\
&< A\big(N + 2, A(N, m)\big) && \text{by property 4} \\
&< A\big(N + 2, A(N + 3, m)\big) && \text{by properties 2 and 4} \\
&= A(N + 3, m') && \text{by the definition of } \quad A \\
&\leq A(N + 4, m) && \text{by property 3}
\end{aligned}
$$

Let $N + 4 = n^{**}$.

PROOF OF LEMMA 6.1 As the lemma must hold for all primitive recursive functions, the proof will be by induction on the formation of primitive recursive functions. Each step of the induction will be treated as a property of A. We will sometimes write $\bar{n}_i$ for $n_1, \ldots, n_k$.

Property 8 For all m, $S(m) < A(1, m)$.

PROOF

$$
\begin{aligned}
S(m) = m' &= A(0, m) && \text{by the definitions of } \quad S \quad \text{and} \quad A \\
&< A(1, m) && \text{by property 4}
\end{aligned}
$$

Therefore, $C = 1$.

Property 9 For all $n_1, \ldots, n_k$, $U_i^k(n_1, \ldots, n_k) < A(0, \sum_1^k n_i)$.

PROOF

$$U_i^k(n_1, \ldots, n_k) = n_i < (\sum n_i) + 1 = A(0, \sum n_i)$$

by the definitions of U_i^k and A. Therefore, $C = 0$.

Property 10 For all m, $Z(m) < A(0, m)$.

PROOF

$$Z(m) = 0 < 1 \leq A(0, m)$$

by the definitions of Z and A. Therefore, $C = 0$.

Property 11 (*Composition*) If $h, g_1, \ldots, g_r$ are prf's and $n^*, n_1^*, \ldots, n_r^*$ are numbers such that $h(n_1, \ldots, n_k) < A(n^*, \sum n_i)$ and, for $i = 1, \ldots, r$,

$$g_j(n_1, \ldots, n_k) < A(n_j^*, \sum n_i)$$

and if f is defined by

$$f(n_1, \ldots, n_k) = h(g_1(n_1, \ldots, n_k), \ldots, g_r(n_1, \ldots, n_k))$$

then there exists a C such that for all $n_1, \ldots, n_k$,

$$f(n_1, \ldots, n_k) < A(C, \sum n_i)$$

PROOF

$f(\bar{n}_i) = h(g_1(\bar{n}_i), \ldots, g_r(\bar{n}_i))$	
$< A(n^*, \sum g_j(\bar{n}_i))$	by (induction) hypothesis
$< A(n^*, \sum A(n_j^*, \sum n_i))$	by (induction) hypothesis and property 2
$< A(n^*, A(n^{**}, \sum n_i))$	for suitable n^{**} by property 7
$\leq A(n^* + n^{**}, A(n^* + n^{**} + 1, \sum n_i))$	by property 4 twice and property 2
$= A(n^* + n^{**} + 1, (\sum n_i) + 1)$	by the definition of A
$\leq A(n^* + n^{**} + 2, \sum n_i)$	by property 3

Therefore, let $C = n^* + n^{**} + 2$.

Property 12 If g_1 and g_2 are prf's and n_1^* and n_2^* are numbers such that for all $n_{-1}, n_0, n_1, \ldots, n_k$,

$$g_1(n_1, \ldots, n_k) < A\left(n_1^*, \sum_1^k n_i\right) \quad \text{and} \quad g_2(n_{-1}, n_0, n_1, \ldots, n_k) < A\left(n_2^*, \sum_{-1}^k n_i\right)$$

and if f is defined by

$$f(0, n_1, \ldots, n_k) = g_1(n_1, \ldots, n_k)$$
$$f(n', n_1, \ldots, n_k) = g_2(f(n, n_1, \ldots, n_k), n, n_1, \ldots, n_k)$$

then there exists a c_1^* such that

$$g_1(\bar{n}_i) + \sum_1^k n_i < A\left(c_1^*, \sum_1^k n_i\right)$$

PROOF We have

$$g_1(\bar{n}_i) + \sum_1^k n_i = g_1(\bar{n}_i) + \sum_{j=1}^k U_j^k(\bar{n}_i)$$

$$\leq A\left(n_1^*, \sum_1^k n_i\right) + \sum_{j=1}^k A\left(0, \sum_1^k n_i\right)$$ by the (induction) hypothesis and properties 9 and 2

$$< A\left(c_1^*, \sum_1^k n_i\right)$$ for suitable c_1^* by property 7

Property 13 Under the hypotheses of property 12, there exists a c_2^* such that

$$g_2(\bar{n}_i) + \sum_{-1}^k n_i < A\left(c_2^*, \sum_{-1}^k n_i\right)$$

where $\bar{n}_i$ is $n_{-1}, n_0, \ldots, n_k$.

Exercise 6.4 Prove property 13.

Property 14 Under the hypotheses of property 12, if $c_0^* = \max(c_1^*, c_2^*) + 1$, then for all $n_0, n_1, \ldots, n_k$,

$$f(n_0, \bar{n}_i) + \sum_0^k n_i < A\left(c_0^*, \sum_0^k n_i\right)$$

PROOF By induction on n_0. For $n_0 = 0$,

$$f(0, \bar{n}_i) + \sum_1^k n_i = g_1(\bar{n}_i) + \sum_1^k n_i$$ by the definition of f

$$< A(c_1^*, \sum n_i)$$ by property 12

$$\leq A(c_0^*, \sum n_i)$$ by the definition of c_0^*

Assume true at $n_0 = p$ and show true at p':

$$f(p', \bar{n}_i) + p' + \sum_1^k n_i$$

$$= g_2(f(p, \bar{n}_i), p, \bar{n}_i) + p' + \sum_1^k n_i$$ by definition of f

$$\leq g_2(f(p, \bar{n}_i), p, \bar{n}_i) + f(p, \bar{n}_i) + p' + \sum_1^k n_i$$ by adding $f(p, \bar{n}_i)$

$$< A(c_2^*, f(p, \bar{n}_i) + p + \sum n_i) + 1$$ by property 13

$< A(c_2^*, A(c_0^*, p + \sum n_i)) + 1$ by (induction) hypothesis and property 2

$\leq A(c_0^* - 1, A(c_0^*, p + \sum n_i)) + 1$ by definition of c_0^*

$\leq A(c_0^*, p' + \sum n_i) + 1$ by definition of A

So, by property 0,

$$f(p', \bar{n}_i) + p' + \sum_1^k n_i < A\left(c_0^*, p' + \sum_1^k n_i\right)$$

Property 15 (*Recursion scheme*) Under the hypothesis of property 12, there exists a C such that for any $n_0, n_1, \ldots, n_k$,

$$f(n_0, n_1, \ldots, n_k) < A\left(C, \sum_0^k n_i\right)$$

PROOF If we let $C = c_0^*$, then $f(p', \bar{n}_i) < A\left(C, p' + \sum_1^k n_i\right)$ from property 14. ▌

Theorem 6.1 The function A is not primitive recursive.

PROOF If A is pr, then the 1-ary function A' defined by $A'(n) = A(n, n)$ is also pr. Considering A' we have, by the lemma, that there exists a C such that for all n, $A'(n) < A(C, n)$. What happens at $n = C$? $A(C, C) = A'(C) < A(C, C)$, a contradiction. ▌

6.3 AN ALTERNATIVE FORMULATION OF THE RECURSIVE FUNCTIONS†

Let $\mathcal{R}$ denote the class of recursive functions and let $\mathcal{R}_1$ denote the unary recursive functions. Let $\mathcal{R}'$ denote the smallest class of functions satisfying the following conditions:

1. The successor function S belongs to $\mathcal{R}'$.
2. The excess over a square function E belongs to $\mathcal{R}'$.
3. If $g, h \in \mathcal{R}'$ and f is defined by

$$f(x) = g(x) + h(x)$$

then $f \in \mathcal{R}'$. (This operation is called *addition of functions*.)

4. If $g, h \in \mathcal{R}'$ and f is defined by

† The material in this section is based on a paper by J. Robinson (1950).

$$f(x) = h(g(x))$$

then $f \in \mathscr{R}'$. (Call this operation "fully restricted composition.")

5. If $f \in \mathscr{R}'$ and is *onto* N, and f^{-1} is defined by

$$f^{-1}(x) = \mu_y[f(y) = x]$$

then $f \in \mathscr{R}'$. (This operation is called *inversion*.)

The main result of this paper is that every recursive function can be obtained from functions in $\mathscr{R}' \cup \{J\}$ by means of composition. A particular advantage in viewing the recursive functions in this way is that since $\mathscr{R}'$ does not include the recursion scheme, every recursive function can be defined in a sort of "closed form"—just one big formula on the right.

Let composition as we have used it so far be called "unrestricted composition." We will introduce two kinds of restrictions, one as in 4 above. Define $\mathscr{J}$ to be the smallest class of functions which includes E, S, Z, U_i^n, for all $i, n \in N$, $i \leq n$, $x + y$, and which is closed under unrestricted composition and inversion. E, the excess over a square function, was defined in exercise 4.21. Let $\mathscr{J}_1$ denote the unary functions in $\mathscr{J}$. Most of the section will be devoted to showing that $\mathscr{R} \subseteq \mathscr{J}$; finally, we will show that $\mathscr{J}_1 \subset \mathscr{R}'$ and so $\mathscr{R}_1 \subseteq \mathscr{R}'$. The main result then follows easily.

Theorem 6.2 If $f \in \mathscr{R}$, then $f \in \mathscr{J}$.

PROOF The proof is, naturally, by induction on the formation of f. Since the initial functions of $\mathscr{R}$ are included in those of $\mathscr{J}$, there is nothing to show at the initial stage. For the rest of the proof, it is necessary to show that inversion can replace μ and that we can somehow manage without the recursion scheme. That is, $\mathscr{J}$ is closed under the usual minimization operation and under the operation of using the recursion scheme. Several pages will be necessary to obtain these results.

Notice that E is onto N; for any n, $E(n^2 + n) = n$ because $(n + 1)^2 > n^2 + n$. So $E^{-1} \in \mathscr{J}$. If n is even, say $2m$, it is clear that $E^{-1}(2m) = m^2 + 2m$. If n is odd, say $2m + 1$, we can see easily that $E^{-1}(2m + 1) = (m + 1)^2 + 2m + 1$. For, by definition, $E^{-1}(2m + 1) = \mu_k[E(k) = 2m + 1]$. Now, $E(k) = 2m + 1$ if and only if there is an n such that $k = n^2 + 2m + 1$ and $(n + 1)^2 > n^2 + 2m + 1$. That we want the least such k implies that we want the least such n. Further, $(n + 1)^2 > n^2 + 2m + 1$ implies $n > m$, and the least such n is just $m + 1$.

There is a slight difference between the functions given here and those defined in chapter IV. In chapter IV, for instance, $x \dot{-} y = 0$ if $y > x$. Here, there is no such requirement; $x - y$ is the usual minus if $x \geq y$, but $x - y$ is "in limbo" for $y > x$. We could say it is not defined; however, there might be confusion with the "not defined" associated with partial recursive functions. In that case, "not defined" is used in a very strong sense; in this case, it would be used in only a very weak sense. We might just say, "weakly not defined." Then we make the convention for $f(n)$ "weakly not defined" that $f(n) \cdot 0 = 0$.

Exercise 6.5 Verify the following:

(a) $x - y = E(E^{-1}(2x + 2y) + 3x + y + 4)$.
(b) $x^2 = E^{-1}(2x) - 2x$.
(c) $\text{sg}\, x = E(S(x^2))$.
(d) $0^x = 1 - \text{sg}\, x$ (This is just $\overline{\text{sg}}\, x$ of chapter IV by the convention that

$$0^x = \begin{cases} 1 & \text{if} \quad x = 0 \\ 0 & \text{if} \quad x > 0.) \end{cases}$$

(e) $a = b$ if and only if $((a - b) + (b - a)) = 0$ and $a \geq b$ if and only if $(((a - b) + b) - a) = 0$. Thus, both $a = b$ and $a \geq b$ have characteristic functions in $\mathscr{J}$.
(f) $\text{pd}(x) = E(\mu_y[E(S(y)) = x]$ (Recall pd is predecessor. Note that it must be verified that $E \circ S$ is onto N.)

(g)
$$E(S(S(E^{-1}(x)))) = \begin{cases} 1 & x \quad \text{even} \\ 0 & x \quad \text{odd} \end{cases}$$

(h) $[x/2] = E(\mu_y[2E(y) + E(S(S(E^{-1}(y)))) = x]$.
[Hint: Show that $2E + ESE^{-1}$ is onto N; that if $x = 2z$ and $x \neq 0$, then the least y satisfying the conditions is $((z - 1)^2 + z)$; and that if $x = 2z + 1$, the least y is $(z^2 + z)$.]
(i) $x^{1/2} = [E(\text{pd}(x))/2] + \text{sg}\, x$ (Remark: This function is defined by

$$x^{1/2} = \begin{cases} x^{1/2} & \text{if} \quad x \quad \text{is a square} \\ \text{does not matter} & \text{if not.)} \end{cases}$$

(j) $[x^{1/2}] = (x - E(x))^{1/2}$.
(k) J, K, and L belong to $\mathscr{J}$.
(l) $x \cdot y = [(((x + y)^2 - x^2) - y^2)/2]$.

Let the set of functions listed in the exercise be called $\mathscr{F}$. It follows from the exercise that $\mathscr{F}$ is a subset of $\mathscr{J}$.

Suppose $f(x_1, x_2) = \mu_y[g(x_1, x_2, y) = 0]$ and define g^* by

$$g^*(z, y) = g(K(z), L(z), y).$$

Then if $f^*(z) = \mu_y[g^*(z, y) = 0]$, $f(x_1, x_2) = f^*(J(x_1, x_2))$. In a similar way, any k-ary function which is defined by means of the μ-operation applied to a $(k+1)$-ary function can be defined by means of the pairing function J, composition, and the μ-operation applied to a function of two variables.

Next, we will show that if $f(x) = \mu_y[g(x, y) = 0]$, then f can be defined by means of composition and inversion using only the functions g, $x + y$, and functions in $\mathscr{F}$. We will use "$\wedge$", "$\vee$", and "$\supset$" as abbreviations for "and" "or" and "implies", respectively. Obviously

$$f(x) = L\big(\mu_z[g(K(z), L(z)) = 0 \wedge K(z) = x]\big)$$

Notice that

$$L\big(\mu_z[0^{g(K(z), L(z))} \cdot K(z) = x]\big) = \begin{cases} f(x) & \text{if } x \neq 0 \\ 0 & \text{if } x = 0 \end{cases}$$

Define h by

$$h(x) = \mu_z[(0^{g(K(z), L(z))} \cdot K(z)) = x]$$

Thus, h is defined by an inversion since, $0^{g(K(z), L(z))} \cdot K(z)$ is onto N. We can then define f by

$$f(x) = L\big(\mu_z[0^{g(K(z), L(z))} \cdot K(z) = x]\big) + 0^x \cdot f(0)$$

Define the function $[x/y]$ to have as its value the largest z such that $y \cdot z \leq x$, that is, $x = y \cdot z + r$ and $0 \leq r < y$. The remainder function R is defined by $R(x, y) = x - y[x/y]$. Let $\mathscr{G}$ be $\mathscr{F} \cup \{[x/y], R(x, y)\}$. We will show that $[x/y]$ and R belong to $\mathscr{J}$. Notice that

$$[x/y] = \mu_z[(y \cdot S(z) > x) \vee y = 0] = \begin{cases} \mu_z[y \cdot S(z) > x] & \text{if } y \neq 0 \\ 0 & \text{if } y = 0 \end{cases}$$

and therefore

$$[x/y] = \big(\mu_z[y \cdot S(z) \geq x \wedge y \cdot S(z) \neq x]\big) \cdot \text{sg}(y)$$

Define the three following functions:

$$\begin{aligned} f_1(x, y, z) &= (((y \cdot S(z) - x) + y \cdot S(z)) - y \cdot S(z)) \\ f_2(x, y, z) &= ((y \cdot S(z) - x) + (x - y \cdot S(z))) \\ f_3(x, y, z) &= \text{sg}\big(f_1(x, y, z)\big) + 0^{f_2(x, y, z)} \end{aligned}$$

So: (i) if $y \cdot S(z) \geq x$, then $f_1(x, y, z) = 0$; (ii) if $y \cdot S(z) = x$, then $f_2(x, y, z) = 0$; and (iii) if $y \cdot S(z) > x$, then $f_3(x, y, z) = 0$. Thus,

$$[x/y] = \big(\mu_z[f_3(x, y, z) = 0]\big) \cdot \text{sg}(y)$$

Following the line developed above, for $v = J(x, y)$, define

$$D(v) = [K(v)/L(v)] = \big(\mu_z[f_3(K(v), L(v), z) = 0]\big) \cdot \text{sg}\big(L(v)\big)$$

With $n = J(v, z)$ and $f_4(v, z) = f_3(K(v), L(v), z)$, we have

$$D(v) = L(\mu_u[0^{f_4(K(u), L(u))} \cdot K(u) = v]) \cdot sg(L(v))$$

Defining $f_5(u) = 0^{f_4(K(u), L(u))} \cdot K(u)$, we see that we are using an inversion; that is,

$$f_5^{-1}(v) = \mu_u[f_5(u) = v]$$

and

$$D(v) = (L(f_5^{-1}(v)))\ sg(L(v)) + 0^v \cdot D(0)$$

Finally,

$$[x/y] = D(J(x, y))$$

Thus, $\mathscr{G} \subseteq \mathscr{J}$.

The last, and most difficult, part of showing that $\mathscr{R} \subseteq \mathscr{J}$ involves showing that we can get along without the recursion scheme. The proof involves some elementary number theory, in particular the Chinese remainder theorem. The reader can consult any text on elementary number theory for the necessary number theory. However, in the appendix in Davis's (1958) work, just exactly the facts necessary for the discussion are presented.† The basic difficulty in eliminating the recursion scheme is that if f is defined using the recursion scheme, then to compute the value $f(n + 1, x_1, \ldots, x_k)$, one must have access to the value $f(n, x_1, \ldots, x_k)$ and hence the value

$$f(n - 1, x_1, \ldots, x_k),$$

etc. That is, to compute $f(n + 1, x_1, \ldots, x_k)$, one must have "available" the sequence of numbers $f(0, x_1, \ldots, x_k)$, $f(1, x_1, \ldots, x_k), \ldots, f(n, x_1, \ldots, x_k)$. (For the rest of the discussion, let $k = 1$.) Naturally, as n increases, the sequence gets longer. All this information must, somehow, be glued, or "coded," together.

The Chinese Remainder Theorem If $a_0, a_1, \ldots, a_k$ is any set of natural numbers and $m_0, m_1, \ldots, m_k$ is a set of numbers such that for all i, j, $i \neq j$, m_i and m_j are relatively prime, then there exists a number u such that $u > a_i$ and

$$u \equiv a_i (\text{mod } m_i) \qquad \text{for all} \quad i, \quad 0 \leq i \leq k$$

† The necessary definitions are: "n divides m" (written $n|m$) was defined in section 4.4, part 5; "n and m are relatively prime" if they have no common divisor greater than 1, that is, $k|n$ and $k|m$ implies that $k = 1$; "k is congruent to n modulo m" [written $k \equiv n(\text{mod } m)$] if either $m|(k - n)$ or $m|(n - k)$.

We will use the theorem and the u whose existence is guaranteed by it to show that for any set of numbers $a_0, a_1, \ldots, a_k$, there is a u such that

$$R(u, 1 + 2Ak!(i+1)) = a_i \qquad \text{for} \quad 0 \le i \le k$$

where $A = \max(a_0, a_1, \ldots, a_k)$.

1. Claim that the members of the set $\{1 + 2Ak!(i+1) | 0 \le i \le k\}$ are pairwise relatively prime. Notice first that if for some i and $t \ne 1$,

$$t | (1 + 2Ak!(i+1)),$$

then $t > k$. For, in general, suppose $p | 1 + pq$; then there is an r such that $pr = 1 + pq$, which implies that $p = 1$ and $(r - q) = 1$. So if for some i, j, $i > j$ and d,

$$d | (1 + 2Ak!(i+1)) \qquad \text{and} \qquad d | (1 + 2Ak!(j+1))$$

then

$$d | [(1 + 2Ak!)(j+1)(i+1) - (1 + 2Ak!)(i+1)(j+1)]$$

But this right side is just $(i - j)$. Since $i, j \le k$, $(i - j) < k$, so $d < k$. It then follows from the first remark that d must equal 1.

2. Let $v = 2Ak!$. Let $m_i = (1 + v(i+1))$ for $0 \le i \le k$. Notice that $a_i < v < m_i$. Let u be the u corresponding to $a_0, a_1, \ldots, a_k$ and $m_0, m_1, \ldots, m_k$ according to the Chinese remainder theorem. Claim that $R(u, m_i) = a_i$. Since $a_i < u$, $a_i < m_i$, and $u \equiv a_i (\text{mod } m_i)$, then for each i, there is a k such that

$$R(a_i, m_i) = a_i \qquad \text{and} \qquad u = km_i + a_i \qquad \text{for} \quad 0 \le a_i \le m_i$$

Substituting one into the other gives

$$u = km_i + R(a_i, m_i)$$

That is,

$$R(u, m_i) = R(a_i, m_i) = a_i$$

3. Letting $w = J(u, v)$ for u and v determined by the numbers $a_0, a_1, \ldots, a_k$, we define

$$T(i, w) = R(K(w), (1 + L(w))S(i))$$

that is, $T(i, w) = a_i$. It is clear that $T \in \mathcal{J}$. Thus, we have shown that for any set of natural numbers $a_0, a_1, \ldots, a_k$, there exist numbers u and v such that $T(i, J(u, v)) = a_i$. This result is due to Gödel and T is known as "Gödel's β function."

4. Suppose that f is defined by the recursion scheme from g and h by

$$f(0, x) = g(x), \qquad f(n+1, x) = h(n, x, f(x))$$

Claim that

$$f(n, x) = T\big(n, \mu_w[T(0, w) = g(x) \\ \wedge \mu_m[T\big(S(m), w\big) \neq h\big(m, x, h(m, x, T(m, w))\big) \vee m = n] = n]\big)$$

It is easy to see that

$$f(n, x) = T\big(n, \mu_w[T(0, w) = g(x) \\ \wedge \quad \text{for all } m\ \big(m < n \supset T(S(m), w) = h(n, x, T(m, w))\big)]\big)$$

The claim will be substantiated by showing that

$$\text{for all } m\ \big(m < n \supset T(S(m), w) = h(n, x, T(m, w))\big) \tag{$*$}$$

is true if and only if

$$\mu_m[T\big(S(m), w\big) \neq h\big(m, x, T(m, w)\big) \vee m = n] = n \tag{$**$}$$

(∗) implies (∗∗) since (∗) is true if and only if

$$\text{for all } m\ \big(T(S(m), w) \neq h(n, x, T(m, w)) \supset m \geq n\big) \tag{$***$}$$

is true, and so the smallest m for which

$$\text{either} \quad T\big(S(m), w\big) \neq h\big(n, x, T(m, w)\big) \quad \text{or} \quad m = n$$

must be n. On the other hand, it is clear that (∗∗) implies (∗∗∗) and therefore implies (∗).

We claim that the proof of the theorem is now complete and summarize: (1) the class of functions $\mathscr{F} \subseteq \mathscr{I}$; (2) any function of more than one variable defined by means of the μ-operation on a function of more than one variable can be defined by means of the μ-operation and composition on a function of two variables and the functions J, K, and L; (3) any function of one variable defined by means of the μ-operation on a function of two variables can be defined by means of inversion [(2) and (3) imply that $\mathscr{I}$ is closed under the minimization operation]; (4) the class of functions $\mathscr{G} \subseteq \mathscr{I}$; and (5) $\mathscr{I}$ is closed under the recursion scheme. ▌

Now we are ready to show that $\mathscr{I}_1 \subseteq \mathscr{R}'$. If f is a function of one variable defined by $f(x) = h(g_1(x), g_2(x), \ldots, g_m(x))$ for some $m \geq 1$, then we will say that f is defined by "partially restricted composition."

Lemma 6.2 If $f \in \mathscr{I}_1$, then the only uses of composition necessary to define f are uses of partially restricted composition.

Exercise 6.6 Prove the lemma.

Theorem 6.3 If $f \in \mathscr{J}_1$, then $f \in \mathscr{R}'$.

PROOF (By induction on the formation of f): If f is U_1^1, then $f(x) = E(E^{-1}(x))$. If f is Z then,

$$f(x) = E(S(E^{-1}(x + x))) = E(S(E^{-1}(E(E^{-1}(x)) + E(E^{-1}(x)))))$$

For the induction step, we must consider not only uses of $x + y$ and partially restricted composition, but also composition by means of substitution into U_i^n and Z. If $f(x) = U_i^n(\ldots, g(x), \ldots)$, where g satisfies the induction hypothesis, then of course we have $f(x)$ without using U_i^n. If g satisfies the induction hypothesis and is onto N and f is defined by $f(x) = \mu_y[g(y) = x]$, then $f \in \mathscr{R}'$. If f is defined by substituting g and h into $x + y$, where g and h satisfy the induction hypothesis, then $f(x) = g(x) + h(x)$ and so $f \in \mathscr{R}'$. If f is defined by partially restricted composition, i.e., $f(x) = h(g_1(x), \ldots, g_m(x))$ for h, $g_1, \ldots, g_m$ satisfying the induction hypothesis, then h must be a unary function and so $m = 1$, i.e., $f(x) = h(g(x))$ and therefore $f \in \mathscr{R}'$. ∎

Corollary If $f \in \mathscr{R}_1$, then $f \in \mathscr{R}'$.

Theorem 6.4 A function $f \in \mathscr{R}$ if and only if it can be obtained from $\mathscr{R}' \cup \{J\}$ by means of composition.

Exercise 6.7 Prove the theorem.

6.4 KALMAR-ELEMENTARY FUNCTIONS AND THE GRZEGORCZYK HIERARCHY†

It is natural when considering a class of functions, that is, rules for carrying out a computation, to wonder "how complicated" it is to perform the computation necessary to obtain the value of any one of those functions for a given argument value. This in turn requires some definition of "measure of complexity." All the classes of functions that we consider in this book satisfy the following general definition. A class $\mathscr{X}$ of functions is *inductively defined* by means of initial functions $f_1, \ldots, f_k$ and operations $O_1, \ldots, O_s$ if $\mathscr{X}$ is the smallest class of functions which contains $f_1, \ldots, f_k$ and is closed under $O_1, \ldots, O_s$. A measure of complexity inherent in the definition is just the number of operations necessary to define a particular function in the class. Numbers can be assigned in various ways, but we use the following: If f belongs to an inductively defined class, then the *order of f* is 0 if f is an

† The theorems and general substance of this section are based directly on Grzegorczyk (1953). However, the various vague remarks about "measure of complexity" should not be blamed on him. The Grzegorczyk paper is very readable, though not so easy to find, and the interested reader is encouraged to read it.

initial function and is $1 + \max[\text{order } g_1, \ldots, \text{order } g_k]$ if f is defined by the use of an operation on functions $g_1, \ldots, g_k$. A measure of the complexity of a function viewed as a Tm computation could be the amount of tape it uses, or the amount of time. Various useful "measures of complexity" have been defined and their properties explored, with the result that there is now a subject called "complexity of computation." In this book, we will consider only three investigations in this area, this section and sections 6.5 and 6.6. At the end of section 6.6, a few references will be listed.

As we know, functions defined using bounded μ are much simpler, by almost any standard, than functions defined using unbounded μ. We also know that the Ackermann function is not primitive recursive because, so to speak, it gets too big, too fast. In an attempt to keep track of the complexity of functions, whatever this may mean, we are thus led to defining functions using operations which are bounded in some manner. In addition to the bounded μ-operation, a few examples of bounded operations are as follows:

1. *Limited recursion*: If g, h, and j belong to some class of functions and f is defined by

$$\begin{aligned} f(0, y) &= g(y) \\ f(x+1, y) &= h(x, y, f(x, y)) \\ f(x, y) &\leq j(x, y) \end{aligned}$$

then we say that *f is defined by limited recursion.* If for any g, h, and j, the function f so defined belongs to the same class of functions, then we way that the class is closed under limited recursion.

2. *Limited summation*: If for any $(k + 1)$-ary function g belonging to some class of functions, the $(k + 1)$-ary function f defined by

$$f(n, x_1, \ldots, x_k) = \sum_{z=0}^{n} g(z, x_1, \ldots, x_k)$$

also belongs, then the class of functions is closed under limited summation and *f is defined by limited summation.*

3. *Limited multiplication*: If for any $(k + 1)$-ary function g belonging to some class of functions, the $(k + 1)$-ary function f defined by

$$f(n, x_1, \ldots, x_k) = \prod_{z=0}^{n} g(z, x_1, \ldots, x_k)$$

also belongs, then the class of functions is closed under limited multiplication, and *f is defined by limited multiplication.*

Another operation that is useful is explicit transformation. A class of functions $\mathscr{X}$ is defined to be closed under *explicit transformation* if it is the case that whenever $g^m \in \mathscr{X}$ and for $k \leq m, f^k$ also belongs to $\mathscr{X}$ if f^k is defined by

$$f^k(x_1, \ldots, x_k) = g^m(\xi_1, \ldots, \xi_m)$$

where for each i, $1 \leq i \leq m$, $\xi_i \in \{x_1, \ldots, x_k\} \cup N$. In chapter VIII, other operations (existential and universal quantification) will be discussed in both bounded and unbounded forms.

Exercise 6.8 Show that if $\mathscr{X}$ is a class of functions whose initial functions include the zero, the successor, and the functions U_i^n and which is closed under composition, then $\mathscr{X}$ is closed under explicit transformation.

The striking effect of using these bounded operations shows up in the following class of functions. The class of *Kalmar-elementary functions* $\mathscr{E}$ is defined to be the smallest class of functions which includes the successor function, $x + y$, $x \dot{-} y$, and is closed under composition, explicit transformation, limited addition, and limited multiplication. Obviously, $\mathscr{E}$ is a subset of the set of primitive recursive functions. The feeling that the class $\mathscr{E}$ constitutes a basic class of functions, in some sense, is reinforced by the following theorem, whose proof, because of its length, will not be given.

Theorem † The following classes of functions are equivalent to the class $\mathscr{E}$:

(a) $\mathscr{E}'$, which is the smallest class of functions that includes the successor, $x \dot{-} y$, x^y, and is closed under composition, explicit transformation, and bounded minimization.

(b) $\mathscr{E}''$, which is the smallest class of functions that includes the successor and x^y and is closed under composition, explicit transformation, and limited recursion.

(c) $\mathscr{E}'''$, which is the smallest class of functions that includes the successor, $x \dot{-} y$, $x \cdot y$, x^y and is closed under composition, explicit transformation, and limited summation.

Before the end of the section, the reader will see that $\mathscr{E}$ is just a small subset of the set of primitive recursive functions.

There are a few theorems about inductively defined classes of functions which are of interest in themselves and will be useful shortly.

Theorem ‡ If $\mathscr{X}$ is inductively defined by means of the operations of composition as well as $O_1, \ldots, O_s$ and if $\mathscr{X}_1$ (the set of unary functions in $\mathscr{X}$) is closed under $O_1, \ldots, O_s$ and $\mathscr{X}$ includes J, K, and L, then $\mathscr{X}_1$ is inductively defined.

† Most of section 2 of Grzegorczyk (1953) is devoted to proving this theorem.

‡ This is Theorem 3.1 of Grzegorczyk (1953, p. 120).

We omit the proof; however, that the theorem is true should not be too surprising in view of the results in section 6.3. Given functions f and g, if for all $x, f(x) \leq g(x)$, we say that *f dominates g* and write $f \leq g$. If $\mathscr{X}$ is a class of functions, then we say that *f increases faster than each function in $\mathscr{X}$* if for each $g \in \mathscr{X}$, there is an n such that for all $x \geq n$, $g(x) < f(x)$.

Theorem 6.5 If $\mathscr{X}$ and $\mathscr{Y}$ are inductively defined by means of the operations of composition and explicit transformation and, besides, at most by limited recursion, and if $\mathscr{X}$ includes nondecreasing functions which dominate the initial functions of $\mathscr{Y}$, then for any k-ary function $f \in \mathscr{Y}$, there is a k-ary function $g \in \mathscr{X}$ such that g is nondecreasing and g dominates f.

Exercise 6.9 Prove the theorem by induction on the formation of f.

Theorem 6.6 If $\mathscr{X}$ and $\mathscr{Y}$ are classes of functions inductively defined by means of substitution and explicit transformation and, besides, at most by limited recursion and if $\mathscr{X}$ includes nondecreasing functions which dominate the initial functions of $\mathscr{Y}$, and if f is any function which increases faster than all functions in $\mathscr{X}_1$, then f increases faster than any function in $\mathscr{Y}_1$.

Exercise 6.10 Prove the theorem.

Let $\mathscr{X}$ be a class of functions. *The universal function for $\mathscr{X}_n$* is an $(n+1)$-ary function F such that $f \in \mathscr{X}_n$ if and only if there is a $t \in N$ such that for all $(x_1, \ldots, x_n)$, $F(t, x_1, \ldots, x_n) = f(x_1, \ldots, x_n)$. Of course, the universal function for Tm's is just the universal Tm. We also know by exercise 4.32 that the universal function for unary primitive recursive functions is not a binary primitive recursive function. Another possible view of the measure of complexity of a class of functions is in terms of the complexity of the universal function for that class. The Ackermann function is an example of the following general theorem.

Theorem 6.7 If $\mathscr{X}$ and $\mathscr{Y}$ include the successor function and are inductively defined by means of operations which include substitution and explicit transformation and if for all $f \in \mathscr{X}_1$, there is a $g \in \mathscr{Y}_1$ such that g dominates f, then $\mathscr{X}$ does not include the universal function for $\mathscr{Y}_1$.

PROOF Let F_y be the universal function for $\mathscr{Y}_1$ and suppose that $F_y \in \mathscr{X}$. Since $S \in \mathscr{X}$ and $\mathscr{X}_1$ is closed under composition and explicit transformation, $S(F_y(x, x)) \in \mathscr{X}_1$. But then there exists a function $g \in \mathscr{Y}_1$ such that for all x, $S(F_y(x, x)) < g(x)$. Since F_y is universal for $\mathscr{Y}_1$, there is an n such that $F_y(n, x) = g(x)$. Then for $x = n$, we have that

$$g(n) = F_y(n, n) < S(F_y(n, n)) = g(n)$$

Contradiction. ▌

The set of functions defined as follows is somewhat reminiscent of the set of functions defined by Ackermann which was mentioned at the beginning of section 6.2:

$$\begin{aligned} f_0(x, y) &= y + 1 \\ f_1(x, y) &= x + y \\ f_2(x, y) &= (x + 1)(y + 1) \\ f_{n+1}(0, y) &= f_n(y + 1, y + 1) \\ f_{n+1}(x + 1, y) &= f_{n+1}(x, f(x, y)) \end{aligned}$$

Lemma 6.3 The functions f_i, $i = 0, 1, 2, \ldots$, satisfy the following properties:

(a) For all $n > 1$, $y < f_n(x, y)$.
(b) For all $n \geq 0$, $f_{n+1}(x, y) < f_{n+1}(x + 1, y)$.
(c) For all $n > 0$, $f_n(x, y) < f_n(x, y + 1)$.
(d) For all $n > 0$, $f_n(x, y) < f_{n+1}(x, y)$.

Exercise 6.11 Prove the lemma. (Hint: Use double induction with n the "outer" and x the "inner" induction.)

Define the class of functions $\mathscr{E}^n$ to be the smallest class of functions that includes, as initial functions, the successor function, U_1^2, U_2^2, and f_n, and is closed under the operations of limited recursion, composition, and explicit transformation. The class of functions $\bigcup_{n \in N} \mathscr{E}^n$ is known as the *Grzegorczyk hierarchy*. That it really is a hierarchy, i.e., that $\mathscr{E}^n \subsetneq \mathscr{E}^{n+1}$, is a theorem to be proved.

Exercise 6.12 Show that E, the excess over a square function, belongs in $\mathscr{E}^3$. (See exercise 4.21 for the definition.)

Exercise 6.13 Let g be defined by

$$g(0, y) = y, \qquad g(x + 1, y) = (g(x, y) + 2)^2$$

Verify (a) that

$$g(x, y) < (y + 2)^{2^{2^{2x}}}$$

and hence g is defined by limited recursion and belongs to $\mathscr{E}$, and (b) that

$$f_3(x, y) = g(2^x, y)$$

Hint: Show, by induction on x, that

$$f_3(x, y) = (\cdots((y + 2)^2 \underbrace{+ 2)^2 + \cdots + 2)^2}_{2^x}$$

and that

$$g(2^{x+1}, y) = (\cdots((g(2^x, y) + 2)^2 \underbrace{+ 2)^2 + \cdots + 2)^2}_{2^x}$$

Exercise 6.14 Verify that $x^y \leq f_3(x, y)$. It then follows by the usual recursion definition of x^y that it belongs to $\mathscr{E}^3$.

Theorem 6.8 The class $\mathscr{E}^3$ is the same as the class $\mathscr{E}$ of Kalmar-elementary functions.

Exercise 6.15 Prove the theorem, given the equivalence of $\mathscr{E}$ and $\mathscr{E}''$.

Exercise 6.16 Show that $\mathscr{E}^0 \subset \mathscr{E}^1 \subset \mathscr{E}^2 \subset \mathscr{E}^3$ and that for all $n \geq 3$, $\mathscr{E}^2 \subset \mathscr{E}^n$. [Hint: You may want to use Lemma 6.3(d).]

Theorem 6.9 For all n, $\mathscr{E}^n \subset \mathscr{E}^{n+1}$.

SKETCH OF PROOF If we can show that for all $i \leq n, f_i \in \mathscr{E}^n$ then the theorem will be proved since, except for f_i and f_j, the initial functions of $\mathscr{E}^i$ are the same as those of $\mathscr{E}^j$ and both classes are closed under the same operations. Exercise 6.16 gives $\mathscr{E}^0 \subset \mathscr{E}^1 \subset \mathscr{E}^2 \subset \mathscr{E}^3$ and $\mathscr{E}^2 \subset \mathscr{E}^n$ for all $n \geq 3$. So, inductively, suppose that for $i < n$ and any $n > 2$, it is true that $f_i \in \mathscr{E}^{n+1}$; it must be shown that $f_{i+1} \in \mathscr{E}^{n+1}$. By definition and Lemma 6.3(d),

$$\begin{aligned} f_{i+1}(0, y) &= f_i(y + 1, y + 1) \\ f_{i+1}(x + 1, y) &= f_{i+1}(x, f_{i+1}(x, y)) \qquad (*) \\ f_{i+1}(x, y) &< f_{n+1}(x, y) \end{aligned}$$

where f_i and $f_{n+1} \in \mathscr{E}^{n+1}$. But, because of (*) this is not a simple limited recursion scheme, and so we cannot conclude that $f_{i+1} \in \mathscr{E}^{n+1}$ simply because $\mathscr{E}^{n+1}$ is closed under limited recursion. In the last section, we saw how to avoid the recursion scheme by what might be called "coding everything up" so as to have the function defined by a single formula. A similar trick will be used here, only in this case it is more complicated and might be called a "double coding." This is required because of (*); notice that in the standard recursion we would have $f_{i+1}(x+1, y) = h(x, f_{i+1}(x, y))$ for some *previously* defined h and hence that h could be used, as in section 6.3. Define, for this discussion, $p(x, y)$ to be the $(J(x, y))$th prime. As in the previous section, there must be a number into which are coded many numbers satisfying certain relationships required by the definition of f_{i+1}; somehow retrievable from that number must also be the value of $f_{i+1}(x, y)$. That is, we must settle on some m which has roughly the following properties:

(a) For any $y \leq m$, the exponent of $p(0, y)$, in the prime decomposition of m, should be $f_i(y+1, y+1)$.

(b) The exponent of $p(x+1, y)$ should be $f_{i+1}(x, f_{i+1}(x, y))$, where $f_{i+1}(x, y)$ is the exponent of $p(x, y)$. But since we cannot use f_{i+1} in the definition, it must also be the case that the exponent of $p(x, y)$ is $f_{i+1}(x-1, f_{i+1}(x-1, y))$, etc., inductively down.

As in the discussion of the history function, we assume that the exponents of primes, corresponding to values of f_{i+1}, have one added. Recall (section 4.4, paragraph 5) that the function $P(n, m)$ is defined to have as its value the exponent of the nth prime in the prime factorization of m. Define

$$g(x, y) = p(x, f_{n+1}(x, y)^{(x+2) \cdot f_{n+1}(x+y)})$$

Using these functions in the conjunction with the outline just above, we define: $f_{i+1}(x, y)$ is to have as its value the smallest $z < f_{n+1}(x, y)$ which satisfies the following conditions: There is an $m \leq g(x, y)$ such that

(i) $z + 1 = P(p(x, y), m)$, and

(ii) for each $v \leq m$, if $P(p(0, v), m) \neq 0$, then $P(p(0, v), m) = (f_i(v+1, v+1) + 1)$, and

(iii) for each t, v, and $w \leq m$, if $t \neq 0$ and $t = P(p(w+1, v), m)$, then

$$t = (P(p(w, P(p(w, v), m)), m) \dot{-} 1)$$

If we were to complete the proof properly, it would be necessary to show that for all $n \geq 3$, the functions p and P are in $\mathscr{E}^n$ and that $\mathscr{E}^n$ is closed under bounded μ, the process of choosing m ("bounded existential quantification"), "and," "less than," etc. The proofs of these facts are omitted. [See §2 and Theorems 4.5 and 4.6 of Grzegorczyk (1953).] A complete proof would also require verifying that the manner in which f_{i+1} is defined just above is, in fact, the f_{i+1} originally given. ▮

Exercise 6.17 The reader who wants to be able to use these "coding" and "double coding" tricks should compare this and section 6.3. Show how to do the coding there in terms of powers of primes and show how to do the double coding here in terms of remainders.

Lemma 6.4 $\bigcup_{n \in N} \mathscr{E}^n \subset \mathscr{P}$, the class of primitive recursive functions.

Exercise 6.18 Prove the lemma. (Hint: Show explicitly for f_0, f_1, f_2 and refer to exercise 4.33.)

For our investigation of $\bigcup_{n \in N} \mathscr{E}^n$, we define another class of functions. Let $\mathscr{W}^n$ be the smallest class of functions that includes J, K, L, $x + y$, x^2, and f_n, and is closed under composition and explicit transformation. By an unproved theorem of this section, $\mathscr{W}_1^n$ is inductively defined and so is the smallest class of functions containing J, K, x^2, and $f_n(K(x), L(x))$ which is closed under composition and summation of functions, i.e., $f, g \in \mathscr{W}_1^n$, then $(f + g) \in \mathscr{W}_1^n$.

Exercise 6.19 (a) Show that the function $g_n \in \mathscr{W}^n$, where g_n is defined by $g_n(x, y) = f_n(x + 1, y + 1)$. (b) Show that for all $n \geq 2$, the initial functions of $\mathscr{E}^n$ are dominated by the function g_n of part (a).

Theorem 6.10 If f is a function of order k and $f \in \mathscr{W}_1^n$ for $n \geq 2$, then $f(x) < f_{n+1}(k, x)$.

PROOF (By induction on k, the order of f) Lemma 6.3 will be used throughout. If $k = 0$,

$$x^2 < (x + 2)^2 = f_3(0, x) \leq f_{n+1}(0, x) \qquad \text{for all} \quad n \geq 2$$

$$K(x),\ L(x) \leq x \leq f_{n+1}(0\ x)$$

$$f_n\big(K(x), L(x)\big) \leq f_n(x, x) < f_n(x + 1, x + 1)$$
$$= f_{n+1}(0, x)$$

For the induction step we must show that if f is of order $k + 1$ then the theorem holds. f being of order $k + 1$ means that it is defined by composition, explicit transformation, or summation of functions by means of functions g and h and that max(order g, order h) $= k$. Without loss of generality, we can assume both have order k. (Why?) By the induction hypothesis,

$$g(x) < f_{n+1}(k, x) \qquad \text{and} \qquad h(x) < f_{n+1}(k, x)$$

If f is defined by $f(x) = g(h(x))$, then

$$f(x) = g\big(h(x)\big) < f_{n+1}\big(k, h(x)\big) < f_{n+1}\big(k, f_{n+1}(k, x)\big) = f_{n+1}(k + 1, x)$$

If f is defined by $f(x) = g(x) + h(x)$, then

$$\begin{aligned} f(x) = g(x) + h(x) &< 2f_{n+1}(k, x) \\ &< (f_{n+1}(k, x) + 1)^2 \\ &= f_3(0, f_{n+1}(k, x)) \\ &< f_{n+1}(k, f_{n+1}(k, x)) \\ &= f_{n+1}(k + 1, x) \quad ▮ \end{aligned}$$

Theorem 6.11 The function $f_{n+1}(x, x)$ increases faster than any function in $\mathscr{E}_1{}^n$.

PROOF (By induction on n) For $n = 0$: Claim that, if $f \in \mathscr{E}_1{}^0$ and f is of order k, then $f(x) < x + 2^k + 1$. If that is so, then since there is an n such that for all $x \geq n$, $x + 2^k + 1 < 2x$, $f_1(x, x) = 2x$ increases faster than any function in $\mathscr{E}_1{}^0$. The claim is proved by induction on k, the only difficulty being a minor one in the case for composition. That is, if f is of order $k + 1$ and defined by composition from g and h, then by the induction hypothesis, $g(x) < x + 2^k + 1$ and $h(x) < x + 2^k + 1$. Therefore, $g(x) \leq x + 2^k$ and so

$$f(x) = h(g(x)) < g(x) + 2^k + 1 \leq x + 2^k + 2^k + 1 = x + 2^{k+1} + 1$$

For $n = 1$: By a similar trick, the reader can verify that if f is of order k and belongs to $\mathscr{E}_1{}^1$, then $f(x) \leq (x + 1)2^{2^k}$. Consequently, since there is an m such that for all $x \geq m$,

$$f(x) < (x + 1)2^{2^k} < (x + 1)^2 = f_2(x, x)$$

the theorem holds.

For $n \geq 2$: If f is of order k and belongs to $\mathscr{E}_1{}^n$, then by the previous theorem, for $x \geq k$, $f(x) < f_{n+1}(x, x)$. That is, f_{n+1} increases faster than any function in $\mathscr{W}_1{}^n$. Then, with $\mathscr{W}_1{}^n$ for $\mathscr{X}_1$ and $\mathscr{E}_1{}^n$ for $\mathscr{Y}_1$, by combining Theorem 6.6 and exercise 6.19, we can conclude that $f_{n+1}(x, x)$ increases faster than all functions in $\mathscr{E}_1{}^n$. ▮

Corollary $\mathscr{E}^n \neq \mathscr{E}^{n+1}$.

Let $\mathscr{P}$ be the class of primitive recursive functions and $\mathscr{P}_1$ those that are unary. As we have seen in section 6.3, one can define the class of recursive functions differently than in the original definition of section 4.6. There is a somewhat similar proof, not given here, of the following theorem.

Theorem † $\mathscr{P}_1$ is equal to the smallest class of functions which includes the successor S and the excess over a square E and is closed under summation

† This is Theorem 3 in R. M. Robinson (1947, p. 940).

of functions, composition, and iteration, i.e., if $h \in \mathscr{P}_1$ and $f(x) = h^x(0)$, then $f \in \mathscr{P}_1$.

Lemma 6.5 If f is a function of order $\leq n$ and $f \in \mathscr{P}_1$, then $f \in \mathscr{E}^{3+n}$.

PROOF (By induction on n): For $n = 0$, by definition and exercise 6.16, the lemma holds. Suppose the lemma holds for functions of order $\leq k$ and show for $k + 1$. If f, of order $k + 1$, is defined by $f(x) = g(x) + h(x)$ or $f(x) = h(g(x))$, where g and h satisfy the induction hypothesis, then $f \in \mathscr{E}^{k+3}$ and therefore is in $\mathscr{E}^{k+1+3}$. This is because, for all n, $x + y \in \mathscr{E}^{n+3}$ and $\mathscr{E}^{n+3}$ is closed under composition. So, what must be shown is that if f, of order $k + 1$, is defined by $f(x) = h^x(0)$ for h of order k and a member of $\mathscr{P}_1$, then $f \in \mathscr{E}^{k+1+3}$. By exercise 6.19 for all n, $f_{n+3}(x + 1, y + 1) \in \mathscr{W}^{n+3}$ dominates the initial functions of $\mathscr{E}^{n+3}$. Thus, $\mathscr{W}^{k+3}$ and $\mathscr{E}^{k+3}$ satisfy the hypothesis of Theorem 6.5 with $\mathscr{W}^{k+3}$ for $\mathscr{X}$ and $\mathscr{E}^{k+3}$ for $\mathscr{Y}$. Consequently, for every $g \in \mathscr{E}^{k+3}$, there is a $g' \in \mathscr{W}^{k+3}$ such that $g \leq g'$; therefore, for the h in question, there is a corresponding h'. Both are unary functions, so by Theorem 6.10,

$$h'(x) \leq f_{k+3+1}(\text{order } h', x)$$

It is an exercise for the reader to verify (by induction on y) that

$$h^y(0) < f_{k+4}(\text{order } h' + y, 0)$$

This bound can then be used to define f by limited recursion, i.e.,

$$\begin{aligned} f(0) &= 0 \\ f(x + 1) &= h(f(x)) \\ f(x) &\leq f_{k+4}\ (\text{order } h' + x, 0) \end{aligned}$$

Thus, f satisfies the requirements for being in $\mathscr{E}^{k+1+3}$. ∎

Theorem 6.12 The class of functions $\mathscr{P}$ is equal to $\bigcup_{n \in N} \mathscr{E}^n$.

PROOF If f^k is a k-ary function in $\mathscr{P}$ and f' is defined by

$$f'(x) = f^k(K^1(x), K^2(x), \ldots, K^k(x))$$

then $f' \in \mathscr{P}_1$. By the lemma, $f' \in \mathscr{E}^{n+3}$, where n is the order of f'. But, for all n, $\mathscr{E}^{n+3}$ includes J and is closed under composition. Therefore, $f^k \in \mathscr{E}^{n+3}$ since

$$f^k(x_1, \ldots, x_k) = f'(J^k(x_1, \ldots, x_k))$$

Thus, $\mathscr{P} \subset \bigcup_{n \in N} \mathscr{E}^n$. The theorem follows by combining this with Lemma 6.4. ∎

Corollary The set of Kalmar-elementary functions is a proper subset of $\mathscr{P}$.

The following theorem, stated without proof, is quite interesting, especially as a relative measure of complexity.

Theorem † For $n > 2$, the class $\mathscr{E}^{n+1}$ includes the universal function for the class $\mathscr{E}_1{}^n$.

There are several further interesting theorems in Grzegorczyk's (1953) paper which have not been touched upon here—in particular, some about relations and about estimations. Once more, the reader is encouraged to read the original paper.

6.5 THE PREDICTABLY COMPUTABLE FUNCTIONS, OR THE RITCHIE HIERARCHY‡

"One sometimes objects to Church's thesis that we do not always know ahead of time how many steps will be required to compute $f(n)$ for a recursive function f. Let us refer to functions which have this (admittedly vague) property as strongly computable. The author's paper can be regarded as a highly suggestive first step toward making this notion mathematically precise."§ In section 6.4, a general notion of measure of complexity was introduced. In this section, a particular definition is given in the general setting of Tm's. For this discussion, we assume that the Tm's work from initial tape inscriptions with binary notation and print out answers in binary. If, in terms of instantaneous descriptions, a computation of some Tm is $\delta_0, \ldots, \delta_t$, then the *amount of tape used* is defined to be $\max_{0 \le i \le t} (|\delta_0|, \ldots, |\delta_t|) - 1$, i.e., the longest piece of tape (from leftmost nonblank to rightmost nonblank) that occurs during the computation. For a Tm T, define the function a_T by: $a_T(n)$ is the amount of tape used by T to compute $T(n)$ if $T(n)$ is defined, and $a_T(n)$ is not defined if $T(n)$ is not defined. Denote by $l(x_1, \ldots, x_n)$ the amount of tape necessary to inscribe $x_1, \ldots, x_n$ on a tape. In this section, we will be interested in functions f for which there is a Tm T such that $T \sim f$ and a_T is subject to certain conditions. Since we do not care which T it is that computes f, provided

† This is Theorem 4.12 of Grzegorczyk (1953, p. 38).

‡ The substance and theorems of this section are based directly on Ritchie (1963). The student is encouraged to read this paper.

§ From the review of Ritchie's paper by Myhill (1964).

that a_T satisfies the conditions, we will simply say "f is Turing computable and a_f satisfies so-and-so."

Theorem 6.13 (a) If h and g are Turing computable and f is defined by $f(x, y) = h(x, g(y))$, then

$$a_f(x, y) = l(x, y) + \max(l(x) + a_g(y), a_h(x, g(y)))$$

where

$$l(x) = \sum_{i=1}^{n} l(x_i) + (n - 1)$$

since there are single blank squares between x_i and x_{i+1} in the tape representation of $x_1, \ldots, x_n$.

(b) If h and g are Tc and f is defined from them by $f(0, y) = g(y)$ and $f(x + 1, y) = h(x, y, f(x, y))$, then

$$a_f(x, y) = 2\left(l(x, y) + \max\left(\max_{z<x}(a_h(z, y, f(z, x)), a_g(x))\right)\right).$$

(c) If g is Tc and f is defined by explicit transformation,† i.e., $f(x_1, \ldots, x_n) = g(\xi_1, \ldots, \xi_k)$, then

$$a_f(x_1, \ldots, x_n) = a_g(\xi_1, \ldots, \xi_k) + l(x_1, \ldots, x_n).$$

Exercise 6.20 Outline a proof of the theorem.

A function is defined to be *finite-automaton computable* if it is Turing computable by a right-moving (only) Turing machine. Let $\mathscr{F}^0$ denote the set of all finite-automaton-computable functions. For each $n \in N$, define the *constant function* C_n by $C_n(x) = n$, for all x.

Theorem 6.14 (a) The successor function S, the constant functions C_n, and the projection functions U_i^n are all in $\mathscr{F}^0$.

(b) If $g, h \in \mathscr{F}^0$ and $f(x, y) = g(x) + h(y)$, then $f \in \mathscr{F}^0$.

(c) If $g, h \in \mathscr{F}^0$ and $f(x) = h(g(x))$, then $f \in \mathscr{F}^0$.

Proof With the possible exception of U_i^n, part (a) is obvious. The machine to compute U_i^n cannot operate like the Tm designed for that purpose in chapter I. However, a right-moving Tm could compute U_i^n by moving from left to right erasing everything except x_i on a tape initially inscribed with $x_1, \ldots, x_i, \ldots, x_n$.

† The definition was given early in section 6.4.

Part (b). Only rough indications of what to do will be given, on the assumption that the reader, with enough time, patience, and inclination, could smooth it all out. Since $g, h \in \mathscr{F}^0$, there are machines M_g and M_h to compute them. Further, as all numbers are assumed to be represented in binary, both machines have the alphabet $\{*, 0, 1\}$. For simplification, we assume both g and h are unary functions. One way to design M_f is to imagine a tape with two tracks, where x, in binary, is written on one and y, in binary, on the other. Thus,

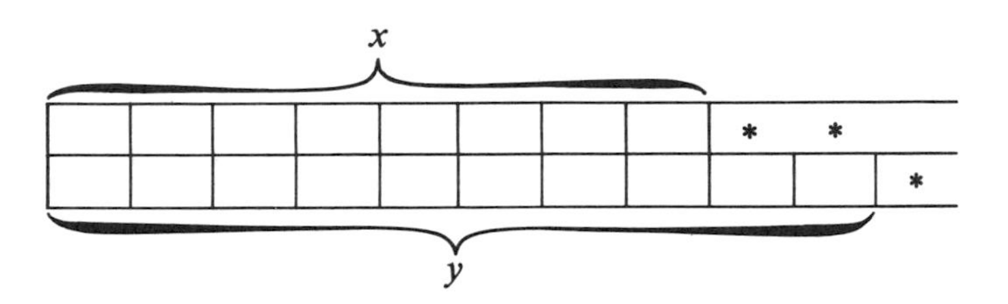

(For finite automata, we make the convention that if M computes from k-tuples, then the initial tape inscription for M is coded on a k-track tape in a manner similar to what has just been described.) As usual, we assume that the tapehead starts at the left side and we therefore assume that M_g and M_h compute g and h, respectively, so as to read left to right from least significance to most significance. We might imagine two tape-heads, one reading the upper track and one the lower, which join to combine their scanned information so as to decide what to print. They print, jointly, the same symbol (or one big symbol over the two tracks of that square). Another way of describing it is to say that there is a single tapehead with two parts, one computing one square of $g(x)$ and the other computing one square of $h(x)$, which adds up the results of that pair of squares, prints 0 or 1 accordingly, and moves right, remembering whether or not to carry 1. To be more specific, suppose M_g has internal states $Z_g = \{q_0, q_1, \ldots, q_m\}$ and M_h has $Z_h = \{q_0', q_1', \ldots, q_{m'}'\}$. Then M_f will have two copies of $Z_g \times Z_h$, one with superscript 0 and one with superscript 1, to indicate no carry and carry, respectively. The alphabet of M_f is $\{*, 0, 1\} \times \{*, 0, 1\} \cup \{*, 0, 1\}$. A typical quintuple of M_f would then be

$$(q_i, q_j')^0(1, 0)(q_k, q_n')^1 1R$$

To see how this works with respect to M_g and M_h, let x and y be variables for elements in the set $\{0, 1\}$ and let a, a', b, b', z be variables for elements of the set $\{*, 1, 0\}$. If $q_i a q_k a' R$ is a quintuple of M_g and $q_j' b q_n' b' R$ is a quintuple of M_h, then

$$(q_i, q_j')^x(a, b)(q_k, q_n')^y zR$$

is a quintuple of M_f, where y and z are functions of a', b', and x in the following way (read in columns):

a'	0	1	1	1	1	
b'	0	0	1	0	1	
x	0	0	0	1	1	etc.
y	0	0	1	0	1	
z	0	1	0	1	1	

There are many details to be attended to, as well as proving that the machine M_f outlined here really computes $g(x) + h(y)$. There are even more gory details involved in figuring out what to do when g and h are functions of more than one variable.

Part (c). The sketch here will be even rougher than above. The idea is somewhat similar, also simpler. There is only a single track on the tape and we design M_f so that where M_g would have read a square and printed a symbol, M_f reads the square, "thinks about" what symbol M_g would print, and "asks" M_h what it would print if it were reading that symbol. M_f then prints what M_h answers. Assuming the same alphabets and internal states for M_g and M_h as in part (b), then the alphabet of M_f is just $\{*, 0, 1\}$ and the set of internal states is just $Z_g \times Z_h$. Thus, if $q_i a q_k bR$ is a quintuple of M_g and $q_j' b q_n' cR$ is a quintuple of M_h, then

$$(q_i, q_j')a(q_k, q_n')cR$$

is a quintuple of M_f. ∎

Corollary $\mathscr{F}^0$ is closed under explicit transformation.

Exercise 6.21 Suppose g and h are any two Turing-computable functions. Obviously, if $f(x, y) = g(x) + h(y)$, then f is Tc. Show that

$$a_f(x, y) \le a_g(x) + a_h(y)$$

Exercise 6.22 (a) Show that there is a Tm T which, with x in binary as its initial tape inscription, will compute and print out 2^x in binary. (b) Show that there is a Tm T satisfying (a) with the property that $a_T(x) \le x + 1$.

Notice that the following inequalities hold:

1. $l(x) \le a_f(x)$.
2. $l(f(x)) \le a_f(x)$.

3. $l(f(x) + g(y)) \leq l(f(x)) + l(g(y)) \leq a_f(x) + a_g(y)$.
4. $f(x) < 2^{l(f(x))}$, since we are dealing with binary representations.

Let $\mathscr{F}^0$ be the class of finite-automaton-computable functions, already established, and for $i > 0$, define f as belonging to $\mathscr{F}^i$ if there is a Turing machine T such that $T \sim f$ and $a_T \leq g$ for some $g \in \mathscr{F}^{i-1}$. $\bigcup_{n \in N} \mathscr{F}_n$ is called the class of *predictably computable functions*; it is also known as the *Ritchie hierarchy*. That $\bigcup_{n \in N} \mathscr{F}^n$ really is a hierarchy remains to be shown.

Lemma 6.6 If g, $h \in \mathscr{F}^i$ and f is defined by $f(x, y) = g(x) + h(y)$, then $f \in \mathscr{F}^i$.

PROOF (By induction on i, with Theorem 6.14 accounting for $i = 0$): For the induction hypothesis, assume that the lemma holds for $i = k$ and show for $k + 1$. Let g, $h \in \mathscr{F}^{k+1}$. By definition, there are g^*, $h^* \in \mathscr{F}^k$ such that

$$a_f(x) + a_g(y) < g^*(x) + f^*(y)$$

If f^* is defined by $f^*(x, y) = g^*(x) + h^*(y)$, then, by the induction hypothesis, $f^* \in \mathscr{F}^k$. By exercise, 6.21, $a_f \leq a_g + a_h$ and therefore $a_f < f^*$. Thus, f satisfies the requirements for belonging to $\mathscr{F}^{k+1}$. ∎

Exercise 6.23 Show (a) for all $k \in N$, there exists a right-moving Tm M_k such that for all x, $M_k(x) = kx$, and (b) for all k and i, if $f \in \mathscr{F}^i$ and $g(x) = kf(x)$, then $g \in \mathscr{F}^i$.

Lemma 6.7 If $f \in \mathscr{F}^0$, then there exists a k such that $a_f(x) \leq l(x) + k$.

Lemma 6.8 For all $n \in N$, $\mathscr{F}^n \subset \mathscr{F}^{n+1}$.

Exercise 6.24 Prove the lemmas.

Lemma 6.9 If $f^n \in \mathscr{F}^0$, then there is a $K \in N$ such that for all $x_1, \ldots, x_n$, $f^n(x_1, \ldots, x_n) < K \max(x_1, \ldots, x_n, 1)$.

PROOF Since we are assuming the convention that a finite automata working from an n-tuple is to have the n-tuple coded up as if on an n-track tape, the length of that coded-up number is the length of $\max(x_1, \ldots, x_n, 1)$. By Lemma 6.7, there is a k such that

$$l(f^n(x_1, \ldots, x_n)) < l(\max(x_1, \ldots, x_n, 1)) + k.$$

It is also true that

$$f^n(x_1, \ldots, x_n) < 2^{f^n(x_1, \ldots, x_n)}$$

and for y in binary, $2^{l(y)} \leq 2y$. Consequently,

$$\begin{aligned} f^n(x_1, \ldots, x_n) &< 2^{l(\max(x_1, \ldots, x_n, 1))+k} \\ &= 2^{l(\max(x_1, \ldots, x_n, 1)} \cdot 2^k \\ &\leq 2^{k+1} \max(x_1, \ldots, x_n, 1) \end{aligned}$$

So, let $K = 2^{k+1}$. ▮

Lemma 6.10 For all $i \in N$, $\mathscr{F}^i$ is closed under explicit transformation.

Proof (By induction on i with the case $i = 0$ accounted for by the corollary to Theorem 6.14): Suppose the lemma holds for $i = k$ and show for $k + 1$. If $g^n \in \mathscr{F}^{k+1}$ and f^m, $m \leq n$, is defined by $f^m(x_1, \ldots, x_m) = g^n(\xi_1, \ldots, \xi_n)$, then by Theorem 6.13(c), $a_f(x_1, \ldots, x_m) = a_g(\xi_1, \ldots, \xi_n) + l(x_1, \ldots, x_m)$. By the definition of $\mathscr{F}^{k+1}$, there exists a $g^* \in \mathscr{F}^k$ such that $a_g \leq g^*$. By the induction hypothesis and Lemma 6.7, the function f^* defined by $f^*(x_1, \ldots, x_m) = g^*(\xi_1, \ldots, \xi_n) + l(x_1, \ldots, x_m)$ is in $\mathscr{F}^k$. But $a_f \leq f^*$, so $f \in \mathscr{F}^{k+1}$. ▮

Lemma 6.11 If $g \in \mathscr{F}^i$ and $h \in \mathscr{F}^j$, then $f \in \mathscr{F}^{i+j}$ for $f(x, y) = h(x, g(y))$.

Proof (By induction on j with i arbitrary): In general, by Theorem 6.13, we have

$$a_f(x, y) = l(x, y) + \max(a_g(y) + l(x), a_h(x, g(y))) \qquad (*)$$

For $j = 0$, by Lemma 6.7, there is a K such that

$$\begin{aligned} a_h(x, g(y)) &\leq l(x) + l(g(y)) + K, && \text{but} \\ &\leq l(x) + a_g(y) + K, && \text{for} \quad g^* \in \mathscr{F}^{i-1} \\ &\leq l(x) + g^*(y) + K \end{aligned}$$

Thus, $a_h(x, g(y))$ for $h \in \mathscr{F}^0$ is bounded by a function in $\mathscr{F}^{i-1}$. Using this and g^* in $(*)$, we have

$$\begin{aligned} a_f(x, y) &\leq l(x, y) + \max(g^*(y) + l(x), l(x) + g^*(y) + K) \\ &= l(x, y) + l(x) + g^*(y) + K \end{aligned}$$

But the right side defines a function belonging to $\mathscr{F}^{i-1}$ and so $f \in \mathscr{F}^{i+0}$. Suppose now that the lemma holds at $j \leq k$ and show for $k + 1$. That is, $h \in \mathscr{F}^{k+1}$ and therefore there is an $h^* \in \mathscr{F}^k$ such that $a_h \leq h^*$. Using these in $(*)$, we obtain

$$a_f(x, y) \leq l(x, y) + \max(g^*(y) + l(x), h^*(x, g(y)))$$

Since $g \in \mathscr{F}^i$ and $h^* \in \mathscr{F}^k$, by the induction hypothesis we have that the function k defined by $k(x, y) = h^*(x, g(y))$ is $\mathscr{F}^{i+k}$. Further, since $\mathscr{F}^i \subset \mathscr{F}^{i+k}$, the function f^* defined by $f^*(x, y) = l(x, y) + g^*(y) + l(x) + h^*(x, g(y))$ is in $\mathscr{F}^{i+k}$. Then $f \in \mathscr{F}^{i+k+1}$ since $a_f \leq f^*$. ▌

To assist in our investigation of $\bigcup_{n \in N} \mathscr{F}^n$ define the following set of functions:

$$f_0(x) = x, \qquad f_{n+1}(x) = 2^{f_n(x)}$$

Notice that for all n, $x < y$ implies $f_n(x) < f_n(y)$ and $x \leq f_n(x)$.

Exercise 6.25 Prove that for all n, $f_n \in \mathscr{F}^n$.

Lemma 6.12 For all $i \geq 1$ and for all $f \in \mathscr{F}^i$, there exists a K such that (a) $f(x) < f_i(K \max(x, 1))$ and (b) $a_f(x) < f_{i-1}(K \max(x, 1))$.

PROOF (By induction on i, of course): If $f \in \mathscr{F}^1$, for the corresponding $f^* \in \mathscr{F}^0$, Lemma 6.9 gives $a_f(x) < f^*(x) < K \max(x, 1)$ and thus

$$f(x) < 2^{l(f(x))} < 2^{a_f(x)} < 2^{K \max(x, 1)} = f_1(K \max(x, 1))$$

Suppose the lemma holds for $i = k$. If $f \in \mathscr{F}^{i+1}$, then there is an $f^* \in \mathscr{F}^i$ such that $a_f < f^* \in \mathscr{F}^i$. So by the induction hypothesis,

$$a_f(x) < f^*(x) < f_i(K^* \max(x, 1))$$

and so

$$f(x) < 2^{l(f(x))} < 2^{a_f(x)} < 2^{f_i(K^* \max(x, 1))} = f_{i+1}(K^* \max(x, 1)) \quad ▌$$

Theorem 6.15 For all n, $\mathscr{F}^n \subsetneq \mathscr{F}^{n+1}$.

PROOF Containment was shown in Lemma 6.8. To see that $\mathscr{F}^i \neq \mathscr{F}^{i+1}$, we will show that $f_{i+1} \notin \mathscr{F}^i$. Suppose $f_{i+1} \in \mathscr{F}^i$. Then there is a k such that for all x,

$$2^{f_i(x)} = f_{i+1}(x) < f_i(k \max(x, 1))$$

But this is false for large enough x. ▌

Lemma 6.13 For any $i, k > 0$, suppose $h \in \mathscr{F}^i$, $j \in \mathscr{F}^k$, and $g \in \mathscr{F}^{i+k}$. If f is defined by means of limited recursion,

$$f(0, x) = g(x)$$
$$f(n + 1, x) = h(n, x, f(n, x))$$
$$f(n, x) \leq j(n, x)$$

then $f \in \mathscr{F}^{i+k}$.

PROOF By Lemma 6.12 and the hypothesis of the lemma,

$$\begin{aligned} a_h(z, x, f(z, x)) &< f_{i-1}(K_h \max(z, x, f(z, x), 1)) \\ &\le f_{i-1}(K_h \max(z, x, j(z, x), 1)) \\ &< f_{i-1}(K_h \max(z, x, f_k(K_j \max(z, x, 1)), 1)) \end{aligned}$$

By Theorem 6.13 and this inequality, we have

$$\begin{aligned} a_f(n, x) &\le 2(\max(\max_{z<n} a_h(z, x, f(z, x)), a_g(x)) + l(n, x)) \\ &\le 2(\max(\max_{z<n} f_{i-1}(K_h \max(z, x, f_k(K_j \max(z, x, 1)), 1)), g^*(x)) + l(n, x)) \\ &= 2(\max(f_{i-1}(K_h \max(n, x, f_k(K_j \max(n, x, 1)), 1)), g^*(x)) + l(n, x)) \end{aligned}$$

because f_n is strictly increasing. Since $x \le f_n(x)$,

$$\begin{aligned} a_f(n, x) &\le 2(\max(f_{i-1}(K_h \cdot f_k(K_j \max(n, x, 1))), g^*(x)) + l(n, x)) \\ &\le 2(f_{i-1}(K_h \cdot f_k(K_j \max(n, x, 1))) + g^*(x) + l(n, x)) \end{aligned}$$

The proof will be complete if we can show that

$$f_{i-1}(K_h \cdot f_k(K_j \max(n, x, 1))) \in \mathscr{F}^{i+k-1}$$

But by exercise 6.23, $K_h \cdot f_k(K_j \max(n, x, 1)) \in \mathscr{F}^k$ and so the theorem follows by Lemma 6.11. ▮

Exercise 6.26 Show that the function $x^y \in \mathscr{F}^2$.

Theorem 6.16 The class of Kalmar-elementary functions is contained in

$$\bigcup_{n \in N} \mathscr{F}^n.$$

PROOF Using the theorem $\mathscr{E} = \mathscr{E}''$ stated in section 6.4, the theorem follows from exercise 6.26 and Lemmas 6.10, 6.11, and 6.13. ▮

Exercise 6.27 Let $f \in \bigcup_{i \in N} \mathscr{F}^i$ and $T_f \sim f$. Let t be the number of steps T_f uses to compute $f(n)$. Show that there exists a function g such that $t < g(n)$.

From section 5.3, we know that for every Tc function (i.e., recursive function) f, there are primitive functions m and q such that

$$f(n) = m(\mu_t(q(t, n) = H), n) \qquad (*)$$

where there is a t such that $q(t, n) = H$. The functions in $\bigcup_{i \in N} \mathscr{F}^i$ are all Tc and thus satisfy (*), but they also satisfy a further restriction about the length of tape necessary for computing the function value. Since t does exist, we know by exercise 6.27 that there is a function which bounds t. Thus, we can use it to represent functions in $\bigcup_{i \in N} \mathscr{F}^i$ by

$$f(x) = m(\mu_{t<g(n)}[q(t, n) = H], n) \qquad (**)$$

The question then arises as to what kind of function that function g is. If it is primitive recursive, then we know that every $f \in \bigcup_{i \in N} \mathscr{F}^i$ is a primitive recursive function. In fact, Ritchie shows something stronger. Recall that (**) is in terms of Tm's operating in unary notation, whereas in this section, all Tm's operate in binary. For Tm's in binary, Ritchie develops a function analogous to (**) in which all functions, including the function bounding t, are Kalmar-elementary. Then, since the Kalmar-elementary functions are closed under bounded μ and composition, the resulting function is Kalmar-elementary. Thus, he proves the following.

Theorem † The class $\bigcup_{i \in N} \mathscr{F}^i$ is equal to the class $\mathscr{E}$.

There are other investigations in Ritchie's (1963) paper, which the reader can look into. These include characterizing each $\mathscr{F}^i$ in terms of number-theoretic functions—without reference to Tm's—and studying relations whose characteristic functions belong to $\mathscr{F}^i$.

6.6 COMPLEXITY OF PARTIAL RECURSIVE FUNCTIONS ‡

As soon as one starts thinking about the notion of complexity in the context of functions, machines, and computations, many ideas come to mind. There have been many specific definitions of complexity; the literature on the subject is already large and growing larger. In this and the previous two sections, only three of the many possible developments are presented. At the end of this section, there is a short bibliography for the reader interested in these ideas. In section 6.4, complexity was considered from the point of view of formulation of functions. Hierarchies of functions were defined with a view toward allowing the functions at each succeeding level to become only slightly more complicated from this point of view. One could also say that, since none are more complicated in formulation than the primitive recursive functions,

† This is Theorem 3 (1963, p. 148).

‡ The material for this section is based directly on two papers: Blum (1967a, pp. 322–336) and Helm and Young (1971).

they are all fairly simple, indeed, primitive. In section 6.5, the Ritchie hierarchy was defined and studied. This hierarchy is an example of the use of complexity with respect to machines. The lowest level of the hierarchy is the class of finite-automaton-computable functions and certainly the finite automata are the simplest, with respect to mode of operation, among machines that do anything interesting. The functions at the $(n + 1)$th level of the hierarchy are those that can be computed by a Turing machine which uses an amount of tape bounded by a function at the nth level. The amount of tape used by a Turing machine is a natural way to view the complexity of a Turing machine computation. It is interesting that the functions computable within this hierarchy are still very primitive, so to speak.

In this section, we will be looking at all Turing machines or all partial recursive functions with complexity in mind. Consider the previously defined set $\mathcal{T}$ of Tm's: $T_0, T_1, T_2, \ldots$. As we know, for any function f which is Turing computable in general, there are infinitely many Tm's which compute f. If T_j is a Tm such that $T_j \sim f$, then we say *j is an index of f*. Although two numbers, say p and q, may be indices for some particular function f, we know that T_p and T_q are not similar because of the way in which the set $\mathcal{T}$ was defined. T_p and T_q are different machines for computing f and, more generally, may be thought of as different programs. One measure of the complexity of a Turing machine is its "size." The most obvious definition of the size of Tm T is simply the number of quintuples of T. Another formulation of complexity is by means of the "amount of tape used by T function" a_T, introduced in the previous section. For a given n, generally, $a_{T_p}(n)$ and $a_{T_q}(n)$ are not equal. We can also define a "number of steps required by a T function" $b_T : b_T(n)$ is equal to the number of steps used by T in computing $T(n)$ for each n. Then, as for a_T, $b_{T_p}(n)$ and $b_{T_q}(n)$ are usually not equal, for a given n. Notice that the amount-of-tape function and the step-counting function share the following properties: (i) they are partial recursive; (ii) $a_T(n)$ is defined if and only if $T(n)$ is defined, and $b_T(n)$ is defined if and only if $T(n)$ is defined; and (iii) there exist total recursive functions A and B such that

$$A(i, n, m) = \begin{cases} 1 & \text{if } \ a_{T_i}(n) = m \\ 0 & \text{otherwise} \end{cases}$$

$$B(i, n, m) = \begin{cases} 1 & \text{if } \ b_{T_i}(n) = m \\ 0 & \text{otherwise} \end{cases}$$

Exercise 6.28 Prove that (iii) is true.

One might well ask, For a given function f, is there a simplest or fastest way to compute f? For example, is there an index i of f such that for all indices j of f, $a_{T_i}(n) \le a_{T_j}(n)$ or $b_{T_i}(n) \le b_{T_j}(n)$? If p and q are indices of f, instead of just wondering if $b_{T_p}(n)$ is less than $b_{T_q}(n)$ for all n, we might ask if they fit into a particular relationship. For example, does $2b_{T_p}(n) < b_{T_q}(n)$ hold for all n, or infinitely many n, or all but a finite number of n, or any n at all? If this relationship is true for all n, or even for all but a finite number of n,† it would be reasonable to say that computing f by means of T_p is "twice as fast" as computing it by means of T_q. Instead of specifying "twice as fast," one might ask if there is some function r such that $b_{T_p}(n) > r(n, b_{T_q}(n))$ for all or almost all n. Such an r would provide a measure on the rate of computing f by means of T_p compared to computing f by means of T_q. When there is such an r, we say that "machine (or program) T_q is an r speed-up of machine (or program) T_p." Another approach is to ask for a function f having the property that for any index i of f, there is an index j of f such that $a_{T_i}(n) > a_{T_j}(n)$, or, $b_{T_i}(n) > 2^{b_{T_j}(n)}$, etc., for all n. Or, given some function r, is there an f with the property that for any index i of f, there is an index j of f such that $b_{T_i}(n) > r(n, b_{T_j}(n))$ for almost all n? Of course, computing f by means of T_j instead of by T_i may be better from one point of view, say length of tape, but worse from another, say number of steps, or size.

The ideas just introduced can be studied in a more general setting. Let $\mathscr{K}$ be the partial recursive functions as usual: $g_0, g_1, g_2, \ldots$. Recall that the *s-m-n* theorem (Theorem 5.12) and the recursion theorem (Theorem 5.13) hold for $\mathscr{K}$ and that there is a universal partial recursive function (Theorem 5.10). To each g_i, we associate as a *complexity function* any partial recursive function C_i satisfying the properties that (i) $C_i(n)$ is defined if and only if $g_i(n)$ is defined and (ii) there is a total recursive function M such that

$$M(i, n, m) = \begin{cases} 1 & \text{if } \; C_i(n) = m \\ 0 & \text{otherwise} \end{cases}$$

M is referred to as a *measure on computation*. Any unary recursive function h is used as a *size function* if for each m, there are at most finitely many (maybe no) n such that $h(n) = m$. A size function will be denoted by the absolute value sign $|\;|$; thus, $|i|$ is "the size of g_i." Any total recursive binary function r will be admitted as a *speed-comparing function*. The main results of this section are that for any speed-comparing function r, there is a total recursive function f, $f\colon N \to \{0, 1\}$, such that if i is an index of f, then there is another index j of f such that

$$C_i(n) > r(n, C_j(n)) \qquad \text{for almost all} \quad n \tag{*}$$

This means that there is a class of functions such that for any program to

† It is customary to say "for almost all n" instead of "for all but a finite number of n."

compute a function in this class, there is another program giving the result faster for almost all n. However, there are drawbacks. The first is obvious; the finitely many n where (*) is not true may be just those n one is interested in computing. The second is that one cannot find in an effective way the j whose existence is guaranteed. Finally, for r used slightly differently than in (*), there is no effective bound on the size of g_j.

Of course, since each C_i is a partial recursive function, there is a k such that $g_k = C_i$. It is of interest to know if there is an effective way to find k for a given i.

Theorem 6.17 There exists a total recursive function D such that $C_i = g_{D(i)}$ for all i.

Exercise 6.29 Prove the theorem. (Hint: Use M.)

Exercise 6.30 Show that to every total recursive function g there corresponds a total recursive function f, f: $N \to \{0, 1\}$, which is so complex that for all T_i such that $T_i \sim f$, $b_{T_i}(n) > g(n)$ for infinitely many n.

Theorem 6.18 † For every total recursive binary function r, there is a total recursive function f, f: $N \to \{0, 1\}$, with the property that for every index i of f, there is another index j of f, such that

$$C_i(n) > r(n, C_j(n)) \qquad \text{for almost all} \quad n$$

PROOF (1) Consider an array of all the values of partial recursive functions as follows:

$$\begin{array}{cccc} g_0(0) & g_0(1) & g_0(2) & \cdots \\ g_1(0) & g_1(1) & g_1(2) & \cdots \\ g_2(0) & g_2(1) & g_2(2) & \cdots \\ \vdots & \vdots & \vdots & \end{array}$$

The general term in the array, $g_u(v)$, lies in the uth row and the vth column. For each pair (u, v), define the set

$$B_{uv} = \{g_k(m) \mid (m < v \text{ and } k \leq m) \quad \text{or} \quad (m \geq v \text{ and } u \leq k \leq m)\}$$

Notice that the members of B_{00} are all the elements of the array that are on or above the diagonal: as u and v increase, the size of B_{uv} decreases. Define $B_{uv}^n = \{g_k(n) \mid g_k(n) \in B_{uv}\}$, i.e., the intersection of B_{uv} and the nth column.

† This is Theorem 6 of Blum (1967b, p. 326), which is sometimes referred to as "Blum's speed-up theorem."

For all given l, u, and v define the partial recursive function $g_{t(u,v,l)}$ and the set $X_{t(u,v,l)}$, inductively on n, as follows:

Let $X^0_{t(u,v,l)} = \varnothing$.

Case 1. If $v < u$, let $g_{t(u,v,l)}(n) = g_{t(u,u,l)}(n)$.

Case 2. If $u \leq v$, and

(a) if

$$n < v \text{ and any of } g_l(0), \ldots, g_l(n) \text{ are not defined,}$$

or if

$$n \geq v \text{ and any of } g_l(0), \ldots, g_l(n-u) \text{ are not defined,}$$

then

$$g_{t(u,v,l)}(n) \text{ is not defined and } X^{n+1}_{t(u,v,l)} = X^n_{t(u,v,l)};$$

(b) otherwise, define

$$D^n_{t(u,v,l)} = \{g_k(n) \mid g_k(n) \in B^n_{uv} \quad \text{and} \quad C_k(n) \leq g_l(n-k)\}$$

If $D^n_{t(u,v,l)} = \varnothing$, define $g_{t(u,v,l)}(n)$ to have the value 0 and $X^{n+1}_{t(u,v,l)}$ to be $X^n_{t(u,v,l)}$. If $D^n_{t(u,v,l)} \neq \varnothing$, look for the smallest k such that $g_k(n) \in D^n_{t(u,v,l)}$ and for no $m < n$ does $g_k(m) \in X^n_{t(u,v,l)}$. If there is no such k, set $g_{t(u,v,l)}(n) = 0$ and $X^{n+1}_{t(u,v,l)} = X^n_{t(u,v,l)}$. If there is such a k, define

$$g_{t(u,v,l)}(n) = \begin{cases} 0 & \text{if } \; g_k(n) \neq 0 \\ 1 & \text{if } \; g_k(n) = 0 \end{cases}$$

and $X^{n+1}_{t(u,v,l)} = X^n_{t(u,v,l)} \cup \{g_k(n)\}$. (End of definition of $g_{t(u,v,l)}$.) Define $X_{t(u,v,l)} = \bigcup_{n \in N} X^n_{t(u,v,l)}$.

(2) Show that there is a total recursive function g_l such that for all u and v,

$$r(n, C_{t(u,v,l)}(n)) \leq g_l(n-u+1) \qquad \text{for almost all} \quad n$$

Define the function h, obviously partial recursive, as follows:

$$h(i, 0) = 0$$
$$h(i, z+1) = \max_{\substack{0 \leq u \leq z \\ 0 \leq v \leq z}} r(z+u, C_{t(u,v,i)}(z+u))$$

By the *s-m-n* theorem (twice), there is a total recursive function s such that for all i and z,

$$g_{s(i)}(z) = h(i, z)$$

By the recursion theorem, there is a number l such that for all z,

$g_l(z) = g_{s(l)}(z)$. We will show, by induction on z, that g_l is total. At $n = 0$, $g_l(0) = g_{s(l)}(0) = h(l, 0) = 0$. Assuming that g_l is defined for $n \leq z$, we will show that it is defined at $z + 1$.

Exercise 6.31 Show that if g_l is defined for $n \leq z$ and $u, v \leq z$, then

$$g_{t(u, v, l)}(z + u)$$

is defined.

By definition,

$$g_l(z + 1) = \max_{\substack{0 \leq u \leq z \\ 0 \leq v \leq z}} r(z + u, C_{t(u, v, l)}(z + u))$$

By the induction hypothesis and the exercise, $g_l(n)$ is defined for $n \leq z$ and if $u, v \leq z$, then $C_{t(u, v, l)}(z + u)$ is defined. Therefore, since r is total, for each u and v less than or equal to z, $r(z + u, C_{t(u, v, l)}(z + u))$ is defined and hence $g_l(z + 1)$ is also. To see that $r(n, C_{t(u, v, l)}(n)) \leq g_l(n - u + 1)$ for almost all n, consider the following: For any u and v, if $n > \max[2u, u + v]$, then $u, v \leq n - u$. For any such n, let $z = n - u$. Thus, for almost all n,

$$\begin{aligned} g_l(n - u + 1) &= \max_{\substack{0 \leq u \leq n-u \\ 0 \leq v \leq n-u}} r(n - u + u, C_{t(u, v, l)}(n - u + u)) \\ &\geq r(n, C_{t(u, v, l)}(n)) \end{aligned}$$

(3) Show that if g_l is total and f is defined to be $g_{t(0, 0, l)}$, then for each index i of f, $C_i(n) > g_l(n - i)$, for almost all n. Suppose not; then, there exists an i such that for infinitely many values $n_1, n_2, \ldots, n_j, \ldots$, $C_i(n_j) \leq g_l(n_j - i)$. Consider each n_j. Since g_l is total and $C_i(n_j) \leq g_l(n_j - i)$, $g_i(n_j) \in X_{t(0, 0, l)}$ unless there is some $n_{j'} < n_j$ such that $g_i(n_{j'}) \in X_{t(0, 0, l)}$. Whichever is the case, it follows that there is an $n \in \{n_1, n_2, \ldots, n_j, \ldots\}$ such that $g_i(n) \in X_{t(0, 0, l)}$. Therefore, on the one hand, by the definition of $g_{t(0, 0, l)}$ and the fact that i is an index for f,

$$g_{t(0, 0, l)}(n) \neq g_i(n) = f(n)$$

and on the other hand, by the definition of f,

$$f(n) = g_{t(0, 0, l)}(n).$$

Contradiction.

(4) Show that if g_l is total, then there exists a total recursive function F such that $g_{t(u, F(u), l)} = g_{t(0, 0, l)}$. Define

$$Y^u_{t(0, 0, l)} = \{g_k(m) \mid g_k(m) \in X_{t(0, 0, l)} \quad \text{and} \quad k < u\}$$

Obviously, $Y^u_{t(0, 0, l)}$ is a finite set and so there is a largest m such that $g_k(m) \in Y^u_{t(0, 0, l)}$. Define $F(u)$ to be 1 plus that m.

Case 1. $n < F(u)$.

Exercise 6.32 Verify the following for $n < F(u)$:

(a) $B^n_{00} = B^n_{uF(u)}$.

(b) $D^n_{t(0,0,t)} = D^n_{t(u,F(u),l)}$.

(c) $X^n_{t(0,0,l)} = X^n_{t(u,F(u),l)}$.

(d) For all n, $n < F(u)$, $g_{t(u,F(u),l)}(n) = g_{t(0,0,l)}(n)$.

Case 2. $n \geq F(u)$. Notice that it follows from the definition of F that if $m \geq F(u)$ and $g_k(m) \in X_{t(0,0,l)}$, then $k \geq u$. Thus, for $n \geq F(u)$, if there is a k such that $g_k(n) \in D^n_{t(0,0,l)} \cap X_{t(0,0,l)}$, $k \geq u$. Define

$$D^{n,u}_{t(0,0,l)} = \{g_k(n) \mid g_k(n) \in D^n_{t(0,0,l)} \quad \text{and} \quad u \leq k\}$$

Exercise 6.33 Show that $D^{n,u}_{t(0,0,l)} = D^n_{t(u,F(u),l)}$.

Thus, if $D^{n,u}_{t(0,0,l)} = D^n_{t(u,F(u),l)} = \varnothing$, $g_{t(u,F(u),l)}(n) = 0$; and if they are not empty, the values of $g_{t(0,0,l)}(n)$ and $g_{t(u,F(u),l)}(n)$ depend on the same set. Assume they are not empty. Let k^* be the smallest k such that $g_{k^*}(n) \in D^{n,u}_{t(0,0,l)}$ and for no $m < n$ does $g_{k^*}(m) \in X_{t(0,0,l)}$, if there is such a k^*. Furthermore, let k^{**} be the smallest k such that $g_{k^{**}}(n) \in D^n_{t(u,F(u),l)}$ and for no $m < n$ does $g_{k^{**}}(m) \in X_{t(u,F(u),l)}$, if there is such a k^{**}.

Exercise 6.34 Prove that k^* exists if and only if k^{**} exists and that if they do exist, they are equal. (Hint: By induction on n and use the fact from Case 1 that $X^{F(u)}_{t(0,0,l)} = X^{F(u)}_{t(0,F(u),l)}$.)

(5) The following summary shows that the theorem is proved.

(a) By step 2, there is a total recursive function g_l with the property that for all u and v,

$$g_l(n - u + 1) \geq r(n, C_{t(u,v,l)}(n)) \qquad \text{for almost all} \quad n$$

(b) Let $f = g_{t(0,0,l)}$, knowing that $g_{t(0,0,l)}$ is a total recursive function with range equal to $\{0, 1\}$. By step 3, for each i which is an index of f, $C_i(n) > g_l(n - i)$ for almost all n.

(c) For i an index of f, let $u = i + 1$ and $v = F(i + 1)$ and combine a and b to obtain $C_i(n) > g_l(n - i) \geq r(n, C_{t(i+1,F(i+1),l)}(n))$ for almost all n. By step 4, $g_{t(i+1,F(i+1),l)} = g_{t(0,0,l)} = f$. So the j promised by the theorem is $t(i + 1, F(i + 1), l)$. ∎

Exercise 6.35 Suppose r is a speed-comparing function such that $r(x, y) \geq y$ and f is the total recursive function associated with r by Theorem 6.19. (a)

Is it true that for any index i of f, there is an infinite sequence of indices of f, $i_0 = i, j_1, \ldots, j_m, j_{m+1}, \ldots$, such that for all m,

$$C_{j_m}(n) > r(n, C_{j_{m+1}}(n)) \qquad \text{for almost all} \quad n?$$

For each m, let $F_m = \{n \mid C_{j_m}(n) \geq r(n, C_{j_{m+1}}(n))\}$. As m increases, what can you say about F_m? (b) Is it possible for the following statement to be true? For every speed-comparing function r, there is a total recursive function f such that for any index i of f, there is an index j of f such that

$$C_i(n) > r(n, C_j(n)) \qquad \text{for } all \quad n$$

Exercise 6.36 Exhibit a total recursive function π which is onto N and has the property that for every $m \in N$, there are infinitely many $n \in N$ such that $\pi(n) = m$.

Exercise 6.37 Show that there is a total recursive function M such that for all n, $g_{M(i,j)}(n) = g_i(g_j(n))$.

Exercise 6.38 (a) Suppose g is a binary total recursive function. Define

$$m(u, v) = \max_{\substack{x \leq u \\ y \leq v}} [g(x, y)] + 1$$

and define $h(u) = m(u, u)$. Certainly h is total recursive. Verify that for all x, y, $g(x, y) < h(x) + h(y)$.

(b) Suppose g is a 4-ary total recursive function. Find a binary total recursive function h such that

$$g(x_1, x_2, x_3, x_4) < h(x_1, x_2) + h(x_3, x_4)$$

***Theorem 6.19*†** Let r be a speed-comparing function and f the function corresponding to it according to Theorem 6.18. If r is sufficiently large, there can be no total recursive function K such that $K(N) = \{i \mid i \text{ is an index of } f\}$.

PROOF Let π be a total recursive function with the property that for every $j \in N$, there are infinitely many $n \in N$ such that $\pi(n) = j$. Define $G(i, n) = g_{g_i(\pi(n))}(n)$ for all i and n. By exercise 6.37 and a few uses of the s-m-n theorem, there is a total recursive function s such that for all i and n,

$$g_{s(i)}(n) = g_{g_i(\pi(n))}(n)$$

Notice that one may think of the value $C_{s(i)}(n)$ as depending just on n and i,

† This is Corollary 1 of Blum (1967a, p. 329).

or as depending on n, i, $C_i(\pi(n))$, and $C_{g_i(\pi(n))}(n)$. Choosing the latter view we see there is a total recursive function g such that

$$C_{s(i)}(n) = g(n, i, C_i(\pi(n)), C_{g_i(\pi(n))}(n))$$

By exercise 6.38, there is a total recursive function h such that

$$C_{s(i)}(n) < h(n, C_{g_i(\pi(n))}(n)) + h(i, C_i(\pi(n))$$

Let r be any total recursive function such that $r(n, m) \geq n + h(n, m)$ and let f be the corresponding function according to Theorem 6.18. The proof of the theorem is by contradiction. Suppose K *is* a total recursive function such that $K(N) = \{i \mid i \text{ is an index of } f\}$. Since K is total recursive, it has an index, say k; i.e., $g_k = K$. Then,

$$g_{s(k)}(n) = g_{g_k(\pi(n))}(n) = g_{K(\pi(n))}(n) = f(n)$$

and so $s(k)$ is an index of f. Of course, for k,

$$C_{s(k)}(n) < h(n, C_{K(\pi(n))}(n)) + h(k, C_k(\pi(n)))$$

Thus, for each j and the infinitely many n for which $\pi(n) = j$, we have

$$\begin{aligned} C_{s(k)}(n) &< h(n, C_{K(j)}(n)) + h(k, C_k(j)) \\ &\leq r(n, C_{K(j)}(n)) - n + h(k, C_k(j)) \end{aligned}$$

For each j, there are infinitely many n such that $\pi(n) = j$ and $n > h(k, C_k(j))$. Thus, for each $j \in N$, there are infinitely many n such that

$$C_{s(k)}(n) \leq r(n, C_{K(j)}(n))$$

Thus $s(k)$ is an index of f for which there is no index of f, i.e., no $K(j)$ for some j, such that

$$C_{s(k)}(n) > r(n, C_{K(j)}(n)) \qquad \text{for almost all} \quad n$$

This contradicts Theorem 6.18. ∎

Notice that if f is a number-theoretic function, to say that $\lim_{n\to\infty} f(n)$ exists is equivalent to saying that there exists an $L \in N$ such that $f(n) = L$ for all but a finite number of n. Given a function f, $f\colon N \to \{0, 1\}$, if B is a function having the property that for all i such that $g_i = f$, $B(i)$ is defined and there exists a j such that $g_j = f$ and $|j| \leq B(i)$ and $C_i(n) \geq C_j(n + 1)$ for almost all n, then we say that B *bounds the size of speed-up of* f. In the lemma about to be stated, we see that there is a special function $g_{s(p)}$ such that if g_p is a partial recursive function which bounds the size of speed-up of a function f, certain relationships between g_p, $g_i = f$, and $g_{s(p)}$ must hold.

Lemma 6.14 There exists a total recursive function s such that for all p, (i) the binary function $g_{s(p)}$ is total, (ii) for all i, $\lim_{n\to\infty} g_{s(p)}(i, n)$ exists, (iii) if g_p is a partial recursive function bounding the size of speed-up of some function f and if $g_i = f$, then there is a $k \le \lim_{n\to\infty} g_{s(p)}(i, n)$ such that $g_k = f$ and $C_i(n) \ge C_k(n + i)$ for almost all n.

PROOF For fixed i and n, consider the set of all $(i + 1)$-termed sequences $(k_0, k_1, \ldots, k_i)$ which have the property that $k_0 = i < k_1 < \ldots < k_i < n$. For fixed p, i, and n, such a sequence is said to be "(p, i, n)-admissible" if (i) $C_p(k_x) \le n$ for $x = 0, 1, \ldots, i$, and (ii) $|k_{x+1}| \le g_p(k_x)$ for $x = 0, 1, \ldots, i - 1$. Clearly, for given p, i, n, the number of (p, i, n)-admissible sequences is finite. Almost equally clear is the fact that the set of (p, i, n)-admissible sequences is contained in the set of $(p, i, n + 1)$-admissible sequences. We claim that for each pair of fixed p and i, there is an n_0 such that for all $n \ge n_0$ the set of (p, i, n)-admissible sequences is equal to the set of (p, i, n_0)-admissible sequences. From condition (ii) for "admissibleness," we see that $|k_1| \le g_p(i)$. Then, since the size function $|\ |$ is finitely-many-to-one, there are only finitely many possible k_1 that can satisfy the condition. Repeating this argument, we see that for each of the finitely many possible k_1, there are only finitely many possible k_2 such that $|k_2| \le g_p(k_1)$, etc. For each p and i, define the function h as follows:

$$h(p, i, n) = \begin{cases} \max\{j \mid j = k_i \text{ where } (k_0, k_1, \ldots, k_i) \text{ is a } (p, i, n)\text{-admissible sequence}\} & \text{if there is such a } j \\ 0 & \text{otherwise} \end{cases}$$

Since h is obviously a partial recursive function, by the s-m-n theorem, there is a total recursive function s such that $g_{s(p)}(i, n) = h(p, i, n)$.

Exercise 6.39 Show (a) that $g_{s(p)}$ is total and (b) that for all p and i, $\lim_{n\to\infty} g_{s(p)}(i, n)$ exists.

To finish the proof of the lemma, notice that if g_p bounds the size of speed-up of some function f and $g_i = f$, then $g_p(i)$ is defined and there is a j such that $g_j = f$, $|j| \le g_p(i)$, and $C_i(n) \ge C_j(n + 1)$ for almost all n. For some i that is an index of f, let $k_0 = i$ and $k_1 = j$ for the j just mentioned. Repeat the analysis with k_1 in place of i to get k_2, etc. Thus, for some n large enough, we obtain a sequence $(k_0, k_1, \ldots, k_i)$ which is (p, i, n)-admissible and for which $C_{k_x}(n) \ge C_{k_{x+1}}(n + 1)$ for all $x < i$ and almost all n. For this sequence, we have that

$$C_i(n) = C_{k_0}(n) \geq C_{k_1}(n+1) \geq C_{k_2}(n+2) \geq \cdots \geq C_{k_i}(n+i)$$
for almost all n

and $k_i \leq \lim_{n\to\infty} g_{s(p)}(i, n)$. ∎

Theorem 6.20 † For every total recursive binary function r with the property that $r(x, y) \geq y$, there exists a total recursive function f, $f: N \to \{0, 1\}$, such that (a) for all i such that $g_i = f$, there exists a j such that $g_j = f$ and

$$C_i(n) > r(n+1, \sum_{m \leq n+1} C_j(m)) \qquad \text{for almost all} \quad n$$

and (b) there is no partial recursive function which bounds the size of speed-up of f.

PROOF The proof is very much like the proof of Theorem 6.18 with, of course, some additions due to the second part of this theorem.

Part I. Define the function $g_{t(u,v,l)}$ and the set $X_{t(u,v,l)}$ simultaneously by induction on n. Let $X^0_{t(u,v,l)} = \varnothing$.

Case 1. If $v < u$, then $g_{t(u,v,l)} = g_{t(u,u,l)}$.
Case 2. If $u \leq v$ and (a) $n < v$, and (1) if not all of $g_l(0), \ldots, g_l(n)$ are defined, then $g_{t(u,v,l)}(n)$ is not defined, or (2) if all of $g_l(0), \ldots, g_l(n)$ are defined, then (i) define

$$A_n = \{i \mid i \leq n, \quad C_i(n) < g_l(n-i), \quad \text{and} \quad i \notin X^n_{t(u,v,l)}\}$$

If $A_n \neq \varnothing$, let i^* be the smallest element in A_n. Define

$$g_{t(u,v,l)}(n) = \begin{cases} 0 & \text{if} \quad g_{i^*}(n) \neq 0 \\ 1 & \text{if} \quad g_{i^*}(n) = 0 \end{cases}$$

$$X^{n+1}_{t(u,v,l)} = X^n_{t(u,v,l)} \cup \{i^*\}$$

If $A_n = \varnothing$, then (ii) define

$$B_n = \{i \mid i \leq n, \quad \text{there is a} \quad j \leq i \quad \text{for which there is a} \quad k \leq g_{s(j)}(i, n) \\ \text{such that} \quad k \notin X^n_{t(u,v,l)} \quad \text{and} \quad C_k(n) \leq C_i(n-i) \leq g_l(n-i)\}$$

If $B_n \neq \varnothing$, let i^* be the smallest element in B_n. Let k^* be the smallest k such that $k \leq g_{s(j)}(i^*, n)$ for some $j \leq i^*$, $k \notin X^n_{t(u,v,l)}$, and $C_k(n) \leq C_{i^*}(n - i^*) \leq g_l(n - i^*)$. Define

$$g_{t(u,v,l)}(n) = \begin{cases} 0 & \text{if} \quad g_{k^*}(n) \neq 0 \\ 1 & \text{if} \quad g_{k^*}(n) = 0 \end{cases}$$

$$X^{n+1}_{t(u,v,l)} = X^n_{t(u,v,l)} \cup \{k^*\}$$

If $B_n = \phi$, define $g_{t(u,v,l)}(n)$ to be 0 and $X^{n+1}_{t(u,v,l)} = X^n_{t(u,v,l)}$.

† This is a theorem due to Helm and Young (1971, p. 22).

(b) Or, if $n \geq v$, and (1) if not all of $g_l(0), \ldots, g_l(n-u)$ are defined, then $g_{t(u, v, l)}(n)$ is not defined; or (2) if all of $g_l(0), \ldots, g_l(n-u)$ are defined, then proceed as in a 2(i) and (ii).

Define $X_{t(u, v, l)} = \bigcup_{n \in N} X^n_{t(u, v, l)}$.

Exercise 6.40 Show that the following are true when $u \leq v$: (a) For $m < v$, $g_{t(0, 0, l)}(m)$ is defined if and only if $g_l(0), \ldots, g_l(m)$ are all defined. (b) For $m \geq v$, $g_{t(u, v, l)}(m)$ is defined if and only if $g_l(0), \ldots, g_l(m-u)$ are defined. (c) If g_l is total, then for all u and v, $g_{t(u, v, l)}$ is total.

Part II. Show that there is an l such that g_l is a monotonic-increasing total recursive function with the properties that for all $u \leq v$,

$$\text{(i)} \qquad g_l(n-u+1) \geq r\Big(n, \sum_{m \leq n} C_{t(u, v, l)}(m)\Big) \qquad \text{for almost all } n$$

and therefore

$$\text{(ii)} \qquad g_l(x-j) \geq r\Big(x+1, \sum_{m \leq x+1} C_{t(j+2, v, l)}(m)\Big) \qquad \text{for almost all } x$$

Define the binary function h as follows:

$$h(i, 0) = 0$$
$$h(i, z+1) = \max\Big\{h(i, z), \max_{u \leq v \leq z}\Big(r\Big(z+u, \sum_{m \leq z+u} C_{t(u, v, i)}(m)\Big)\Big)\Big\}$$

It is clear that h is partial recursive and therefore by the *s-m-n* theorem there is a total recursive function s' such that for all i and z, $g_{s'(i)}(z) = h(i, z)$. By the recursion theorem, there is an l such that $g_l = g_{s'(l)}$.

Exercise 6.41 Show that the function g_l so defined has the required properties (i) and (ii).

Exercise 6.42 Show that there is a function F such that for all u and l, $g_{t(u, F(u), l)} = g_{t(0, 0, l)}$. [Hint: For each i, define the set $K_i = \{k \mid \text{there is a } j \leq i$ for which there is a $k \leq \lim_{n \to \infty} g_{s(j)}(i, n)\}$. Notice that the only difference in the definitions of $g_{t(0, 0, l)}$ and $g_{t(u, v, l)}$ is when $n \geq v$; in computing $g_{t(u, v, l)}(n)$, no $i \leq u$ and no member of K_i for $i \leq u$ is added to $X_{t(u, v, l)}$.]

Exercise 6.43 Show that if $g_i = g_{t(0, 0, l)}$, then

$$C_i(n) > g_l(n-i) \qquad \text{for almost all } n$$

Exercise 6.44 Show that for all total recursive functions r, such that $r(x, y) \geq y$, there is a total recursive function f, $f\colon N \to \{0, 1\}$, such that if

$g_i = f$, there is a j such that $g_j = f$ and $C_i(n) > r\left(n+1, \sum_{m \leq n+1} C_j(m)\right)$ for almost all n (i.e., first part of the theorem).

Part III. For the l estabished in Part II, we will see that there are infinitely many i such that $g_i = g_{t(0,0,l)}$ and $C_i(n) \leq g_l(n)$ for almost all n. Recall that the function r has the property that $r(x, y) \geq y$. Let i_0 be some index of $g_{t(0,0,l)}$, i.e., $g_{i_0} = g_{t(0,0,l)}$. Then, by exercises 6.43 and 6.44, there is an i_1 such that $g_{i_1} = g_{t(0,0,l)}$ and

$$\begin{aligned} C_{i_0}(n) &> g_l(n - i_0) \\ &\geq r\left(n+1, \sum_{m \leq n+1} C_{i_1}(m)\right) \\ &\geq \sum_{m \leq n+1} C_{i_1}(m) \\ &\geq C_{i_1}(n) \qquad \text{for almost all} \quad n. \end{aligned}$$

Since $C_{i_0}(n) > C_{i_1}(n)$, for almost all n, $i_0 \neq i_1$. Since g_l is monotonic increasing, $g_l(n) \geq g_l(n - i_0) \geq C_{i_1}(n)$. The same argument can be repeated taking i_1 for i_0 to get i_2, etc., to obtain the required infinite set.

Part IV. For the l established in Part II, we will show that no partial recursive function is a bound for the size of speed-up of $g_{t(0,0,l)}$. Suppose the contrary: there is a partial recursive function g_b which bounds the size of speed-up of $g_{t(0,0,l)}$. Since for all p and i, $\lim_{n \to \infty} g_{s(p)}(i, n)$ exists, if

$$k \leq \lim_{n \to \infty} g_{s(b)}(i, n)$$

for some i, then $k \leq g_{s(b)}(i, n)$ for that i and almost all n. By Part III, we know there are infinitely many i such that $i > b$ and $g_i = g_{t(0,0,l)}$ and $C_i(n) \leq g_l(n)$ for almost all n; let i_0 be one such i, say the smallest. By Lemma 6.14 and the assumption that g_b bounds the size of speed-up of $g_{t(0,0,l)}$, since $g_{i_0} = g_{t(0,0,l)}$ there is a $k \leq \lim_{n \to \infty} g_{s(b)}(i_0, n)$ and $g_k = g_{t(0,0,l)}$ and $C_{i_0}(n) \geq C_k(n + i_0)$ for almost all n. Let k_0 be the smallest such k. Thus, there is an $n_0 \geq i_0$ such that for all $n \geq n_0$, $k_0 \leq g_{s(b)}(i_0, n)$ and $C_{k_0}(n) \leq C_{i_0}(n - i_0) \leq g_l(n - i_0)$. Claim that $k_0 \in X_{t(0,0,l)}$ and therefore by the definition of $g_{t(0,0,l)}$, there is an n for which $g_{t(0,0,l)}(n) \neq g_{k_0}(n)$; this would contradict the fact that $g_{k_0} = g_{t(0,0,l)}$. If k_0 is added to $X_{t(0,0,l)}$ under case 2, b2(i), then done; so, suppose it is not. For the n_0 mentioned above, consider the sets $B_{n_0}, B_{n_0+1}, B_{n_0+2}, \ldots$, defined under case 2, b2(ii). Notice that i_0 belongs to each unless k_0 is added to $X_{t(0,0,l)}$ at some n. We will show that eventually there is an $n^* \geq n_0$ where i_0 is the smallest element of B_{n^*}. It then follows that

$$k_0 \in X_{t(0,0,l)}^{n^*+1}.$$

Suppose i^* is the smallest element in B_{n_0} and let k^* be the smallest k such that for some $j < i^*$, $k \leq g_{s(j)}(i^*, n_0)$ and $C_k(n_0) \leq C_{i^*}(n_0 - i_0) \leq g_l(n_0 - i_0)$ and $k \notin X_{t(0,0,l)}^{n_0}$. Then, $X_{t(0,0,l)}^{n_0+1} = X_{t(0,0,l)}^{n_0} \cup \{k^*\}$. Then, $i^* \in B_{n_0+1}$ only if there is a $j < i^*$ and there is a $k \leq g_{s(j)}(i^*, n_0 + 1)$ such that $C_k(n_0 + 1) \leq C_{i^*}(n_0 + 1 - i_0) \leq g_l(n_0 + 1 - i_0)$ and $k \notin X_{t(0,0,l)}^{n_0+1}$. Notice that there are only finitely many j less than i^* and for each pair j, i^*, only finitely many $k \leq g_{s(j)}(i^*, n)$ no matter how big n gets, because $\lim_{n \to \infty} g_{s(j)}(i^*, n)$ exists. Therefore, for some $n' > n_0$, $i^* \notin B_{n'}$. Suppose i' is the smallest i in $B_{n'}$, repeat the argument, etc. So, eventually there is an $n^* \geq n_0$ such that i_0 is the smallest element of B_{n^*}. ∎

Here, we outline briefly the approaches to complexity which are more or less connected with the subjects presented in this book. There are a few papers which might be said to constitute the beginning of the study of complexity. In addition to Grzegorczyk (1953), Ritchie (1963), and Blum (1967a), there are Hartmanis and Stearns (1964, 1965), Myhill (1960), Rabin (1960, 1963), and Yamada (1962). The basic ideas in these papers have led into, roughly speaking, four directions. (However, it must be emphasized that even these divisions do not exhaust the subject of complexity.):

1. Hierarchies of functions within the recursive functions. (These are usually referred to as "subhierarchies.")
2. The study of complexity defined in terms of specific computing devices.
3. Combinations of 1 and 2 [e.g., Ritchie (1963)].
4. Complexity of partial recursive functions.

For the first, the reader should also see Axt (1963). The subject matter of 2 has only been suggested in these pages, e.g., size, amount of tape, number of steps for Turing machines. However, a great deal of work is being done and the reader can consult Arbib (1969, pp. 229–248) and Hopcroft and Ullman (1969a, chapters 10, 11) to get an idea of what it is about and for further bibliography. He can also look into Hartmanis (1968), Hopcroft and Ullman (1968, 1969b), and watch the appropriate journals for the latest developments. For subhierarchies in terms of machines, the reader can consult Cleave (1963) and Cobham (1965). The student interested in complexity of partial recursive functions should read Blum (1967a), which includes interesting general comments as well as theorems not presented here. The student can also read Arbib (1966, 1969, pp. 248–270), Blum (1967b, 1969), and Young (1969a, b), and generally watch the appropriate journals.

Chapter VII

RECURSIVELY ENUMERABLE SETS†

7.1 DEFINITIONS AND BASIC THEOREMS

A set S (of natural numbers) *is recursive* (rec.) if there exists a recursive function f such that

$$f(n) = \begin{cases} 1 & \text{if} \quad n \in S \\ 0 & \text{if} \quad n \notin S \end{cases}$$

That is, S has a recursive characteristic function. This function is sometimes referred to as *the function which decides S*. Some examples of recursive sets are: (i) any set of one element; (ii) any finite set; (iii) the empty set; (iv) N; (v) for fixed k, $\{kn \mid n \in N\}$; (vi) the set of primes; (vii) a set of natural numbers is recursive if and only if it is Turing decidable.

Any set S of natural numbers which is the image of a unary partial recursive function is a *recursively enumerable set* (r.e. set). That is, a set S is r.e. if there is a partial recursive function f such that $f(N) = S$. (Of course, f may not be defined on all of N, but we shall indicate the image of f in this way.) The function f whose image is S will be referred to as the *function which generates S* or *the function which enumerates S*. Some examples of r.e. sets are: (i) the set of even positive integers; (ii) the set $\{2\}$; (iii) any finite set.

Exercise 7.1 Show that the set of Gödel numbers of the elements of $\mathscr{K}$ is an r.e. set. Is it a recursive set? These questions should be answered in terms of *any* Gödel numbering, not just for an onto N Gödel numbering, since in that case the answer is obvious. (You may wish to use Church's thesis.)

Exercise 7.2 Do you think that all subsets of N are r.e.?

† Most of the material in this chapter is based on Post (1944).

Usually, we will be considering sets of numbers which represent, by means of Gödel numbering, sets of objects such as Tm's, partial-recursively defined functions, formulas, sentences, etc. Once it has been shown that the set of objects to be considered can be Gödel numbered and that the set of numbers is r.e. or recursive, we can then say that the set of objects itself is r.e. or recursive. In the usual notation, we take the complement of S to be $\bar{S} = N - S$.

Theorem 7.1 A set S is recursive if and only if both S and $\bar{S}$ are r.e.

PROOF Suppose that S is recursive; then, there is a recursive function f such that

$$f(n) = \begin{cases} 1 & \text{if } n \in S \\ 0 & \text{if } n \notin S \end{cases}$$

S can be generated by defining the function g as follows: $g(0)$ is to be the smallest i such that $f(i) = 1$, $g(1)$ the next smallest, etc. This is a sufficient explanation for a proof by Church's thesis. However, if S is infinite, it is easy to write down g explicitly:

$$g(0) = \mu_m[f(m) = 1]$$
$$g(n + 1) = \mu_m[(sg(m \dot{-} g(n)) \cdot f(m)) = 1]$$

$\bar{S}$ is r.e. by similar arguments. That is, h generates $\bar{S}$ if $h(0)$ is the smallest i such that $f(i) = 0$, $h(1)$ is the next smallest i such that $f(i) = 0$, etc. If both S and $\bar{S}$ are r.e., then suppose that g generates S and h generates $\bar{S}$. We show that S is recursive by using the extended Church thesis. Suppose we want to know about an arbitrary n whether or not it belongs to S. Then, we can compute $g(0)$, $h(0)$, $g(1)$, $h(1)$, $g(2)$, ... Eventually, since $g(N) \cup h(N) = N$, n will be computed. If it is computed by g, then $n \in S$; if by h, it belongs to $\bar{S}$. To give the proof without using the extended Church thesis is considerably more difficult at this point. ▌

One might well be wary of this use of the extended Church thesis since it says "to compute" a certain sequence of numbers and wait until the n in question turns up; for, we are computing with *partial* recursive functions and so if the waiting time includes the time to compute first $g(0)$, and then $h(0)$, etc., maybe n will not turn up eventually. For example, perhaps $h(0) = n$, but if $g(0)$ is not defined, we would be waiting forever for $g(0)$ to get computed. The proof of the next theorem gives a method for designing a Tm to carry out this kind of procedure. Thus, we will be reassured that the above argument is not as dubious as it may appear.

Exercise 7.3 If S is a finite set, show that there is a recursive function that generates it.

***Theorem** 7.2* If S is an r.e. set, then there is a recursive function that generates it.

PROOF If S is finite, then the theorem follows by exercise 7.3. Assume S is infinite.

Part I. We consider first an intuitive sketch. Since S is r.e., it is generated by a partial recursive function g, and so there exists a Tm T_g such that $T_g \sim g$. We shall work with T_g (infinitely many copies are available) and set up a procedure involving time. We will be defining a total effectively computable function f which will generate S. At time $t = 0$, turn on T_g, reading a tape with initial tape inscription 0. At $t = 1$, T_g reading 0 will perform its first print-move operation, while another copy of T_g, reading a tape with initial tape inscription 1, is turned on. At $t = 2$, T_g on 0 is performing its second operation while T_g on 1 is performing its first and T_g on 2 is being turned on. Etc. Let $f(0)$ be the first, with respect to time, answer computed by a copy of T_g. Then, $f(1)$ is the next, and so on. Admittedly, at some t, it is possible that several (say k) copies of T_g will simultaneously halt, but as there can be only finitely many, at a given t, this presents no problem. If at $t - 1$, $f(m)$ was established, then assign to $f(m + 1), \ldots, f(m + k)$ the values computed at time t. Thus, f is a total effectively computable function whose image is exactly the range of g; that is, f generates S.

Part II. To verify that f, or a function similar to f, is really a recursive function, we outline here how a Tm T_f could be designed to generate S. T_f will work on a tape with two distinct halves. On the right side, T_f will be simulating (as in the universal Tm) what T_g does to different initial tape inscriptions. On the left side will be written only the initial tape inscription and eventually the answer. Suppose we wish to compute T_f on n. T_f is to read the tape

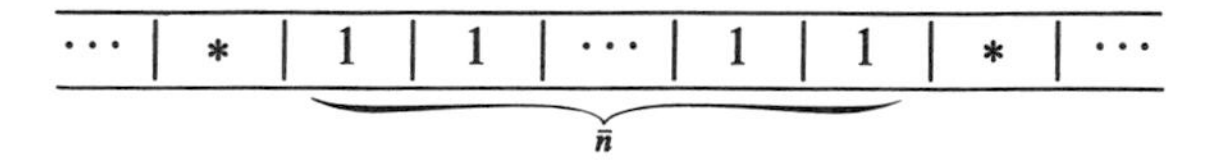

and then write on the right side $\rho_{T_g}\rho_0$, in a manner like the universal Tm, followed by some number of blanks and a special marker E, which is to indicate that there is nothing written on the tape to the right of E. Then, the machine, in the manner of the universal machine, simulates one print-move operation of T_g reading 0. Then it moves to the right and, erasing E, writes out ρ_{T_g} and ρ_1, skips so many blanks, and prints E. Then, it simulates one print-move operation of T_g on 0 and one of T_g on 1, goes to the right so as to write out ρ_{T_g} and ρ_2. And so on. One difficulty is that the extra blanks left for operating space may not be sufficient. If this happens, then everything to the right of the place that needs more space will be copied out beyond the present E, finally erasing that E and printing a new one in the appropriate place. The first time one of these simulated versions of T_g halts, T_f leaves a marker M there and goes to the left side of the tape to erase the leftmost 1 of $\bar{n}$. T_f checks to see if any 1's remain and if so it returns to the right, removes M, and continues as before. Of course, the machine that has halted stays halted, by definition. Each time a new halt occurs, there is a trip to the left to erase another 1 and to check if there are any 1's remaining. Once there are none, the machine copies onto the left the answer given by the simulation with the marker M by it, erases everything else, and halts. ▌

Corollary If S is an infinite recursive set, then there is a recursive one–one function which generates it in order of magnitude.

Exercise 7.4 Prove the corollary.

Exercise 7.5 Define a recursive characteristic function of S if both S and $\bar{S}$ are r.e. That is, prove Theorem 7.1 without using Church's thesis.

Exercise 7.6 Is the following true? There exists an r.e. set S with $\bar{S}$ not r.e. if and only if there exists a set which is r.e. but not recursive.

Theorem 7.3 There exists a recursively enumerable set which is not recursive.

PROOF Keeping in mind exercise 7.6, we see that we can construct this set if we can construct an r.e. set whose compliment is not r.e. By exercise 7.1, we know that the set of Gödel numbers of the elements of $\mathscr{K}$ is an r.e. set, so we may think of an enumeration of $\mathscr{K}$: $g_0, g_1, \ldots, g_i, \ldots$. We want a set S with the properties (a) there exists a j such that $g_j(N) = S$ and (b) for *no* k is $g_k(N) = \bar{S}$. A plausible approach is to try to construct S so that $\bar{S}$ will differ from every r.e. set by at least one element. This should make us look for some sort of diagonal argument. We define S to have the natural number n as an element if the nth partial recursive function g_n generates the natural number n. That is, $n \in S$ if and only if there exists an i such that $g_n(i) = n$. We show that $\bar{S}$ is not r.e., by showing that for each i, $g_i(N) \neq \bar{S}$. For, if $i \in g_i(N)$, then $i \in S$, so $g_i(N) \neq \bar{S}$; and if $i \notin g_i(N)$, then $i \notin S$, so $i \in \bar{S}$ and thus $g_i(N) \neq \bar{S}$. To see that S is r.e., consider an array of the function values of the partial recursive functions:

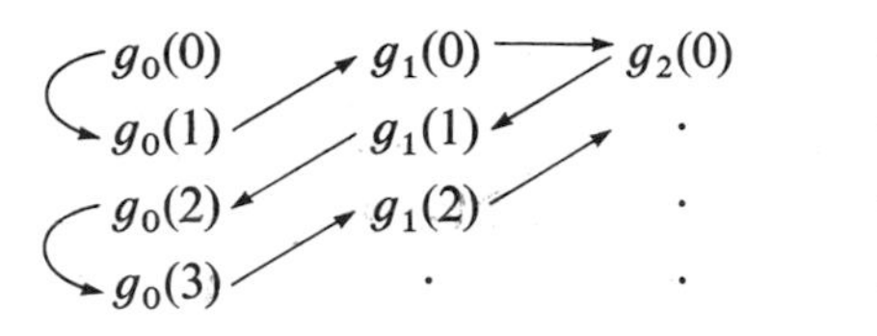

By searching through them in a zigzag procedure, we see that we can generate the elements of S in an effective manner. By the extended Church thesis, there is a partial recursive function that generates S, so S is r.e. ▮

Exercise 7.7 One might well object to the zigzag procedure of the previous proof on the grounds that we do not know, for a given i and j, whether or not $g_i(j)$ is defined, so "searching through" is rather vague. That is, do we really have an effective procedure for generating S? Give a more careful proof that S is r.e. (Hint: Use methods similar to those used in the proof of Theorem 7.2. The function will be quite different from the one in the proof of Theorem 7.3. The exercise is rather difficult.)

Exercise 7.8 (a) If S is a subset of a rec. set, is S rec.?
(b) If S is a subset of an r.e. set, is S r.e.?
(c) If S is an infinite r.e. set, does it contain an infinite rec. subset?
(d) If S is a subset of a rec. set, is S r.e.?
(e) If S is r.e., is $\bar{S}$ r.e.?
(f) If S is rec., is $\bar{S}$ rec.?
(g) If S is not r.e., can $\bar{S}$ be finite?
(h) If R and S are rec. (r.e.), is $R \cup S$ rec. (r.e.)? Is $R \cap S$ rec. (r.e.)?
(i) If each member of $\{R_i \mid i \in S$, for S r.e.$\}$ is r.e., is $\bigcup_{i \in S} R_i$ r.e.?
(j) If each member of $\{R_i \mid i \in S$, S any subset of $N\}$ is r.e., is $\bigcup_{i \in S} R_i$ r.e.?
(k) Formulate questions similar to these. Try to answer them.

An r.e. set was defined to be the image of a partial recursive function, and as we know from Theorem 7.2, every r.e. set can be generated by a recursive function. Another, and common, definition is that an r.e. set is the domain of definition of a partial recursive function. We now show that this is equivalent to the original definition.

Theorem 7.4 A set S is r.e. if and only if it is the domain of a partial recursive function.

PROOF *Part I.* Suppose that S is r.e. (a) If $S = \varnothing$, then the partial recursive function that is not defined anywhere has S for its domain. (b) If $S \neq \varnothing$, let S be generated by the recursive function f. We want a partial recursive function g such that, for some constant, say 1, for all n, $g(n) = 1$ if and only if there is an i such that $f(i) = n$, and g is not defined elsewhere. We show there is such a function g by defining the Tm T_g as follows. With n written on its tape, T_g, in universal Tm fashion, writes ρ_{T_f} and ρ_0 on its tape, computes $T_f(0)$, and compares it to n. If $T_f(0) = n$, then T_g prints 1, erases everything else, and halts. If $T_f(0) \neq n$, then T_g computes $T_f(1)$, etc. Thus,

$$g(n) = \begin{cases} 1 & \text{if there is an } i \text{ such that } f(i) = n \\ \text{undefined} & \text{otherwise} \end{cases}$$

Part II. Suppose S is the domain of a partial recursive function g. If S is empty or finite, it is recursive and therefore there is a recursive function which generates it. So suppose S is infinite. Let T_g be the Tm that computes g, i.e., $T_g \sim g$ and T_g accepts S. We will design a Tm T_h so that T_h enumerates S, i.e., h is a (partial) recursive function such that $h \sim T_h$ and $h(N) = S$. Let n, in unary, be the initial tape inscription. T_h is to operate in the same fashion as the Tm T_f in the proof of Theorem 7.2, with one exception. That is as follows. When the initial tape inscription has been erased, there is one pair, ρ_{T_g} and ρ_i, with a marker to indicate that (the simulation of) T_g has just halted on i. T_h is to print out that i, erase everything else, and halt. ▮

Corollary A subset S of N is r.e. if and only if it is Turing acceptable.

Exercise 7.9 Without using Theorem 7.3, prove that there is an r.e. set which is not rec.

At this stage, the student might be confused as to whether to decide that a set is recursive because of what he knows about the generating function or what he knows about the set itself. For example, if the set generated is finite, it is a recursive set no matter how peculiar a function may be given to generate it. Consider the following:

(i) $$f(n) = \begin{cases} 1 & \text{if it rains during the } n\text{th day from today} \\ 0 & \text{if it does not rain during the } n\text{th day from today} \end{cases}$$

(ii) $$g(n) = \begin{cases} 1 & \text{if Fermat's theorem is true} \\ 0 & \text{if Fermat's theorem is false} \end{cases}$$

In the first case, the image of f is just $\{0, 1\}$ and so is recursive. In the second case, the image of g is either $\{0\}$ or $\{1\}$. If Fermat's theorem is false, we can use, for example, the prf $f(n) = Z(n)$, and if it is true, we can use the prf $h(n) = S(Z(n))$ to generate the set. The only problem is that *we* do not know which function to use, but whichever, it is recursive and its image is a recursive set. Similar remarks can be made concerning r.e. sets. After all, whatever the function may be that is given as the generating function, the question is whether or not there is a partial recursive function that generates the same set. Suppose that the image of a certain function h is given as, or is obviously, $\{2\}$. No matter what else is true of h, its image $\{2\}$ is a recursive set. On the other hand, we could be given an extremely complicated partial-recursively defined function with no information and about which nothing is obvious. If we start computing values and keep getting 2, we can conclude nothing about whether the image is $\{2\}$ and hence recursive. Some of the exercises just below are to help the reader understand this distinction, and some others give properties of functions which guarantee that a set generated by a function with such a property is recursive.

Exercise 7.10 Prove that if f is a strictly increasing recursive function, then the image of f is a recursive set.

Exercise 7.11 (a) Suppose that S is generated by a recursive function f, with the property that for finitely many natural numbers $m \in S$, there exists $n_1 < n_2 < \cdots < n_{k(m)}$ such that $f(n_1) = f(n_2) \cdots = f(n_{k(m)}) = m$, for $m \leq f(n_1 + 1)$ and $f(n_1 - 1) < f(n_1)$. But other than this finite number of exceptions, f is strictly increasing. (i) Draw the graph of an example of such a function. (ii) Can S be finite? (iii) Is S recursive? (b) Suppose we allow infinitely many m to have finitely many repetitions. (i) Can S be finite? (ii) Is S recursive? (c) Suppose we allow finitely many m to have infinitely many repetitions, i.e., for finitely many m, there exists a sequence $n_1 < n_2 < \cdots$ such that $m = f(n_1) = f(n_2) = \cdots$. (i) Can S be finite? (ii) Is S recursive?

Exercise 7.12 Prove that if f is a monotonic increasing, *partial* recursive function, then the image of f is not necessarily a recursive set. (Hint: Find a monotonic increasing, partial recursive function defined in terms of some function whose domain is an r.e., not rec., set.)

Exercise 7.13 Let f be a recursive function with the property that there exists an n_0 such that for all $n_2 > n_1 > n_0$, $f(n_2) \leq f(n_1)$. (a) Is the image of f a recursive set? (b) What do you think about n_0, is it important that it be specifically given, or can it merely be known to exist? (c) If we take f to be a partial recursive function with the property that there exists an n_0 such that for all $n_2 > n_1 > n_0$, $f(n_2) \leq f(n_1)$ whenever both exist, is the image of f a recursive set?

Exercise 7.14 Suppose S is an r.e. set with recursive generating function f and there exists an n_0 such that for all $n_1, n_2, n_0 < n_1 < n_2, f(n_1) < f(n_2)$; is S a recursive set?

Exercise 7.15 (a) Suppose that f is a one–one recursive function and S is a recursive set contained in $f(N)$. Is $f^{-1}(S)$ a recursive set? (b) Suppose f is a many–one recursive function and S is a recursive set in $f(N)$. Define $g(x) = \mu_y [f(y) = x]$. Is $g(S)$ a recursive set?

Exercise 7.16 If f is any primitive recursive function, is the image of $\tilde{f}$, its history function, a recursive set?

Exercise 7.17 Exercise 4.39 was to show that the function $f(n, x) = \mu_y [n^y \geq x]$ is a partial recursive function. For a fixed x, is the image of $f(n, x)$ a rec. set?

Exercise 7.18 Suppose that both R and S are nonempty and recall the definition of the cross-product of two sets R and S: $R \times S = \{(r, s)\} | r \in R$ and $s \in S\}$. (a) If R and S are both recursive, show that $J(R \times S)$ is recursive. (b) If R and $J(R \times S)$ are recursive, is S also recursive? (c) If R is r.e. and $J(R \times S)$ is recursive, is S recursive? (d) If R and S are r.e., show that $J(R \times S)$ is also r.e.

Exercise 7.19 If T is the set $J(R \times S)$ for sets R and S, define the *diagonal set of* T, D_T, to be $\{i | J(i, i) \in T\}$. (a) Show that if R and S are r.e., then D_T is r.e. (b) Show that the function d defined by: for all i, $g_{d(i)}(N) = D_T$, where $g_i(N) = T$, is a recursive function. (Hint: This is very easy by Church's thesis; but figuring out how to design a Tm to compute d is a very instructive exercise.)

Exercise 7.20 Is each of the following true or false? Prove carefully.

(a) If f is a one–one recursive function such that the image of f is a recursive set and if g is a recursive function such that for all $n \in N$, $g(n) \leq f(n)$, then the image of g is a recursive set.

(b) If f is a one–one recursive function such that the image of f is a recursive set and if g is a recursive function such that for all $n \in N$, $g(n) \geq f(n)$, then the image of g is a recursive set.

Exercise 7.21 Show that *if* "there exists a recursive function F such that for all m, n,

$$F(m, n) = \begin{cases} 1 & \text{if } m = \mu_y[y \in g_n(N)] \\ 0 & \text{otherwise} \end{cases} \text{"}$$

is true, then "there exists a recursive function f such that $g_{f(n)}(N) = g_n(N) - \{\mu_y [y \in g_n(N)]\}$" *cannot be true.*

We take a moment to review the form of "diagonal arguments" for certain classes of functions. Let $\mathscr{F}$ be either the class of primitive recursively defined functions, the class of recursively defined functions, or the class of partial-recursively defined functions. Let $S_{\mathscr{F}}$ be the set of Gödel numbers of the functions in the class $\mathscr{F}$. (Thus for each $\mathscr{F}$, $S_{\mathscr{F}} \subseteq S_{\mathscr{K}}$.) Let $U_{\mathscr{F}}$ be the universal function for the class $\mathscr{F}$, that is, for all $n \in S_{\mathscr{F}}$, and for all $m \in N$, $U_{\mathscr{F}}(n, m) = g_n(m)$ for g_n in $\mathscr{F}$. Define the diagonal function of $\mathscr{F}$, $D_{\mathscr{F}}$, by: for all $n \in S_{\mathscr{F}}$, $D_{\mathscr{F}}(n) = U_{\mathscr{F}}(n, n)$. Define the successor to the diagonal function, $D_{\mathscr{F}}'$, by: for all $n \in S_{\mathscr{F}}$, $D_{\mathscr{F}}'(n) = D_{\mathscr{F}}(n) + 1$. Notice the following: If $S_{\mathscr{F}}$ is r.e., then $U_{\mathscr{F}}$ is effectively computable, and hence also $D_{\mathscr{F}}$ and $D_{\mathscr{F}}'$. If $S_{\mathscr{F}}$ is r.e. and $\mathscr{F}$ includes all effectively computable functions, then there are numbers u, $d(u)$, and $d'(u)$ in $S_{\mathscr{F}}$ such that $g_u = U_{\mathscr{F}}$, $g_{d(u)} = D_{\mathscr{F}}$, $g_{d'(u)} = D_{\mathscr{F}}'$. If $U_{\mathscr{F}}(n, m)$ is defined for all $n \in S_{\mathscr{F}}$ and $m \in N$, then $D_{\mathscr{F}}$ and $D_{\mathscr{F}}'$ are defined for all $n \in S_{\mathscr{F}}$. Consequently, if the three conditions: (i) $S_{\mathscr{F}}$ is r.e., (ii) $\mathscr{F}$ includes all effectively computable functions, and (iii) $U_{\mathscr{F}}(n, m)$ is defined for all $n \in S_{\mathscr{F}}$ and all $m \in N$, all hold, then we can derive the contradiction that $g_{d'(u)}(d'(u)) = g_{d'(u)}(d'(u)) + 1$.

Theorem 7.5 The set of Gödel numbers of the recursively defined functions is not a recursively enumerable set.

Exercise 7.22 Prove the theorem.

For $\mathscr{F}$ the class of primitive recursively defined functions, certainly $S_{\mathscr{F}}$ is r.e. and $U_{\mathscr{F}}$ is effectively computable. However, if one assumes that the primitive recursively defined functions include all total effectively computable functions it follows that $U_{\mathscr{F}}$ is primitive recursive and thus a contradiction is derived. In case $\mathscr{F}$ is the class of recursively defined functions, by Church's thesis, we believe $\mathscr{F}$ to include all the total effectively computable functions. Thus, it is the assumption that $S_{\mathscr{F}}$ is r.e. that leads to difficulties. For $\mathscr{F} = \mathscr{K}$, we know that $S_{\mathscr{F}}$ is r.e. and believe that $\mathscr{F}$ includes all effectively computable functions. However, we do not arrive at a contradiction via the usual diagonal argument because it need not be true that $U_{\mathscr{F}}(n, m)$ is defined for all $n \in S_{\mathscr{F}}$ and $m \in N$.

Exercise 7.23 Use some sort of diagonal argument to define a function which is not partial recursive.

In order to "diagonalize out" of the class of partial recursive functions, it is necessary use a function "stronger" than simply $D_{\mathscr{K}}'$. The general form is

$$f(i) = \begin{cases} \text{Do} \quad A & \text{if} \quad g_i(i) \quad \text{is defined} \\ \text{Do} \quad B & \text{if} \quad g_i(i) \quad \text{is not defined} \end{cases}$$

It is the instruction "Do B if $g_i(i)$ is not defined" which leads out of the partial recursive functions. Recall the warning about using the extended Church thesis.

Exercise 7.24 Prove that there is a recursive set which is not a type 1 language. (Hint: See section 2.5. As we did for Tm's, prf's, etc., establish a standard set of symbols for grammars and languages. Gödel number all type 1 grammars; Gödel number all terminal words. Diagonalize.)

7.2 THE COMPLETE SET K

The *decision problem for a set of natural numbers S* is to decide of an arbitrary $n \in N$ whether or not $n \in S$. Clearly, by the definitions of solvable decision problem and recursive set and the equivalence of Tc functions and recursive functions, a set S has a solvable decision problem if and only if it is a recursive set.

Since the set of Gödel numbers of $\mathscr{K}$ is r.e., we have the usual enumeration of $\mathscr{K}$: $g_0, g_1, \ldots$. Thus, to say, "A set S is r.e." is equivalent to saying, "There exists an i such that $g_i(N) = S$." Since S is generated by g_i, to ask, "Does $n \in S$?" is equivalent to asking, "Is there a j such that $g_i(j) = n$?" For a fixed i, the set $\{(n, g_i)?\,|\,n \in N\}$ is thus a way of representing the decision problem for S. If we let i vary over all of N, the set (of symbols)

$$E' = \{(n, g_i)?\,|\,n, i \in N\}$$

is a way of representing the decision problem for all r.e. sets. We can enumerate the set E' in the usual zigzag manner. Instead of the set E' of pairs—natural number, partial recursively defined function—consider the set of ordered pairs of natural numbers

$$E = \{(n, i)\,|\,n, i \in N\}$$

or simply the set $\{J(n, i)\,|\,n, i \in N\}$, which is, in fact, N.

We shall consider certain properties of the set E'. First, we designate by K' the subset of E' containing just those pairs (n, g_i) for which the answer is "yes, n is generated by g_i":

$$K' = \{(n, g_i)\,|\,\text{there is a } j,\ g_i(j) = n\}$$

Is K' an r.e. set? To show that it is, we give an outline of a function (or Tm) for generating it. Consider the usual zig-zag arrangement

$$\begin{array}{ll} T_{g_0}(0) & T_{g_0}(1) \\ T_{g_1}(0) & \end{array}$$

which we interpret as follows. At time $t = 0$, turn on Tm T_{g_0} with initial tape inscription 0; at $t = 1$, turn on Tm T_{g_1} with initial tape inscription 0 while T_{g_0} reading 0 performs its first print-move operation; etc. Whenever a T_{g_i} computing from some j halts, write down the pair $(T_{g_i}(j), g_i)$. In this fashion, K' is generated.

Exercise 7.25 Let $K = \{J(n, i) \mid \text{there exists a } j, g_i(j) = n\}$. Is K r.e.? In view of Theorem 7.3, do you think that K is a recursive set?

Exercise 7.26 Show that the set $\{J(m, n) \mid \text{there is a } j, g_n(j) = m \text{ and } m \geq 2n\}$ is a recursively enumerable set.

The second aspect of our considerations of the set E' is an investigation of $\overline{K}'$ (i.e., $\overline{K}' = E' - K'$). Let F' be *any* r.e. subset of K'. We shall see that F' cannot coincide with $\overline{K}'$ (and so K' is not recursive) and, further, that F' itself leads us to discover elements of $\overline{K}'$ which are not contained in F'. Since $F' \subset \overline{K}'$, for any pair (n, g_i), if $(n, g_i) \in F'$, then g_i does not generate n. Further, since F' is itself an r.e. set, it has a recursive generating function. Using the function that generates F', we generate a new set S as follows. As each pair (n, g_i) of F' is generated, if $n = i$, put i into S. Thus, $S = \{i \mid (i, g_i) \in F'\}$. Notice that S is a set of numbers, not a set of pairs. It is not a subset of E'. Certainly, S is an r.e. set, so it has a generating function g_s. Where is the pair (s, g_s)?

1. On the one hand, if $(s, g_s) \in F'$ then (by the definition of F') s *is not generated by* g_s, which happens if and only if (by definition of S) $s \in S$, i.e., s *is generated by* g_s. Contradiction; so $(s, g_s) \notin F'$.

2. On the other hand, $(s, g_s) \in K'$ if and only if (by definition of K') s *is generated by* g_s, but if this happens then (by definition of S) $s \notin S$, i.e., s *is not generated by* g_s. Contradiction; so $(s, g_s) \notin K'$.

Therefore, $(s, g_s) \in \overline{K}'$ but $(s, g_s) \notin F'$.

The set K is usually known as *the complete set K*. As we have seen, the set K seems to hold the key to all decision problems.

7.3 ONE–ONE (MANY–ONE) REDUCIBILITIES

In exercise 7.18, the reader was asked to consider the recursiveness of the set $J(R \times S)$ by referring to the recursiveness of the sets R and S. Since $n \in J(R \times S)$ if and only if $K(n) \in R$ and $L(n) \in S$, we see that since R and S are recursive, $J(R \times S)$ is also recursive. It is often the case, as we already have seen on several occasions, that we can say something about the decision problem for one set by referring to some other set. In section 1.7, we showed how the decision problem for the set of Diophantine equations can be referred to the "truth problem" for first-order arithmetic. In that case, the Diophantine equation, say $x^3 + 4x^3y^2 - z^7 = 5$, has a solution if and only if the sentence $(\exists x)(\exists y)(\exists z)(x^3 + 4x^3y^2 - z^7 = 5)$ is true. In chapter II, there was a series of decision problems each related to the other in such a way that the unsolvability of the halting problem implied the unsolvability of the ambiguity problem for context-free grammars.

We say that *the decision problem for a set S_1 is one–one (many–one) reducible to the decision problem for a set S_2* if there exists a one–one (many–one) recursive function f such that $n \in S_1$ if and only if $f(n) \in S_2$.

Exercise 7.27 Let S_1 and S_2 be r.e. sets such that S_1 is one–one reducible to S_2. What can you say about the truth or falsity of each of the following: (a) If S_1 is recursive, then S_2 is recursive. (b) If S_1 is not recursive, then S_2 is not recursive.

At the end of section 7.2, we said that the complete set K seems to hold the key to all decision problems. That idea is formalized and proved by the following theorem.

***Theorem** 7.6* The decision problem for every r.e. set is one–one reducible to the decision problem for K.

PROOF If S is an r.e. set, then for some i, g_i generates it. But then $n \in S$ if and only if $(n, g_i) \in K'$ if and only if $J(n, i) \in K$. Recall that J is a one–one recursive function. ∎

Exercise 7.28 An r.e. set can be defined as either (i) the image of a partial recursive function or (ii) the domain of a partial recursive function. The complete set K was developed in terms of (i). (a) Carry through a similar development in terms of (ii), call the resulting set H. (b) Let D_H and $D_{\bar{H}}$ be the diagonal sets of H and $\bar{H}$, respectively. For the pair (i, j), suppose j refers to the Gödel number of a Tm (rather than a partial-recursively defined function as such). Write out a proof of the unsolvability of the halting problem using Gödel numbers instead of ρ_T's and ρ_τ's, and compare it, step-by-step, to a proof of the nonrecursiveness of H.

A *set S is productive* if there is a recursive function p such that if $g_i(N) \subset S$, then $p(i) \in (S - g_i(N))$. Such a function p is called a *productive function*. A *set C is creative* if C is r.e. and $\bar{C}$ is productive.

***Theorem** 7.7* Creative sets exist.

PROOF We show that K is creative by producing the required productive function. This is done by following through the development of K in the previous section. Let F be any r.e. subset of $\bar{K}$; then, there is an f such that $g_f(N) = F$. Let D_F be the diagonal set of F. There is a recursive function d (see exercise 7.19) such that $g_{d(f)}(N) = D_F$. Claim that the productive function is g_p defined by $g_p(i) = J(d(i), d(i))$, for all i. ▮

Exercise 7.29 Verify that if $g_i(N) \subset \bar{K}$, then $g_p(i) \notin g_i(N)$ but $g_p(i) \in \bar{K}$.

Exercise 7.30 Design a Tm that computes g_p.

Exercise 7.31 Show that if C is a creative set, then $\bar{C}$ contains an infinite r.e. subset.

Exercise 7.32 Prove that if T is a set and R is a recursive set and $R \cap T$ is productive, then T is productive.

An r.e. set S is *simple* if $\bar{S}$ is infinite but contains no infinite r.e. subset.

Theorem 7.8 Simple sets exist.

PROOF If we can construct S in an effective way so that (1) it intersects every infinite r.e. set and (2) $\bar{S}$ is infinite, then S will be simple. Let

$$R = \{J(m, n) \mid \text{there is a } j, g_n(j) = m \text{ and } m \geq 2n\}$$

By exercise 7.26, R is r.e.; let g be the recursive function which generates R. Define h by

$$h(0) = g(0)$$

$$h(k+1) = \begin{cases} g(\mu_i[L(g(i)) \neq L(h(0)) \\ \text{and} \ldots \text{and } L(g(i)) \neq L(h(k))]) & \text{if there is such an } i \\ \text{undefined} & \text{otherwise} \end{cases}$$

Certainly h is a partial recursive function. The idea behind h is that it takes the next element $J(m, n) \in R$ so that for no element $J(m', n')$ already in the image of h is $n = n'$. Define S to be the set generated by the function $f = K \circ h$. (1) Claim that the image of f intersects every infinite r.e. set. Any r.e. set is generated by g_n for some n and if, further, it is an infinite r.e. set, then it has infinitely many elements which are greater than or equal to $2n$. For all those values m, the element $J(m, n) \in R$. By the definition of h, for each n exactly one such element is in the image of h. By the definition of f, for each n, exactly one value $\geq 2n$ which is in the image of g_n belongs to S. (2) Claim that for all $n > 0$, $\bar{S}$ contains at least n elements and hence is infinite. The fact that for all n, $f(n) \geq 2n$ implies that for all n, the set $A_n = \{0, 1, \ldots, 2n\}$ contains, as elements of S, at most the set $B_n = \{f(0), \ldots, f(n)\}$. Thus, the set A_n contains at least n elements which belong to $\bar{S}$. ∎

Theorem 7.9 No creative set is one–one reducible to a simple set.

Exercise 7.33 Prove the theorem.

Most of the topics considered in this chapter were developed by Post (1944), as already mentioned. In particular, Post defined, as an abstract set, the complete set K and showed that all r.e. sets are one–one reducible to it. Since all r.e. sets are one–one reducible to K and the decision problem for K is unsolvable, K may be thought of as being a set with the highest "degree of unsolvability." Post first asked if K is one–one reducible to all r.e. sets and we have seen by Theorem 7.9 that they are not. (It is a theorem in his paper.) However, in the same paper, he extended the notion of reducible to more and more complicated formalisms which are, nevertheless, effective. In all, five such notions were defined, the most complicated being known as Turing reducibility. By the end of his paper, many questions had been asked and answered, but his question as to whether K is Turing reducible to each r.e. set was still unanswered. Subsequently, the answer was found to be "no" by Friedberg (1957) in the USA and Muchnik (1956) in the USSR. In the next to last section of the next chapter, we will consider briefly Turing reducibility, and in the last section, state and prove Friedberg's theorem. There are many interesting studies which are direct or indirect outgrowths of Post's paper. The reader who wishes to learn more about these matters can get a good start by reading Rogers (1967) from cover to cover.

Chapter VIII

RECURSIVE AND RECURSIVELY ENUMERABLE RELATIONS

8.1 DEFINITIONS

Recall that any subset S^k of N^k is an n-ary number-theoretic relation. In section 4.4, such a relation was defined to be primitive recursive if there is a k-ary primitive recursive function f such that

$$f(n_0, \ldots, n_{k-1}) = \begin{cases} 1 & \text{if } (n_0, \ldots, n_{k-1}) \in S^k \\ 0 & \text{if not} \end{cases}$$

We can extend this idea in the obvious way as follows: If $R^k \subset N^k$ and g is a k-ary recursive function such that

$$g(n_0, \ldots, n_{k-1}) = \begin{cases} 1 & \text{if } (n_0, \ldots, n_{k-1}) \in R^k \\ 0 & \text{if not} \end{cases}$$

then R^k is a *recursive relation.* Since a set of natural numbers is just a 1-ary relation, the definition of a recursive relation is the natural extension of the definition of a recursive set. Noting that an r.e. set is the domain of definition of a 1-ary partial recursive function, in an analogous way, we make the natural definition of an r.e. relation. If $R^k \subset N^k$ and there exists a k-ary partial recursive function h such that $h(n_0, \ldots, n_{k-1})$ is defined if and only if $(n_0, \ldots, n_{k-1}) \in R^k$, then R^k is an *r.e. relation.* One should expect recursive and r.e. relations to be related in the same way that recursive and r.e. sets are. It is obvious from the definitions that if R^k is a recursive relation, then it is an r.e. relation. The analog of Theorem 7.1 will be given in Theorem 8.5.

Keeping in mind that relations are just sets of k-tuples, we can form new relations from given ones; e.g., if S^k and R^k are k-ary relations, then (i) $S^k \cup R^k$ is a k-ary relation, (ii) $S^k \cap R^k$ is a k-ary relation, and (iii) $\overline{S^k} = N^k - S^k$ is a k-ary relation.

Exercise 8.1 Let R^k, S^k be (a) primitive recursive relations, (b) recursive relations, (c) recursively enumerable relations. For (a) show that $R^k \cup S^k$, $R^k \cap S^k$, $\bar{R}^k$, and D_R are primitive recursive relations. For (b) show that $R^k \cup S^k$, $R^k \cap S^k$, $\bar{R}^k$, and D_R are recursive relations. And for (c), show that $R^k \cup S^k$, $R^k \cap S^k$, and D_R are recursively enumerable relations. [Define D_R by $n \in D_R$ if and only if $(n, \ldots, n) \in R^k$. D_R is the *diagonal relation* of R^k.]

There are other interesting ways of defining new relations from given ones. Suppose R^{k+1} is some number-theoretic relation; then, it makes sense to speak of the set of k-tuples

$$(n_1, \ldots, n_k) \quad \text{such that for some} \quad n_0, \quad (n_0, n_1, \ldots, n_k) \in R^{k+1}$$

But this set is just a subset of N^k and hence is a k-ary number-theoretic relation. We write

$$S^k(n_1, \ldots, n_k) \leftrightarrow (\exists n_0) R^{k+1}(n_0, n_1, \ldots, n_k)$$

to define the relation S^k (in terms of the relation R^{k+1}) by

$$(n_1, \ldots, n_k) \in S^k \qquad \text{if and only if} \qquad \text{there exists an } n_0 \in N$$
$$\text{such that} \quad (n_0, n_1, \ldots, n_k) \in R^{k+1}$$

The symbol $\exists$ is known as the *existential quantifier* and the symbol $\forall$ is known as the *universal quantifier*. Using the latter, we can define another k-ary relation Q^k (in terms of the relation R^{k+1}) by

$$Q^k(n_1, \ldots, n_k) \leftrightarrow (\forall n_0) R^{k+1}(n_0, n_1, \ldots, n_k)$$

by which we mean

$$(n_1, \ldots, n_k) \in Q^k \qquad \text{if and only if} \qquad \text{for every } n_0 \in N,$$
$$(n_0, n_1, \ldots, n_k) \in R^{k+1}$$

Exercise 8.2 Let $R_1{}^2 = \{(n, 7n) \mid \text{all } n \in N\}$, $R_2{}^2 = \{(2n, 3n) \mid \text{all } n \in N\}$, and $R_1{}^3 = \{(4n, 6n) \mid \text{all } n \in N\}$. Define new relations from these (or others you make up) using unions, intersections, complements, and existential and universal quantifiers, and describe the sets so defined.

Exercise 8.3 Given the relation R^{k+1}, define a sort-of projection, S^k, as follows: for a fixed value of n, $S^k(n_1, \ldots, n_k) \leftrightarrow R^{k+1}(n, n_1, \ldots, n_k)$. Show that if R^{k+1} is a primitive recursive relation, then S^k is also.

By recalling the definition of the bounded μ-operation, one can easily imagine what is meant by the *bounded quantifiers* and so define new relations (in terms of a given relation R^{k+1}) by

$$S^k(n_1, \ldots, n_k) \leftrightarrow (\exists n_0)_{n_0 < n} R^{k+1}(n_0, n_1, \ldots, n_k) \tag{1}$$

$$Q^k(n_1, \ldots, n_k) \leftrightarrow (\forall n_0)_{n_0 < n} R^{k+1}(n_0, n_1, \ldots, n_k) \tag{2}$$

Thus, for fixed n, by (1), we mean

$(n_1, \ldots, n_k) \in S^k$ if and only if there exists an $n_0 < n$ such that $(n_0, n_1, \ldots, n_k) \in R^{k+1}$

and by (2)

$(n_1, \ldots, n_k) \in Q^k$ if and only if for all $n_0 < n$, $(n_0, n_1, \ldots, n_k) \in R^{k+1}$

Exercise 8.4 Show that if R^{k+1} is a primitive recursive (or recursive) relation, then S^k and Q^k as defined in (1) and (2) above are primitive recursive (or recursive) relations.

Exercise 8.5 If $n \in N$, then there is a unique k, $n = 2^{m_0}3^{m_1} \cdots p_k^{m_k}$, where $m_k > 0$. Show that the function l defined by $l(n) = k$ is a primitive recursive function.

When relations are defined by using unbounded quantifiers on recursive relations, very interesting things happen. We will be looking into this shortly. A particularly important fact that we wish to show in this chapter is that the set of k-ary r.e. relations coincides with the set of $(k + 1)$-ary recursive relations prefixed by an existential quantifier. En route to showing this, we shall consider an especially interesting primitive recursive predicate.

8.2 THE T PREDICATE

If S^k is a relation, we generally write $(n_0, \ldots, n_{k-1}) \in S^k$, but it is equally customary to write $S^k(n_0, \ldots, n_{k-1})$ as we have been doing for definining a relation. The latter case is more likely to be referred to as a "predicate" (instead of a relation), by which is meant

$S^k(n_0, \ldots, n_{k-1})$ is true if and only if $(n_0, \ldots, n_{k-1}) \in S^k$

For this chapter, we will generally use the notion (on the right) of a relation. However, since the subject about to be discussed is used extensively in the literature and always in the form of a predicate, we will do so here. This predicate, known as the *Kleene T-predicate*, or simply the T-predicate, is quite amazing for the amount of information it packs into a small space. We write

$T(z, n, y)$ to mean, "z is the Gödel number of a Tm T, n is a natural number, and y is the Gödel number of the computation performed by T when given a tape representation of n as an initial tape inscription." This is equivalent to saying that T is the 3-ary number-theoretic relation defined by

$$T = \{(z, n, y) \mid z \text{ is the Gödel number of a Tm } T, n \text{ is a natural number, and } y \text{ is the Gödel number of the computation performed by } T \text{ when given a tape representation of } n \text{ as an initial tape inscription}\}$$

Although one might well imagine what is meant by the phrase "y is the Gödel number of a computation ...," we must make it precise. (The reader is advised to review chapter 2, section 1; exercise 3.10; and chapter 5, section 3.) Suppose that, as we have often done, we restrict our attention to Tm's operating from a half-infinite tape and with alphabet $\{*, 1\}$. If, as we have been assuming, we use a Gödel numbering of $\mathscr{T}$ which is onto N, then every $z \in N$ is the Gödel number of some Tm. Further, if τ is an initial tape inscription, then τ is the binary representation of some number which we can call n, and if τ' is the answer computed by such a Tm, τ' is the binary representation of a number which we can call m. In these terms, Theorem 2.1 can be restated:

Theorem 2.1 (Alternate) For all z, n, and m, $T_z(n) = m$ if and only if there is a computation of T_z, $\delta_0, \delta_1, \ldots, \delta_t$, where $\delta_0 \equiv q_0\tau, \ldots, \delta_t \equiv \tau' q_H *$, τ is the binary representation of n and τ' is the binary representation of m.

Since $\delta_0, \delta_1, \ldots, \delta_t$ is a finite sequence of words on a finite alphabet, we can give a Gödel number to such a sequence, e.g.,

$$2^{\#\delta_0} 3^{\#\delta_1} \cdots p_t^{\#\delta_t} \qquad (*)$$

That number is then the number y which is "the Gödel number of the computation...." To say that the Tm T_z halts on a tape with initial tape inscription n is to say that $(\exists y)T(z, n, y)$ is true.

Theorem 8.1 T is a recursive relation.

Exercise 8.6 Prove the theorem.

Exercise 8.7 What do you think is true about the binary relation defined from T by prefixing it with $(\exists y)$?

In fact, T is a primitive recursive relation. For a complete and careful proof, the reader is invited to consult Davis (1958, pp. 56–62). However, we will prove that a relation almost the same as T is primitive recursive. The following will be based on the construction used in the proof of Theorem 5.4. In that construction, for the Tm T_z, there are primitive recursive functions q_{T_z}, m_{T_z}, k_{T_z}, and s_{T_z} which depend on time t and the natural number n represented by the initial tape inscription of T_z. By that construction, if $T_z(n)$ is defined, it equals $m_{Tz}(\mu_t[q_{Tz}(t, n) = H], n)$. Define the relation T^* by

$$T^* = \{(z, n, t) \mid q_{T_z}(t, n) = H\}$$

T^* is a primitive recursive relation because the function $=(q_{T_z}(t, n), H)$ is its characteristic function and is primitive recursive. It is instructive to consider the relationship between the relations T and T^*. If $(z, n, y) \in T$, then $(z, n, l(y)) \in T^*$, assuming that the Gödel number of a computation is given as suggested in $(*)$ above and that l is the function of exercise 8.5. Given $(z, n, t) \in T^*$ it is not so simple to find the corresponding y, though obviously there is one. Consequently, we have the following: there is a t (and hence a least one) such that $q_{T_z}(t, n) = H$ [i.e., $(z, n, t) \in T^*$] if and only if there is a y (and hence a least one) such that $T(z, n, y)$ is true [i.e., $(z, n, y) \in T$]. If there is a y such that $T(z, n, y)$, then the answer $T_z(n)$ is coded into y. It is in the power of the highest prime divisor of y. Recall that $P(n, y)$ is the power of the nth prime in the prime decomposition of y, then $P(l(y), y)$ is the power of the highest prime and so if y is the Gödel number of a computation, $P(l(y), y)$ is the Gödel number of the terminal instantaneous description. As the terminal instantaneous description is of the form $\tau' q_H *$, one can retrieve the τ'. We define $U(y)$ to be the number for which τ' is the binary representation. That is,

$$U(\mu_y T(z, n, y)) = m_{T_z}(\mu_t[q_{T_z}(t, n) = H], n)$$

[The proof that U is a primitive recursive function can be found in Davis (1958) as above.] We complete the comparison of T and T^* by referring to Theorems 5.1 and 5.4. That is, by Theorem 5.1, if g is a unary function in $\mathscr{K}$, then there exists a Tm T_z such that $g \sim T_z$, and so by the construction of Theorem 5.4, there exists a unary function g' in $\mathscr{K}$ such that

$$g(n) = g'(n) = m_{T_z}(\mu_t[q_{T_z}(t, n) = H], n) = U(\mu_y T(z, n, y))$$

The T predicate can be defined in the obvious way to have several argument values to correspond to functions of several variables. Thus, from the discussion above, we have the following theorem which is just Theorem 5.7 in slightly different form.

Theorem 8.2 (Kleene Normal Form Theorem) If g is a k-ary function in $\mathscr{K}$, then there exists a z such that for all k-tuples in N^k,

$$g(n_0, \ldots, n_{k-1}) = U(\mu_y T(z, n_0, \ldots, n_{k-1}, y))$$

In chapter VII, we found that there are three equivalent definitions of an r.e. set. The following theorem shows that there is a fourth equivalent definition.

Theorem 8.3 If S is a nonempty r.e. set, then there is a primitive recursive function f such that $f(N) = S$.

PROOF Since S is an r.e. set, there is a partial recursive function g such that $g(n)$ is defined if and only if $n \in S$. By the normal form theorem, there exists a z_0 such that $g(n) = U(\mu_y T(z_0, n, y))$, which means that $n \in S$ if and only if $g(n)$ is defined if and only if $(\exists y)T(z_0, n, y)$. Since T is a primitive recursive relation, it has a primitive recursive characteristic function C such that

$$C(z, n, y) = \begin{cases} 1 & \text{if } \; T(z, n, y) \; \text{ is true} \\ 0 & \text{if false} \end{cases}$$

As we are interested in a fixed z_0, we define C_{z_0} by

$$C_{z_0}(n, y) = 1 \qquad \text{if and only if} \qquad C(z_0, n, y) = 1$$

and so C_{z_0} is primitive recursive by exercise 8.3.

We shall define the function f whose range is S as follows. Suppose for the moment that we know the value of f at 0. Then, take some recursive enumeration of the ordered pairs of natural numbers, $O_1, O_2, O_3, \ldots$, and let the components of O_i be n_i, m_i. To compute the value of f at 1, compute $C_{z_0}(n_1, m_1)$. If it is 1 [i.e., $T(z_0, n_1, m_1)$ is true, equivalently, $g(n_1)$ is defined], define $f(1) = n_1$; if it is 0, define $f(1) = f(0)$. Inductively, for O_i, compute $C_{z_0}(n_i, m_i)$. If it is 1, $f(i) = n_i$; if it is 0, $f(i) = f(i-1)$. To ascertain what $f(0)$ is, since $S \neq \varnothing$, one can set up the usual scheme with regard to time, using T_g so that eventually a copy of T_g halts on some initial tape inscription n_0. To write down the definition of f formally requires the use of J, K, and L. It is a good exercise for the reader. ▮

This is a very strong theorem, especially from the computing point of view, for it shows that if a set can be generated in the loosest sense of effectively (say, by a partial recursive function), no matter how complex that function may be, there is a function no more complicated than a primitive recursive function which generates it, though, of course, not necessarily in the same order. Thus, as far as effective enumerations are concerned, we can restrict our attention to primitive recursive functions. As surprising as this fact may be, even more surprising is the following.

***Theorem*†** For every recursively enumerable set S, there is a function in $\mathscr{E}^0$ of the Grzegroczyk hierarchy which generates S.

One might then wonder if primitive recursive functions, or perhaps even just functions in $\mathscr{E}^0$, are not sufficient for dealing with recursive sets. That is, knowing that every r.e. set can be generated by a primitive recursive function, indeed by a function in $\mathscr{E}^0$, we should ask if every recursive set has a primitive recursive characteristic function, or better still, a characteristic function in $\mathscr{E}^0$ for some n.

Exercise 8.8 Show that there exists a recursive set which does not have a primitive recursive characteristic function.

8.3 QUANTIFYING ON RECURSIVE AND R.E. RELATIONS

Consider a binary relation which is the graph of a unary function. If we ask if n_1 belongs to the image of f, it is the same as asking if there exists an n_0 such that (n_0, n_1) belongs to the relation which is its graph. Similarly, to ask if f is defined for some n_0 is the same as asking if there exists an n_1 such that (n_0, n_1) belongs to the graph. Usually, it has been the custom to specify the graph of f as a set of ordered pairs the first component of which is the argument value and the second component the function value. However, we may equally well take the graph to be the set of ordered pairs listing the argument value second and its function value first: e.g., $\{(f(x), x)\}$. By allowing either ordering, the expression $(\exists n_0)R_f^{\,2}(n_0, n_1)$ may be interpretated by either (i) n_1 is in the range of f, using the standard order, or (ii) f is defined for n_1, using the reverse order. The graphs of k-ary functions are, of course, $(k+1)$-ary relations and we may define them so that the first k components are the argument values and the $(k+1)$th is the function value, or so that the first component is the function value and the rest are the argument values. Thus, if f is a k-ary function and R_f^{k+1} is its graph, then to say $(\exists n_0)R_f^{k+1}(n_0, n_1, \ldots, n_k)$ is equivalent to saying $f(n_1, \ldots, n_k)$ is defined and has the value n_0. If f is a partial recursive function, then the previous statement also may be interpreted as saying that $(n_1, \ldots, n_k)$ belongs to the r.e. relation which is the domain of definition of f. The above remarks may be useful for the reader as he studies this section.

Exercise 8.9 (a) If f is a k-ary recursive (partial recursive) function and

† This is Theorem 5.3 in Grzegroczyk (1953, p. 44). The reader is encouraged to look up the proof. Although a little long, in terms of theorems which it requires, the proof is not difficult.

R_f^{k+1} is its graph, is R_f^{k+1} a recursive (r.e.) relation? (b) If G is the graph of a function f and G is recursive, is f a partial recursive function?

Theorem 8.4 (a) If R^{k+1} is a recursive relation, then the relation Q^k defined by

$$Q^k(x_1, \ldots, x_k) \leftrightarrow (\exists x_0) R^{k+1}(x_0, x_1, \ldots, x_k)$$

is an r.e. relation.

(b) If Q^k is an r.e. relation, then there exists a recursive relation R^{k+1} such that

$$Q^k(x_1, \ldots, x_k) \leftrightarrow (\exists x_0) R^{k+1}(x_0, x_1, \ldots, x_k)$$

PROOF (a) We must find a partial recursive function whose domain coincides with Q^k. Since R^{k+1} is a recursive relation, it has a recursive characteristic function C:

$$C(x_0, x_1, \ldots, x_k) = \begin{cases} 1 & \text{if } (x_0, x_1, \ldots, x_k) \in R^{k+1} \\ 0 & \text{if not} \end{cases}$$

The function f defined by

$$f(x_1, \ldots, x_k) = \begin{cases} \mu_{x_0}[C(x_0, x_1, \ldots, x_k) = 1] & \text{if there is such an } x_0 \\ \text{undefined} & \text{otherwise} \end{cases}$$

is a partial recursive function. That its domain coincides with Q^k can be seen as follows: $f(x_1, \ldots, x_m)$ is defined if and only if there is an x_0 such that $C(x_0, x_1, \ldots, x_k) = 1$, which, since C is the characteristic function of R^{k+1}, is true if and only if $(\exists x_0) R^{k+1}(x_0, x_1, \ldots, x_k)$ is true; and that, by the definition of Q^k, is true if and only if $(x_1, \ldots, x_k) \in Q^k$.

(b) Q^k r.e. means there is a partial recursive function f such that

$$(x_1, \ldots, x_k) \in Q^k \quad \text{if and only if} \quad f(x_1, \ldots, x_k) \text{ is defined}$$

By the normal form theorem, there is a z_0 such that

$$f(x_1, \ldots, x_k) = U(\mu_y T(z_0, x_1, \ldots, x_k, y))$$

But that means $f(x_1, \ldots, x_k)$ is defined if and only if $(\exists y) T(z_0, x_1, \ldots, x_k, y)$. So, there is a z_0 such that

$$Q^k(x_1, \ldots, x_k) \leftrightarrow (\exists y) T(z_0, x_1, \ldots, x_k, y)$$

where T is a (primitive) recursive relation. ∎

Corollary (Alternative form of part 2 of Theorem 8.4, known as the Kleene Enumeration Theorem) If Q^k is a recursively enumerable relation, then there is a number z_0 such that $Q^k(x_1, \ldots, x_k) \leftrightarrow (\exists y) T(z_0, x_1, \ldots, x_k, y)$.

Exercise 8.10 Show that this corollary is essentially the same as Theorem 5.10, which is also referred to as the Kleene enumeration theorem.

Exercise 8.11 If R^k is a recursive relation, show that both R^k and $\bar{R}^k$ are r.e. relations.

Exercise 8.12 If R^{k+1} is a recursive relation and f^k is a recursive function, show that the relation Q^k defined by

$$Q^k(x_1, \ldots, x_k) \leftrightarrow R^{k+1}(f^k(x_1, \ldots, x_k), x_1, \ldots, x_k)$$

is a recursive relation.

Theorem 8.5 R^k is a recursive relation if and only if both R^k and $\bar{R}^k$ are r.e. relations.

PROOF If R^k and $\overline{R^k}$ are r.e. relations, then by Theorem 8.4, there are recursive relations P^{k+1} and Q^{k+1} such that

$$R^k(x_1, \ldots, x_k) \leftrightarrow (\exists x_0)P^{k+1}(x_0, x_1, \ldots, x_k)$$
$$\overline{R^k(x_1, \ldots, x_k)} \leftrightarrow (\exists x_0)Q^{k+1}(x_0, x_1, \ldots, x_k)$$

Since P^{k+1} and Q^{k+1} are recursive, $(P^{k+1} \cup Q^{k+1})$ is also recursive. The function f defined by

$$f^k(x_1, \ldots, x_k) = \mu_{x_0}(P^{k+1}(x_0, x_1, \ldots, x_k) \cup Q^{k+1}(x_0, x_1, \ldots, x_k))$$

is recursive because for each $(x_1, \ldots, x_1)$, there is an x_0 such that either $(x_0, x_1, \ldots, x_k) \in P^{k+1}$ or $(x_0, x_1, \ldots, x_k) \in Q^{k+1}$, but not both. By the exercise, $S^k(x_1, \ldots, x_k) \leftrightarrow P^{k+1}(f(x_1, \ldots, x_k), x_1, \ldots, x_k)$ is a recursive relation. But $S^k(x_1, \ldots, x_k) \leftrightarrow R^k(x_1, \ldots, x_k)$, so with exercise 8.9, the theorem is proved. ▌

Theorem 8.6 The relation $(\exists y)T(z, n, y)$ is r.e. but not recursive.

Exercise 8.13 Theorem 8.6 is, of course, just the halting problem again, in slightly different clothing. Prove it directly, without reference to the halting problem per se.

Exercise 8.14 Is $\{(z, n) \mid (\exists y)T(z, n, y)\}$ creative?

Exercise 8.15 (a) Show that if $S^k(n_1, \ldots, n_k) \leftrightarrow (\exists n_0)R^{k+1}(n_0, n_1, \ldots, n_k)$, then

$$\overline{S^k(n_1, \ldots, n_k)} \leftrightarrow (\forall n_0)\overline{R^{k+1}(n_0, n_1, \ldots, n_k)}$$

(b) Show that if $S^k(n_1, \ldots, n_k) \leftrightarrow (\forall n_0)R^{k+1}(n_0, n_1, \ldots, n_k)$, then

$$\overline{S^k(n_1, \ldots, n_k)} \leftrightarrow (\exists n_0)\overline{R^{k+1}(n_0, n_1, \ldots, n_k)}$$

Exercise 8.16 Show that the relation Q^k defined by

$$Q^k(n_1, \ldots, n_k) \leftrightarrow (\forall n_0)R^{k+1}(n_0, n_1, \ldots, n_k)$$

for R^{k+1} recursive, is not necessarily an r.e. relation.

Exercise 8.17 Show that a relation P^k can be defined by both

$$P^k(n_1, \ldots, n_k) \leftrightarrow (\exists n_0)R^{k+1}(n_0, n_1, \ldots, n_k)$$

and

$$P^k(n_1, \ldots, n_k) \leftrightarrow (\forall n_0)S^{k+1}(n_0, n_1, \ldots, n_k)$$

for R^{k+1} and S^{k+1} recursive relations if and only if P^k is a recursive relation.

We have seen by Theorems 8.4 and 8.6 that a relation defined by prefixing a recursive relation with an existential quantifier is certainly an r.e. relation and may not be a recursive relation. What if we prefix a recursive relation by a universal quantifier? Is the result recursive, r.e., or something else? Or, what if we prefix an r.e. relation with an existential quantifier, or a universal quantifier? Etc.

Theorem 8.7 If R^k is an r.e. relation and Q^{k-1} is defined by

$$Q^{k-1}(x_2, \ldots, x_k) \leftrightarrow (\exists x_1)R^k(x_1, x_2, \ldots, x_k)$$

then Q^{k-1} is an r.e. relation.

PROOF Since R^k is r.e., there is a recursive relation P^{k+1} such that

$$R^k(x_1, \ldots, x_k) \leftrightarrow (\exists x_0)P^{k+1}(x_0, x_1, \ldots, x_k)$$

and so

$$(\exists x_1)R^k(x_1, \ldots, x_k) \leftrightarrow (\exists x_1)(\exists x_0)P^{k+1}(x_0, x_1, \ldots, x_k)$$

We will use the pairing function J and the associated functions K and L. Since P^{k+1} is recursive, by exercise 8.10, the relation S^k defined by

$$S^k(x, x_2, \ldots, x_k) \leftrightarrow P^{k+1}(K(x), L(x), x_2, \ldots, x_k)$$

is recursive, so $(\exists x)S^k(x, x_2, \ldots, x_k)$ is an r.e. relation. We claim that

$$Q^{k-1}(x_2, \ldots, x_k) \leftrightarrow (\exists x)S^k(x, x_2, \ldots, x_k)$$

and thus Q^{k-1} is r.e. The verification of the claim is fairly obvious and is left to the reader. ▮

The trick of using the pairing function to get one existential quantifier to do the work of two or more is known as *collapsing quantifiers*. The same idea can be extended to collapse k existential quantifiers into one existential quantifier.

Exercise 8.18 Show that if the relation Q^{k-1} is defined by

$$Q^{k-1}(x_2, \ldots, x_k) \leftrightarrow (\forall x_0)(\forall x_1)R^{k+1}(x_0, x_1, x_2, \ldots, x_k)$$

for R^{k+1} a recursive relation then there exists a recursive relation P^k such that

$$Q^{k-1}(x_2, \ldots, x_k) \leftrightarrow (\forall x)P^k(x, x_2, \ldots, x_k)$$

Suppose we take the set of all recursive relations (predicates) and start piling up quantifiers in front, collapsing like ones. We can easily see what is meant by the $\exists\forall$-relations (predicates), the $\forall\exists\forall\exists$-relations (predicates), etc. One might well ask how these all fit together. We have just seen, for example, that the intersection of the $\forall$-relations and the $\exists$-relations is the set of recursive relations, that there are $\forall$-relations which are not r.e. and hence not among the $\exists$-relations.

Theorem 8.8 (1) The set of $\forall$-relations is contained in (a) the set of $\forall\exists$-relations and (b) the set of $\exists\forall$-relations. (2) The set of $\exists$-relations is contained in (a) the set of $\exists\forall$-relations and (b) the set of $\forall\exists$-relations.

PROOF (1) To say that a relation P^k belongs to the $\forall$-relations means of course, that there is a recursive relation R^{k+1} such that

$$P^k(x_1, \ldots, x_k) \leftrightarrow (\forall x_0)R^{k+1}(x_0, x_1, \ldots, x_k)$$

To show (1a), we must also show that there is also a recursive relation Q^{k+2} such that

$$P^k(x_1, \ldots, x_k) \leftrightarrow (\forall x_0)(\exists x_{-1})Q^{k+2}(x_{-1}, x_0, x_1, \ldots, x_k)$$

We define Q^{k+2} as follows:

$$Q^{k+2}(x_{-1}, x_0, \ldots, x_k) \leftrightarrow [R^{k+1}(x_0, \ldots, x_k) \quad \text{and} \quad x_{-1} = x_{-1}]$$

Since $x_{-1} = x_{-1}$ is true for all $x_{-1} \in N$, we have

$$R^{k+1}(x_0, \ldots, x_k) \qquad \text{if and only if} \qquad (\exists x_{-1})Q^{k+2}(x_{-1}, x_0, \ldots, x_k) \quad (*)$$

(This is a very common way of introducing an extra variable without changing anything essential.) Prefixing both sides of (*) by $(\forall x_0)$, we have

$$P^k(x_1, \ldots, x_k) \leftrightarrow (\forall x_0)R^{k+1}(x_0, x_1, \ldots, x_k)$$
$$\leftrightarrow (\forall x_0)(\exists x_{-1})Q^{k+2}(x_{-1}, x_0, x_1, \ldots, x_k)$$

For (1b), we do almost the same thing. Using the same definition of Q^{k+2}, we see that for any x_{-1}, $(x_0, x_1, \ldots, x_k) \in R^{k+1}$ if and only if $(x_{-1}, x_0, x_1, \ldots, x_k) \in Q^k$. Thus, $(\forall x_0)R^{k+1}(x_0, \ldots, x_k)$ if and only if $(\forall x_0)Q^{k+2}(x_{-1}, x_0, \ldots, x_k)$. Further, since there are $x_{-1} \in N$ such that $x_{-1} = x_{-1}$, certainly

$$P^k(x_1, \ldots, x_k) \leftrightarrow (\forall x_0)R^{k+1}(x_0, x_1, \ldots, x_k)$$
$$\leftrightarrow (\exists x_{-1})(\forall x_0)Q^{k+2}(x_{-1}, x_0, x_1, \ldots, x_k)$$

The proof of part 2 is left to the reader. ▌

Corollary The set of $\forall\exists$-relations contains some relations which are not r.e. and hence not $\exists$-relations.

PROOF There are $\forall$-relations which are not r.e. and by the theorem, these are contained in the $\forall\exists$-relations. ▌

Suppose that all relations (predicates) are written in collapsed quantifier form. The pattern already begun is continued in a hierarchy of number-theoretic relations known as the *Kleene–Mostowski hierarchy* or *arithmetic hierarchy*. A standard way of referring to the sets of relations beginning with an existential quantifier is by $\Sigma_1{}^0, \Sigma_2{}^0, \Sigma_3{}^0, \ldots$, where $\Sigma_1{}^0$ is the set of $\exists$-relations and $\Sigma_i{}^0$ is the set of

$$\underbrace{\exists\forall\cdots}_{i \text{ quantifiers}}\text{-relations}$$

In the same way, $\Pi_1{}^0, \Pi_2{}^0, \Pi_3{}^0, \ldots$ designate the different sets of relations beginning with a universal quantifier. Then, $\Sigma_i{}^0 \cap \Pi_i{}^0$ is called Δ_i. In terms of this notation, $\Delta_1{}^0$ is the set of recursive relations and $\Sigma_1{}^0$ is the set of r.e. relations. Some of the facts (theorems) about the Kleene–Mostowski hierarchy are:

1. For all $i, j, j > i$: $\Pi_i{}^0 \subsetneq \Sigma_j{}^0$; $\Pi_i{}^0 \subsetneq \Pi_j{}^0$; $\Sigma_i{}^0 \subsetneq \Pi_j{}^0$; $\Sigma_i{}^0 \subsetneq \Sigma_j{}^0$.
2. For all i, $\Pi_i{}^0 \cup \Sigma_i{}^0 \subsetneq \Pi_{i+1}^0 \cap \Sigma_{i+1}^0$.

For the proofs of the statements and further information about the hierarchy, the reader can consult Rogers (1967, section 14.1), Davis (1958, chapter 9), or Hermes (1965, section 29). The following diagram indicates how the hierarchy looks:

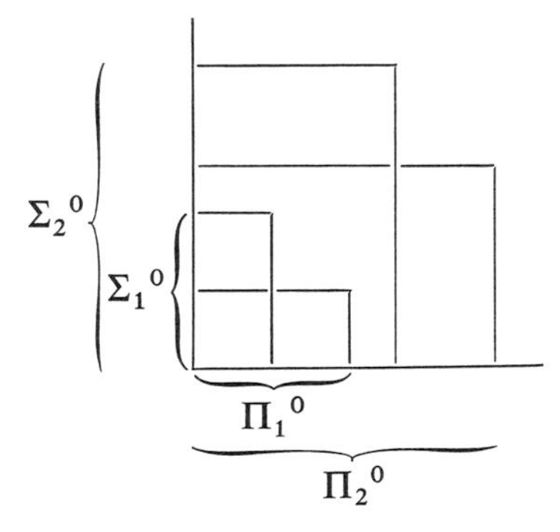

8.4 TURING REDUCIBILITY, OR ORACLES

It was mentioned earlier that five effective procedures have been defined for saying that the decision problem for S_1 is reducible to the decision problem for S_2. The simplest of these procedures is one–one reducibility, which we have already discussed. The most complex is usually called Turing reducibility, and we discuss it here. [For a careful discussion of all five procedures, the reader should consult Rogers (1967, chapters 6–9).] Roughly, a *set* S_1 *is Turing reducible to set* S_2 if there is a Tm which can compute the characteristic function of S_1 provided it "has access" to the characteristic function of S_2. (It is not to be concluded from this that S_2 has a recursive characteristic function. Only that if S_2 does have a recursive characteristic function, then S_1 also has one, just by the definition of Turing reducible.) We might imagine "has access" as follows: in one corner of the room is the Tm to compute the characteristic function of S_1 and in another corner there is an oracle who answers questions about S_2. There is also a Tm operator, i.e., a human being. Having been given a tape with n on it, the Tm starts to compute an answer as to whether or not n belongs to S_1. Every so often during the computation, the Tm rings a bell and stops. The operator reads a number m on a certain section of the tape and goes to the oracle and asks, humbly, "Tell me, oh oracle, does m belong to S_2?" The oracle gives an answer "yes" or "no" and the operator, returning to the Tm, punches into its tape, by hand, the answer given by the oracle and pushes a "restart" button, at which point the Tm continues its computation. Notice (i) for a given n, the Tm stops only a finite number of times to ask a question, (ii) the number of times q depends on n and is not uniform or even uniformly bounded for all n, and (iii) if the Tm asks about the membership of $m_1, m_2, \ldots, m_q$ in S_2

during a computation from n, the form of each question, $m_{i+1} \in S_2$?, may depend on the answers supplied for the previous questions, $m_k \in S_2$?, for $k = 1, \ldots, i$.

There are several ways of formalizing the oracle-asking procedure. [The reader may consult Davis (1958), Kleene (1967), and Rogers (1967) for several versions.] Another one is the following. Let S be a set. We define an *S-relative Tm* T^S to be a Tm which has in addition to the usual style of quintuples, one quintuple of the form $q_S M_S q_i$—R. Such a quintuple is interpreted by: if T^S is in internal state q_S reading the special marker M_S, it has access to the set S in such a way that it will print Y or N according to whether or not the number which is between the M_S being read and the next occurrence of M_S to the left is in S or is not in S. (We could allow it to print I, for incomplete information, in case there is no M_S to the left). For an *S-relative Tm* T^S, we have the corresponding kind of *S-relative T-predicate*,

$$T^S = \{(z, n, y) \mid z \text{ is the Gödel number of an } S\text{-relative Tm } T_z^S, n \text{ is a natural number, and } y \text{ is the Gödel number of the computation of } T_z^S \text{ from a tape with an initial tape inscription representing } n\}$$

It is important to notice that for any $(z, n, y) \in T^S$, if any information as to whether or not certain $m_1, \ldots, m_q$ are in S is used during the computation, the $m_1, \ldots, m_q$ are less than y. A *function is S-recursive* if there is an *S-relative Tm* that computes it. That is, a function f is S-recursive if and only if there exists a z such that

$$f(x) = U(\mu_y T^S(z, x, y))$$

A set (relation) is recursive in S if it has an S-recursive characteristic function.

Exercise 8.19 Show that if S_1 is one-one reducible to S_2, then S_1 is recursive in S_2.

Exercise 8.20 Show that if a set S is recursive, then it is recursive in R for all sets R.

Exercise 8.21 Show that if there exist two r.e. sets S and R such that S is not recursive in R and R is not recursive in S, then (a) neither R nor S is recursive, and (b) there exists a set Q such that the complete set K is not recursive in Q.

Exercise 8.22 Suppose f is the characteristic function of a set A and g is the characteristic function of a set B. Show that if for all $e \in N$ there exists a number $x(e) \in N$ such that

$$f(x(e)) \neq U(\mu_y T^B(e, x(e), y))$$

then A is not recursive in B.

8.5 FRIEDBERG'S THEOREM†

This section will be devoted to proving the following theorem.

Theorem 8.9 There exist two r.e. sets neither of which is recursive in the other.

In order to prove the theorem, we will construct sets S_1 and S_2 by way of constructing their characteristic functions f_1 and f_2. The aim of the construction is to make sure that S_1 is r.e. but not recursive in S_2 and that S_2 is r.e. but not recursive in S_1. The r.e.-ness will follow from the construction by Church's thesis. In an effort to keep each set from being recursive in the other, we will construct f_1 and f_2 and functions x_1 and x_2 so that for all $e \in N$,

$$\text{(i)} \quad f_1(x_1(e)) = 1 \quad \text{if and only if} \quad U(\mu_y T^{S_2}(e, x_1(e), y)) = 0$$
$$\text{(ii)} \quad f_2(x_2(e)) = 1 \quad \text{if and only if} \quad U(\mu_y T^{S_1}(e, x_2(e), y)) = 0$$

(See exercise 8.22.) Since f_i is to be the characteristic function of S_i, it is not essentially different to write T^{f_i} instead of T^{S_i}, and so we will do so henceforth.

The construction of the necessary functions will be in stages a, for $a = 0$, $1, \ldots$. The functions, as they are defined at stage a, will be called $f_1{}^a, f_2{}^a, x_1{}^a$, and $x_2{}^a$. We will think of a as being written in the form $a = 2n + 1$ or $2n + 2$. For each a, let e_a be the number of prime divisors of n. For any given number e, as a increases through N, the number e will recur infinitely often as e_a. The functions are defined as follows:

At stage $a = 0$:

$$f_1{}^0(x) = f_2{}^0(x) = 0 \qquad \text{for all} \quad x \in N$$
$$x_1{}^0(e) = x_2{}^0(e) = 2^e \qquad \text{for all} \quad e \in N$$

At stage $a = 2n + 1$:

Case 1.1 If $f_1^{a-1}(x_1^{a-1}(e_a)) = 0$ and

$$(\exists y < a)[T^{f_2{}^{a-1}}(e_a, x_1^{a-1}(e_a), y) \quad \text{and} \quad U(y) = 0]$$

then set $f_1{}^a(x_1^{a-1}(e^a)) = 1$ and $x_2{}^a(e) = 2^e(2a + 1)$ for all $e \geq e_a$. For all other

† The proof given here is taken directly from Friedberg (1957), though we have filled in between the lines a bit. Not only is the result, i.e., that the theorem is true, important, but the method developed here to prove it. Because of the importance of both, there have been several subsequent analyses and generalizations of the method, now called the *priority method.* The two main places for learning about these generalizations are Sacks (1963) and Rogers (1967).

functions and arguments, the function values do not change from stage $a-1$.

Case 1.2 If the above condition does not hold, everything stays the same.

At stage $a = 2n+2$: (This is just as above except reversing the roles of 1 and 2 *and* changing $x_1{}^a$ for all e strictly greater than e_a. It is all written out as follows.)

Case 2.1 If $f_2^{a-1}(x_2^{a-1}(e_a)) = 0$ and

$$(\exists y < a)[T^{f_1{}^{a-1}}(e_a, x_2^{a-1}(e_a), y) \quad \text{and} \quad U(y) = 0]$$

then set $f_2{}^a(x_2^{a-1}(e_a)) = 1$ and $x_1{}^a(e) = 2^e(2a+1)$ for all $e > e_a$. Everything else stays the same.

Case 2.2 If the above condition does not hold, everything stays the same.

Notice that both $f_1{}^0$ and $f_2{}^0$ start with the value 0 everywhere. The value for different x *may*, at some stage $a > 0$, be changed to 1, but once 1, the value for that argument is never changed again. With that in mind, we define, for $i = 1$ and 2,

$$f_i(x) = \begin{cases} 1 & \text{if there exists an } a \text{ such that } f_i{}^a(x) = 1 \\ 0 & \text{otherwise} \end{cases}$$

(In fact, by setting $f_i{}^a(x)$ equal to 1 at some stage a, we are saying that x is to be in S_i, and once in, we do not take it out.)

The following little discussion may be helpful in understanding the proof. There are at least two useful ways (not essentially different) to think about the predicates $T^{f_i{}^a}$, $i = 1, 2$.

1. One may imagine that we are defining an infinite sequence of sets, $S_i{}^0, S_i{}^1, \ldots, S_i{}^a, \ldots$ (for which $f_i{}^0, f_i{}^1, \ldots, f_i{}^a, \ldots$ are the characteristic functions) with the property that for all a, $S_i{}^a \subset S_i^{a+1}$ (S_i^{a+1} contains at most one more element than $S_i{}^a$) and that $\bigcup_{a \in N} S_i{}^a = S_i$ and similarly, $\lim_{a \to \infty} f_i{}^a = f_i$.

2. It is well known that oracles, at least those of Greek times, were uncooperative and ambiguous. With that in mind, one may imagine that the oracle associated with the predicate T^{f_i} is likely to answer the question, "Is n in S_i?" by sometimes saying, "I'm saying 'no' today, but come back tomorrow and ask again." But, if one day the oracle answers "yes" to the question, then for that same n, it will henceforth always answer "yes."

Lemma 8.1 For all e, $x_1{}^a(e)$ changes only finitely many times as a increases through N.

PROOF Assume false. Let $\bar{e}$ be the smallest e such that $x_1{}^a(\bar{e})$ changes infinitely often.

1. Under what circumstances does $x_1{}^a(\bar{e})$ change? Only if the hypotheses of case 2.1 are satisfied, i.e., both $f_2^{a-1}(x_2^{a-1}(e_a)) = 0$ and

$$(\exists y < a)[T^{f_1{}^{a-1}}(e_a, x_2^{a-1}(e_a), y) \quad \text{and} \quad U(y) = 0]$$

Then, in that case, at a, we have

$$f_2{}^a(x_2^{a-1}(e_a)) = 1 \qquad \text{and} \qquad x_1{}^a(e) = 2^e(2a+1) \qquad \text{for all} \quad e > e_a$$

2. Since this is the cause of $x_1{}^a(\bar{e})$ changing, it means that $\bar{e} > e_a$. Further, if $x_1{}^a(\bar{e})$ is to change infinitely often, the hypotheses of case 2.1 must be satisfied at infinitely many a for which $e_a < \bar{e}$. As there are only finitely many numbers less than $\bar{e}$, there must be at least one, call it e_{a*}, for which there are infinitely many a such that $e_a = e_{a*}$ and for which the hypotheses of case 2.1 are satisfied; thus

$$x_1{}^a(e) = 2^e(2a+1) \qquad \text{for all} \quad e > e_a = e_{a*}$$

3. But now we must ask how it can happen that the hypotheses of case 2.1 are satisfied infinitely often. Since the value of $f_2{}^a$, once changed to 1 for a particular argument, is never changed again, it must be that the argument is changing infinitely often. That is, if $f_2{}^a(x_2^{a-1}(e_{a*}))$ is to be set equal to 1 infinitely often, it must be that $x_2{}^a(e_{a*})$ is changed infinitely often.

4. But how does $x_2{}^a(e_{a*})$ get changed infinitely often? $x_2{}^a$ is changed when the hypotheses of case 1.1 are satisfied for a such that $e_a \leq e_{a*}$. As in step 1, notice that for $x_2{}^a(e_{a*})$ to change infinitely often, there must be a number $e_{a**} \leq e_{a*}$ so that for infinitely many a, $e_a = e_{a**}$, and by satisfying the hypotheses of case 1.1, the result is that both

$$\begin{aligned} x_2{}^a(e) &= 2^e(2a+1) \qquad \text{for all} \quad e \geq e_{a**} \\ f_1{}^a(x_1^{a-1}(e_{a**})) &= 1 \end{aligned} \tag{**}$$

5. But, as in step 3, the conditions of case 1.1 provoking (**) can happen infinitely often only if $x_1{}^a(e_{a**})$ is changed infinitely often.

6. But $e_{a**} \leq e_{a*} < \bar{e}$. And the assumption was that $\bar{e}$ is the smallest number on which $x_1{}^a$ changes infinitely often. Contradiction. ▮

One can show in a similar manner that for all e, $x_2{}^a(e)$ changes only finitely many times as a increases through N. Consequently, for each e, define $z_i(e)$ to be the final value of $x_i{}^a(e)$, for $i = 1$ and 2.

Lemma 8.2 For a fixed e, let $z_1(e)$ be the final value of $x_1{}^a(e)$ as a increases. Then,

$$(\exists y)[T^{f_2}(e, z_1(e), y) \text{ and } U(y) = 0] \quad \text{if and only if} \quad f_1(z_1(e)) = 1.$$

PROOF *Part I.* Given that $(\exists y)[T^{f_2}(e, z_1(e), y)$ and $U(y) = 0]$, we must show that there is an a such that $f_1{}^a(z_1(e)) = 1$. (Then done, by definition of f_1.) Looking at the left side, we notice the following facts:

1. There exists an a_e such that $x_1{}^a(e) = z_1(e)$ for all $a \geq a_e$ by Lemma 8.1.
2. By the definition of the predicate T^{f_2}, any information about the set defined by f_2 that is used in a computation is coded into the Gödel number of the computation y. Thus, there are only finitely many u such that $f_2(u)$ is used. In each case, by the definition of f_2, there is an a_u such that $f_2{}^a(u) = f_2^{a_u}(u)$ for all $a \geq a_u$.
3. There are infinitely many $a > y$.

In view of these three items, we can take some a greater than all a_u's and a_e and y; call it a^*. Thus, for all $a \geq a^*$,

$$(\exists y < a)[T^{f_2{}^a}(e, z_1(e), y) \quad \text{and} \quad U(y) = 0] \qquad (*)$$

Consider $f_1^{a^*}(z_1(e))$. If it is 1, then we are done. Suppose it is not 1. Then, let a^{**} be the next stage such that $e_{a^{**}} = e$. Since $a^{**} > a^*$, we still have $(*)$, i.e.,

$$(\exists y < a^{**})[T^{f_2{}^{a^{**}-1}}(e, z_1(e), y) \quad \text{and} \quad U(y) = 0]$$

and, further, $f_1^{a^{**}-1}(z_1(e)) = 0$. But that is to say that the hypotheses of case 1.1 are satisfied at a^{**}, so set $f_1^{a^{**}}(z_1(e)) = 1$, and so, done.

Part II. Since for every e, $x_1{}^a(e)$ has a final value $z_1(e)$, the discussion will be for fixed $\bar{e}$ and corresponding $z_1(\bar{e})$. (Wish to show that the right side implies the left side.) Since, by the assumption, $f_1(z_1(\bar{e})) = 1$, by the definition of f_1, there was a stage a at which $f_1{}^a(z_1(\bar{e}))$ was set equal to one. Call that stage a^*. This occurred because the hypotheses of case 1.1 were satisfied at stage a^*; thus, $f_1^{a^*-1}(x_1^{a^*-1}(e_{a^*})) = 0$ and

$$(\exists y < a^*)[T^{f_2{}^{a^*-1}}(e_{a^*}, x_1^{a^*-1}(e_{a^*}), y) \quad \text{and} \quad U(y) = 0] \qquad (**)$$

where, by definition of a^*, $x_1^{a^*-1}(e_{a^*}) = z_1(\bar{e})$. Consider the two following items about the functions x_1 and x_2:

1. Claim that $z_1(\bar{e}) = x_1^{a^*-1}(e_{a^*})$ implies that $\bar{e} = e_{a^*}$. This is true because functions x_1 and x_2 are one–one and so, in general, $2^{e'}(2a' + 1) = 2^e(2a + 1)$ implies $e = e'$ and $a = a'$.
2. For $\bar{e} = e_{a^*}$, if, for some $a > a^* - 1$, $x_1{}^a(e_{a^*})$ changes its value, then $z_1(\bar{e})$ cannot be the final value of x_1 on e because (for fixed e) x_1 and x_2 do not decrease as a increases.

With these items in mind, we can rewrite $(**)$ above as $f_1^{a^*-1}(z_1(\bar{e})) = 0$,

$$(\exists y < a^*)[T^{f_2{}^{a^*-1}}(\bar{e}, z_1(\bar{e}), y) \quad \text{and} \quad U(y) = 0] \qquad (***)$$

and so at a^*, $f_1^{a^*}(z_1(\bar{e})) = 1$ and $x_2^{a^*}(e) = 2^e(2a^* + 1)$ for all $e \geq e_{a^*} = \bar{e}$.

Now, the question is, does (***) stay true for all $a > a^* - 1$? If so, we are done, because it is the same as saying

$$(\exists y)[T^{f_2}(\bar{e}, z_1(\bar{e}), y) \quad \text{and} \quad U(y) = 0]$$

So, suppose not. At some $a^{**} > a^*$,

$$(\exists y < a^{**})[T^{f_2^{a^{**}-1}}(\bar{e}, z_1(\bar{e}), y) \quad \text{and} \quad U(y) = 0] \qquad (****)$$

is false. That means, since nothing else is changing, that the information provided by $f_2{}^a$ is changing. That can happen only if there is some argument value u such that $f_2^{a^*-1}(u) = 0$ was used in (***) and $f_2^{a^{**}-1}(u) = 1$ is used in (****). Thus, the value at u changed sometime between a^* and a^{**}. Suppose at $\bar{a}$. Further, by the meaning of the $f_2{}^a$-relative T-predicate $T^{f_2{}^a}$, any changes in $f_2{}^a$ (for $a^* \le a \le a^{**}$) that might have changed (***) from being true to (****) being false had to be on argument values less than y. That is, the value u just discussed is less than the y of (***), i.e., $u < y < a^*$. Let w be the smallest such u. Thus, we have

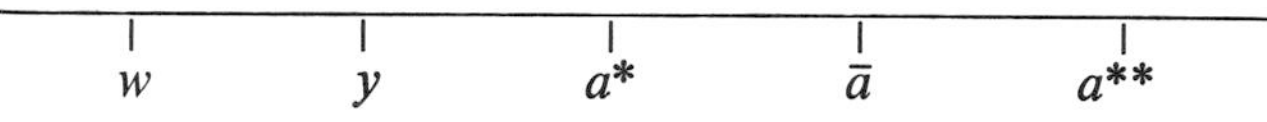

So we must consider how it happened that $f_2{}^{\bar{a}}(w)$ was set equal to 1. At $\bar{a}$, the hypotheses of case 2.1 were satisfied, i.e.,

$$f_2^{\bar{a}-1}(w) = 0 \qquad \text{and} \qquad (\exists y < \bar{a})[T^{f_1^{\bar{a}-1}}(e_{\bar{a}}, w, y) \quad \text{and} \quad U(y) = 0]$$

Consequently, set $f_2{}^{\bar{a}}(w) = 1$ *and* $x_1{}^{\bar{a}}(e) = 2^e(2\bar{a} + 1)$ for all $e > e_{\bar{a}}$. But since $\bar{a} > a^*$, $x_1{}^a(\bar{e}) = z_1(\bar{e})$, so it must be that $e_{\bar{a}} \ge e_{a^*} = \bar{e}$. *However*, $w = x_2^{\bar{a}-1}(e_{\bar{a}})$, by virtue of the fact that we are in case 2.1. But, back at a^* [when $f_1^{a^*}(z_1(\bar{e}))$ was set equal to 1], we also set $x_2^{a^*}(e) = 2^e(2a^* + 1)$ for all $e \ge e_{a^*}$, thus including $e_{\bar{a}}$. Since x_2 is nondecreasing with respect to a, we have

$$w = x_2^{\bar{a}-1}(e_{\bar{a}}) \ge x_2^{a^*}(e_{\bar{a}}) = 2^{e_{\bar{a}}}(2a^* + 1) > a^* > y$$

But this contradicts w being less than y. So (***) stays true for all $a > a^* - 1$. ∎

One must state and prove a lemma similar to Lemma 8.2 which relates T^{f_1} and f_2 in the same fashion. From Lemma 8.2 and its twin, we have by exercise 8.22 that S_1 and S_2 are not Turing reducible to (recursive in) each other. If we can show that each is r.e., then the theorem will be proved.

Notice that for each a, $S_1{}^a$ and $S_2{}^a$ are finite and so recursive. Thus, the predicates, for each a, $T^{f_1{}^a}$ and $T^{f_2{}^a}$ are recursive predicates. Also, at each stage a, one need make only $2a + 1$ computations in order to decide how to proceed, e.g., $f_2^{a-1}(x_2^{a-1}(e_a))$ and $T^{f_1^{a-1}}(e_a, x_2^{a-1}(e_a), y)$ and $U(y)$ for $y = 0$,

$1, \ldots, a-1$. So, by Church's thesis, the function defined as $F_1(a, n) = f_1{}^a(n)$ is a recursive function. From this, we define g_1 by

$$g_1(n) = \begin{cases} \mu_a[F_1(a, n) = 1] & \text{if there is such an } a \\ \text{undefined} & \text{otherwise} \end{cases}$$

g_1 is a partial recursive function.

Exercise 8.23 Show that the domain of g_1 is just S_1 and so S_1 is r.e.

PART II

Mathematical Logic

Chapter IX

THE PROPOSITIONAL CALCULUS AS AN EXAMPLE OF A RECURSIVELY AXIOMATIZABLE THEORY

9.1 DEFINITION OF A RECURSIVELY AXIOMATIZABLE THEORY

A *formal language* L is specified by first giving a countably infinite set of symbols and then choosing some particular recursive subset F of the set of all words on those symbols. We call F the set of *formulas* of L. In order to have a logical theory T from F, we choose some subset A which we call the *axioms* of T and we specify a recursive set of relations, on elements of F, of the form $\{\mathbf{A}_1, \mathbf{A}_2, \ldots, \mathbf{A}_n\} \rightarrow \mathbf{A}_{n+1}$ (where all the **A**'s belong to F) which we call the *rules of inference*. (We say "$\mathbf{A}_{n+1}$ follows from $\mathbf{A}_1, \mathbf{A}_2, \ldots, \mathbf{A}_n$" or "from $\mathbf{A}_1, \mathbf{A}_2, \ldots, \mathbf{A}_n$, infer $\mathbf{A}_{n+1}$.") We then define, inductively, the subset D of F of *formulas derivable from the axioms by means of the rules of inference*: If $\mathbf{A} \in$ A, then $\mathbf{A} \in$ D, and if $\{\mathbf{A}_1, \mathbf{A}_2, \ldots, \mathbf{A}_n\} \rightarrow \mathbf{A}_{n+1}$ is a rule of inference and if $\mathbf{A}_i \in$ D for $1 \leq i \leq n$, then $\mathbf{A}_{n+1} \in$ D. If $\mathbf{A} \in$ D, then **A** is defined to be a *theorem* of T. Thus, to say that a formula **A** is a theorem of T is equivalent to saying that it has a *derivation* or *proof* or *is provable* in T, i.e., that there is a finite sequence of formulas in F, $\mathbf{B}_1, \ldots, \mathbf{B}_k$ such that each $\mathbf{B}_i$ is either (i) an axiom or (ii) $\mathbf{B}_i$ follows from $\mathbf{B}_{j_1}, \ldots, \mathbf{B}_{j_t}$ (also in the sequence for $j_i < i$) and $\{\mathbf{B}_{j_1}, \ldots, \mathbf{B}_{j_t}\} \rightarrow \mathbf{B}_i$ is a rule of inference, and (iii) $\mathbf{B}_k$ is **A**. The *length of* (or *number of steps in*) *a proof* is k. To indicate that **A** is provable in a theory T, we write $\vdash_{\mathsf{T}} \mathbf{A}$, omitting T when there can be no confusion. The *theory* T is usually identified with its set of theorems. T is a *recursively axiomatizable theory* if the set of formulas which are its axioms is a recursive set. T is *finitely axiomatizable* if the set of axioms is finite. (This definition will be amended slightly for first-order theories in chapter XI.)

For a given language, we can have many theories. *Two theories are equal* if they are exactly the same set of theorems. An interesting kind of question that one may reasonably ask is: If T_1 is a theory, is there a recursively axiomatizable theory T_2 such that T_1 equals T_2? We will be very much concerned with two such questions later. We define a theory T_1 to be a *subtheory* of a theory T_2 if $T_1 \subseteq T_2$. In that case, we also say that T_2 is an *extension* of T_1.

To get a better idea of what all these definitions really mean, we next consider a very simple example of a recursively axiomatizable theory.

9.2 A FORMULATION OF THE PROPOSITIONAL CALCULUS

For our first example, we will consider the recursively axiomatizable theory usually known as the *propositional calculus* (or *sentential calculus*) and we give this particular formulation of it the name P_0.

I. The language $L(P_0)$ is given by†:

1. The following set of symbols: the *propositional variables* are $\mathbf{p_1}, \mathbf{p_2}, \mathbf{p_3}, \ldots$; the *propositional connectives* are $\sim$, $\supset$; the parentheses **(,)**.

2. The set of formulas $F(P_0)$ is defined inductively by (a) every propositional variable belongs to $F(P_0)$; (b) if **A** and **B** belong to $F(P_0)$, then $\sim\mathbf{A}$ belongs to $F(P_0)$ and $(\mathbf{A} \supset \mathbf{B})$ belongs to $F(P_0)$.

II. The axioms $A(P_0)$ are, for every **A**, **B**, and **C** in $F(P_0)$,

$$(\mathbf{A} \supset (\mathbf{B} \supset \mathbf{C})) \qquad \text{axiom 1}$$
$$((\mathbf{A} \supset (\mathbf{B} \supset \mathbf{C})) \supset ((\mathbf{A} \supset \mathbf{B}) \supset (\mathbf{A} \supset \mathbf{C}))) \qquad \text{axiom 2}$$
$$((\sim\mathbf{A} \supset \sim\mathbf{B}) \supset (\mathbf{B} \supset \mathbf{A})) \qquad \text{axiom 3}$$

III. The rule of inference of P_0 is

$$\{\mathbf{A}, (\mathbf{A} \supset \mathbf{B})\} \to \mathbf{B}$$

(This rule of inference is known as *modus ponens* (mp).)

A, **B**, and **C** are variables for formulas. Thus, each axiom is an *axiom scheme*, because there are infinitely many formulas in $F(P_0)$ that fit into that scheme. For example, $(\mathbf{p_1} \supset (\mathbf{p_2} \supset \mathbf{p_1}))$, and $((\mathbf{p_1} \supset \mathbf{p_2}) \supset ((\mathbf{p_2} \supset \mathbf{p_1}) \supset (\mathbf{p_1} \supset \mathbf{p_2})))$ each fit into $(\mathbf{A} \supset (\mathbf{B} \supset \mathbf{A}))$. A formula which fits into a scheme is an instance of the scheme. (Since " fits into a scheme " is not defined formally, neither is "instance.") Thus, the two examples are each instances of the first axiom scheme. In general, we will not belabor the distinction between

† A formal language whose only symbols are propositional variables, propositional connectives, and parentheses will be called a *propositional language* and any logical theory based on that language, a *propositional theory*.

formula and scheme, but from time to time, it is necessary to take into account the distinction. For example, P_0 does not have a finite set of axioms, but since it has a finite set of axiom schemes, it is certainly recursively axiomatizable.

Exercise 9.1 (a) Devise a Gödel numbering for the formulas of $F(P_0)$. (b) Is the set $\{n \in N \mid n$ is the Gödel number of a formula in $F(P_0)\}$ a recursive set?

Exercise 9.2 (a) Extend your Gödel numbering so as to give Gödel numbers to all the proofs in P_0. (b) Is the set $\{n \mid n$ is the Gödel number of a proof in $P_0\}$ a recursive set? (c) Is the set $\{n \mid n$ is the Gödel number of a theorem of $P_0\}$ a recursively enumerable set? Is it a recursive set? (Use Church's thesis.)

Exercise 9.3 Use Church's thesis to show that if T is a recursively axiomatizable theory, then $\{n \mid n$ is the Gödel number of a theorem of T$\}$ is a recursively enumerable set.

A *logical theory is consistent* if and only if there is no formula **A** such that both **A** and $\sim$**A** are theorems; i.e., not both $\vdash$**A** and $\vdash\sim$**A**. A *logical theory is complete* if for every pair of formulas **A** and $\sim$**A**, one of them is a theorem. (This is a temporary definition which will be modified in a later chapter.) The *decision problem for a logical theory* T is the problem of deciding of an arbitrary formula in the language of T whether or not it is a theorem of T. This is the same as asking if the set $\{n \mid n$ is the Gödel number of a theorem in T$\}$ is a recursive set. We will see before the end of the chapter that the decision problem for P_0 is solvable. First, however, we shall consider some other important properties of P_0 and those properties will help us get the solvability result. In order to do any of these things, we must see what some of the theorems of P_0 look like.

Exercise 9.4 Prove that every recursively axiomatizable theory which is consistent and complete is decidable.

9.3 SOME THEOREMS IN P_0 AND THE DEDUCTION THEOREM

(A ⊃ A) is a very simple formula and we would like to see if $\vdash_{P_0}$ **(A ⊃ A)**. At the moment, the only thing we can do is actually produce the proof. We have no other way of finding out. We give the formulas that constitute the proof in a column, with explanations to the right:

$((\mathbf{A} \supset ((\mathbf{B} \supset \mathbf{A}) \supset \mathbf{A})) \supset ((\mathbf{A} \supset (B \supset \mathbf{A})) \supset (\mathbf{A} \supset \mathbf{A})))$	axiom 2
$(\mathbf{A} \supset ((\mathbf{B} \supset \mathbf{A}) \supset \mathbf{A}))$	axiom 1
$((\mathbf{A} \supset (\mathbf{B} \supset \mathbf{A})) \supset (\mathbf{A} \supset \mathbf{A}))$	first two steps and mp
$(\mathbf{A} \supset (\mathbf{B} \supset \mathbf{A}))$	axiom 1
$(\mathbf{A} \supset \mathbf{A})$	third and fourth steps and mp

If it is this complicated to prove that $(\mathbf{A} \supset \mathbf{A})$ is a theorem of P_0, what horrible things await us when we wish to prove

$$((\mathbf{B} \supset \mathbf{A}) \supset ((\sim \mathbf{B} \supset \mathbf{A}) \supset \mathbf{A}))!$$

There must be an easier way.

If T is an axiomatizable theory on formulas F, we define, inductively, the subset $\mathsf{D}_{\{B_1, \ldots, B_t\}}$ of F of *formulas derivable from the axioms and the hypotheses* $\mathbf{B}_1, \ldots, \mathbf{B}_t$ *by means of the rules of inference by*: If $\mathbf{A} \in \{\mathbf{B}_1, \ldots, \mathbf{B}_t\}$, then $\mathbf{A} \in \mathsf{D}_{\{B_1, \ldots, B_t\}}$ and if $\{\mathbf{A}_1, \ldots, \mathbf{A}_k\} \to \mathbf{A}_{k+1}$ is a rule of inference and $\mathbf{A}_1, \ldots, \mathbf{A}_k$ belong to $\mathsf{D}_{\{B_1, \ldots, B_t\}}$, then $\mathbf{A}_k$ belongs to $\mathsf{D}_{\{B_1, \ldots, B_t\}}$. If $\mathbf{A} \in \mathsf{D}_{\{B_1, \ldots, B_t\}}$, then there is a *derivation of* $\mathbf{A}$ *relative to hypotheses* $\{\mathbf{B}_1, \ldots, \mathbf{B}_t\}$, or, *a proof of* $\mathbf{A}$ *from hypotheses*: i.e., there is a finite sequence of formulas in F, $\mathbf{C}_1, \ldots, \mathbf{C}_n$, such that each $\mathbf{C}_i$ is either (1) an axiom or a member of $\{\mathbf{B}_1, \ldots, \mathbf{B}_t\}$ or (2) $\mathbf{C}_i$ follows from $\mathbf{C}_{r_1}, \ldots, \mathbf{C}_{r_s}$ (also in the sequence for $r_j < i$) and $\{\mathbf{C}_{r_1}, \ldots, \mathbf{C}_{r_s}\} \to \mathbf{C}_i$ is a rule of inference, and (3) $\mathbf{A}$ is $\mathbf{C}_n$. In this case, we write $\mathbf{B}_1, \ldots, \mathbf{B}_t \vdash \mathbf{A}$.

Exercise 9.5 (a) Can you prove $\mathbf{A} \vdash_{\mathsf{P}0} \mathbf{A}$?! (b) Considering the ease with which that can be done, and the complications of getting $\vdash_{\mathsf{P}0} (\mathbf{A} \supset \mathbf{A})$, what would you like to have be true?

In proving $(\mathbf{A} \supset \mathbf{A})$ is a theorem in P_0, we wrote down a certain sequence of formulas, just as required by the definition of proof. It is convenient to identify provable formulas by beginning them with $\vdash$. Proofs will usually appear in this form from now on.

We point out a possible confusion concerning the word "theorem." We have already proved several theorems in these pages and now we are also speaking of "theorems in P_0." In a few lines, we will be speaking of "theorems about P_0." The latter are often referred to as "metatheorems." English is our language for talking about a logical theory and the formal language of the theory is a subset of English. Thus, we may refer to English as a "metalanguage." Indeed, since the main aim of logic is to set up mathematics in a formal language and then choose some appropriate formulas as axioms so that it will be a recursively axiomatizable theory, the theorems of mathematics become the formal theorems in the theory. The theorems we prove about the theory, such as whether or not it has a solvable decision problem, are metatheorems. It is for this reason that the study of logic is sometimes referred to as metamathematics. The first metatheorem concerning P_0 is known as the *deduction theorem.*

Theorem 9.1 If $\mathbf{B}_1, \ldots, \mathbf{B}_{k-1}, \mathbf{B}_k \vdash_{P_0} \mathbf{C}$, then $\mathbf{B}_1, \ldots, \mathbf{B}_{k-1} \vdash_{P_0} (\mathbf{B}_k \supset \mathbf{C})$.

PROOF To assert "$\mathbf{B}_1, \ldots, \mathbf{B}_k \vdash_{P_0} \mathbf{C}$" is short-hand for asserting the existence of a finite sequence of formulas constituting a proof of **C** from hypotheses as defined just above. Hence, the proof of this theorem will be by induction on the number of steps in the proof from hypothesis, i.e., the length n of the sequence. (1) Suppose $n = 1$. That means that **C** is $\mathbf{C}_1$ and that it must be either an axiom or a hypothesis.

(a) If **C** is an axiom, then

$\mathbf{B}_1, \ldots, \mathbf{B}_{k-1} \vdash_{P_0} \mathbf{C}$	
$\mathbf{B}_1, \ldots, \mathbf{B}_{k-1} \vdash_{P_0} (\mathbf{C} \supset (\mathbf{B}_k \supset \mathbf{C}))$	axiom 1
$\mathbf{B}_1, \ldots, \mathbf{B}_{k-1} \vdash_{P_0} (\mathbf{B}_k \supset \mathbf{C})$	first two steps and mp

(b) Suppose **C** is a hypothesis. If it is $\mathbf{B}_j$ for $j \neq k$, then

$\mathbf{B}_1, \ldots, \mathbf{B}_{k-1} \vdash_{P_0} (\mathbf{B}_j \supset (\mathbf{B}_k \supset \mathbf{B}_j))$	axiom 1
$\mathbf{B}_1, \ldots, \mathbf{B}_{k-1} \vdash_{P_0} \mathbf{B}_j$	a hypothesis
$\mathbf{B}_1, \ldots, \mathbf{B}_{k-1} \vdash_{P_0} (\mathbf{B}_k \supset \mathbf{B}_j)$	mp

If it is $\mathbf{B}_k$, then using our proof that $\vdash_{P_0}(\mathbf{A} \supset \mathbf{A})$,

$$\mathbf{B}_1, \ldots, \mathbf{B}_{k-1} \vdash_{P_0} (\mathbf{B}_k \supset \mathbf{B}_k)$$

(2) For the induction hypothesis, we suppose that the theorem is true for all proofs from hypotheses with fewer than n steps and try to show that it is also true for proofs of length n. (a) & (b). If **C** is an axiom or a hypothesis, then the proof is just as given above. (c) Suppose that **C** results from earlier steps by a use of the rule of inference *modus ponens.* If that were the case, then at some steps i and j less than n, we have

$\mathbf{B}_1, \ldots, \mathbf{B}_k \vdash_{P_0} \mathbf{C}_i$	and
$\mathbf{B}_1, \ldots, \mathbf{B}_k \vdash_{P_0} (\mathbf{C}_i \supset \mathbf{C})$	so that the nth step is
$\mathbf{B}_1, \ldots, \mathbf{B}_k \vdash_{P_0} \mathbf{C}$	

Since the ith and jth steps satisfy the induction hypothesis, we have, by the induction hypothesis,

$$\mathbf{B}_1, \ldots, \mathbf{B}_{k-1} \vdash_{P_0} (\mathbf{B}_k \supset \mathbf{C}_i)$$
$$\mathbf{B}_1, \ldots, \mathbf{B}_{k-1} \vdash_{P_0} (\mathbf{B}_k \supset (\mathbf{C}_i \supset \mathbf{C}))$$

but we also know that

$$\mathbf{B}_1, \ldots, \mathbf{B}_{k-1} \vdash_{P0} ((\mathbf{B}_k \supset (\mathbf{C}_i \supset \mathbf{C})) \supset ((\mathbf{B}_k \supset \mathbf{C}_i) \supset (\mathbf{B}_k \supset \mathbf{C}))) \qquad \text{axiom 2}$$

and so we can conclude, by using *modus ponens* twice, that

$$\mathbf{B}_1, \ldots, \mathbf{B}_{k-1} \vdash_{P0} (\mathbf{B}_k \supset \mathbf{C}) \quad \blacksquare$$

Exercise 9.6 What is the converse to the deduction theorem? Is it true?

Exercise 9.7 (a) Prove that if $\mathbf{B}_1, \ldots, \mathbf{B}_k \vdash \mathbf{C}$, then $\mathbf{B}_1, \ldots, \mathbf{B}_k, \mathbf{B}_{k+1}, \ldots, \mathbf{B}_{k+j} \vdash \mathbf{C}$. (b) If $\mathbf{B}_1, \ldots, \mathbf{B}_k \vdash \mathbf{C}$, can we change the order of the hypotheses and still deduce $\mathbf{C}$?

Both for practice and because we shall need them, we shall see if the following formulas are theorems of P_0.

Lemma 9.1 The following formulas are theorems of P_0:

1. $((\mathbf{A} \supset \mathbf{B}) \supset ((\mathbf{B} \supset \mathbf{C}) \supset (\mathbf{A} \supset \mathbf{C})))$.
2. $(\mathbf{B} \supset ((\mathbf{B} \supset \mathbf{C}) \supset \mathbf{C}))$.
3. $(\sim\mathbf{B} \supset (\mathbf{B} \supset \mathbf{C}))$.
4. $(\sim\sim\mathbf{B} \supset \mathbf{B})$.
5. $(\mathbf{B} \supset \sim\sim\mathbf{B})$.
6. $((\mathbf{A} \supset \mathbf{B}) \supset (\sim\mathbf{B} \supset \sim\mathbf{A}))$.
7. $(\mathbf{B} \supset (\sim\mathbf{C} \supset \sim(\mathbf{B} \supset \mathbf{C})))$.
8. $((\mathbf{B} \supset \mathbf{A}) \supset ((\sim\mathbf{B} \supset \mathbf{A}) \supset \mathbf{A}))$.

PROOF The proofs of some are given, the easier ones are left as exercises. We point out that the order is important, as the proofs of some depend on others previous in the list.

1. $(\mathbf{A} \supset \mathbf{B}), (\mathbf{B} \supset \mathbf{C}) \vdash (\mathbf{A} \supset (\mathbf{B} \supset \mathbf{C})) \supset ((\mathbf{A} \supset \mathbf{B}) \supset (\mathbf{A} \supset \mathbf{C}))$ axiom 2
   $(\mathbf{A} \supset \mathbf{B}), (\mathbf{B} \supset \mathbf{C}) \vdash (\mathbf{B} \supset \mathbf{C}) \supset (\mathbf{A} \supset (\mathbf{B} \supset \mathbf{C}))$ axiom 1
   $(\mathbf{A} \supset \mathbf{B}), (\mathbf{B} \supset \mathbf{C}) \vdash (\mathbf{B} \supset \mathbf{C})$ hypothesis
   $(\mathbf{A} \supset \mathbf{B}), (\mathbf{B} \supset \mathbf{C}) \vdash (\mathbf{A} \supset (\mathbf{B} \supset \mathbf{C}))$ steps 2, 3 and mp
   $(\mathbf{A} \supset \mathbf{B}), (\mathbf{B} \supset \mathbf{C}) \vdash ((\mathbf{A} \supset \mathbf{B}) \supset (\mathbf{A} \supset \mathbf{C}))$ 1, 4 and mp
   $(\mathbf{A} \supset \mathbf{B}), (\mathbf{B} \supset \mathbf{C}) \vdash (\mathbf{A} \supset \mathbf{B})$ hypothesis
   $(\mathbf{A} \supset \mathbf{B}), (\mathbf{B} \supset \mathbf{C}) \vdash (\mathbf{A} \supset \mathbf{C})$ 5, 6 and mp
   $\vdash((\mathbf{A} \supset \mathbf{B}) \supset ((\mathbf{B} \supset \mathbf{C}) \supset (\mathbf{A} \supset \mathbf{C})))$ deduction theorem twice

4. $\sim\sim\mathbf{B} \vdash \sim\sim\mathbf{B} \supset (\sim\sim\sim\sim\mathbf{B} \supset \sim\sim\mathbf{B})$ axiom 1
   $\sim\sim\mathbf{B} \vdash \sim\sim\mathbf{B}$ hypothesis
   $\sim\sim\mathbf{B} \vdash (\sim\sim\sim\sim\mathbf{B} \supset \sim\sim\mathbf{B}) \supset (\sim\mathbf{B} \supset \sim\sim\sim\mathbf{B})$ axiom 3
   $\sim\sim\mathbf{B} \vdash (\sim\mathbf{B} \supset \sim\sim\sim\mathbf{B})$ mp twice
   $\sim\sim\mathbf{B} \vdash (\sim\mathbf{B} \supset \sim\sim\sim\mathbf{B}) \supset (\sim\sim\mathbf{B} \supset \mathbf{B})$ axiom 3

	$\sim\sim B \vdash (\sim\sim B \supset B)$	mp
	$\sim\sim B \vdash B$	mp
	$\vdash (\sim\sim B \supset B)$	deduction theorem
6.	$(A \supset B) \vdash (\sim\sim A \supset \sim\sim B) \supset (\sim B \supset \sim A)$	axiom 3
	$(A \supset B) \vdash (\sim\sim A \supset A)$	number 4
	$(A \supset B) \vdash (A \supset B)$	hypothesis
	$(A \supset B) \vdash (B \supset \sim\sim B)$	number 5
	$(A \supset B) \vdash (\sim\sim A \supset A) \supset ((A \supset B) \supset (\sim\sim A \supset \sim\sim B))$	number 1
	$(A \supset B) \vdash (\sim\sim A \supset \sim\sim B)$	mp twice
	$(A \supset B) \vdash (\sim B \supset \sim A)$	steps 1, 6 and mp
	$\vdash ((A \supset B) \supset (\sim B \supset \sim A))$	deduction theorem
8.	$(B \supset A), (\sim B \supset A), \sim A \vdash (B \supset A) \supset (\sim A \supset \sim B)$	number 6
	$(B \supset A), (\sim B \supset A), \sim A \vdash (B \supset A)$	hypothesis
	$(B \supset A), (\sim B \supset A), \sim A \vdash (\sim A \supset \sim B)$	mp
	$(B \supset A), (\sim B \supset A), \sim A \vdash \sim A$	hypothesis
	$(B \supset A), (\sim B \supset A), \sim A \vdash \sim B$	mp
	$(B \supset A), (\sim B \supset A), \sim A \vdash (\sim B \supset A) \supset (\sim A \supset \sim\sim B)$	number 6
	$(B \supset A), (\sim B \supset A), \sim A \vdash (\sim B \supset A)$	hypothesis
	$(B \supset A), (\sim B \supset A), \sim A \vdash \sim\sim B$	mp twice
	$(B \supset A), (\sim B \supset A), \sim A \vdash \sim\sim B \supset (\sim B \supset \sim(B \supset A))$	number 3
	$(B \supset A), (\sim B \supset A), \sim A \vdash \sim(B \supset A)$	mp twice
	$(B \supset A), (\sim B \supset A) \vdash \sim A \supset \sim(B \supset A)$	deduction theorem
	$(B \supset A), (\sim B \supset A) \vdash (\sim A \supset \sim(B \supset A)) \supset ((B \supset A) \supset A)$	axiom 3
	$(B \supset A), (\sim B \supset A) \vdash A$	steps 2, 10, 12 and mp twice
	$\vdash (B \supset A) \supset ((\sim B \supset A) \supset A)$	deduction theorem ▮

9.4 MORE ABOUT PROPOSITIONAL LANGUAGES; TRUTH VALUES

Consider the propositional languages $\mathrm{L}(\mathrm{P}_1)$ with symbols†

$$\{p_1, p_2, \ldots\} \cup \{(,)\} \cup \{\wedge, \vee, \sim, \supset, \equiv\}$$

† Read "$\wedge$" as "and" and in $(A \wedge B)$, refer to "A" and "B" as conjuncts. Read "$\vee$" as "or" and in $(A \vee B)$, refer to "A" and "B" as disjuncts. Read "$\supset$" as "implies" and in $(A \supset B)$ refer to "A" as the antecedent (or premise) and to "B" as the consequent (or conclusion). Read "$\equiv$" as "equivalent" and read "$\sim$" as "not."

and where the formulas $F(P_1)$ are defined by: (i) every propositional variable in $F(P_1)$ and (ii) if **A** and **B** are in $F(P_1)$, then $(\mathbf{A} \wedge \mathbf{B})$, $(\mathbf{A} \vee \mathbf{B})$, $(\mathbf{A} \supset \mathbf{B})$, $(\mathbf{A} \equiv \mathbf{B})$, and $\sim\mathbf{A}$ are in $F(P_1)$.

A formula of a formal language is *atomic* if no propositional connectives occur in it. Otherwise, a formula is said to be *compound.* Thus, atomic formulas of any propositional language are just the propositional variables. Notice that in the compound formula $(\mathbf{A} \vee \mathbf{B})$, the disjuncts **A** and **B** themselves may be compound or atomic, e.g., $((\mathbf{p}_1 \supset \mathbf{p}_2) \vee \sim\mathbf{p}_1)$ and $(\mathbf{p}_1 \vee \mathbf{p}_2)$.

We may think of the elements of $F(P_1)$ as a set of formal but meaningless symbols (ink marks on paper), just as we have treated P_0 as a meaningless game. But it is clear, by the very fact that we read " $\wedge$ " as "and," etc., that sometimes we do wish to interpret these formulas so as to make them correspond to mathematical and everyday language. So we associate in the most natural way sentences which are conjunctions with the formula $(\mathbf{A} \wedge \mathbf{B})$, where what formulas **A** and **B** are depends on the two sentences making up the conjunction. We associate sentences which are conditional with $(\mathbf{A} \supset \mathbf{B})$, etc. Just following in the same way (inductively down), we associate with propositional variables, sentences which contain no connectives or sentences whose connectives are subordinate to one or more quantifiers. By the latter, we mean that to divide the sentence at the connective into two sentences would destroy the meaning of the sentence. For example, the sentence "for all x, if $x \in P$ then $x \in Q$" is the definition of the set P being included in the set Q; but it is quite different to say "if for all x $x \in P$, then $x \in Q$," which we can associate with $(\mathbf{p}_1 \supset \mathbf{p}_2)$ and which has no standard mathematical meaning. In such an association, $\mathbf{p}_1$ is "for all x $x \in P$" and $\mathbf{p}_2$ is "$x \in Q$."

The property of sentences which interests us at this point is truth (or falsity). Ordinarily, we say that a disjunction is true if at least one of its disjuncts is true; otherwise, it is false. We shall assign *truth values* (tv) $\mathfrak{t}$ or $\mathfrak{f}$ to the formulas of $F(P_0)$ in an analogous way. It is easy enough to figure out how to make the assignment to formulas of $F(P_1)$ which are compound.

1. If **C** is of the form $(\mathbf{A} \wedge \mathbf{B})$, then **C** is assigned the truth value $\mathfrak{t}$ if and only if **A** and **B** are both assigned the tv $\mathfrak{t}$.
2. If **C** is of the form $(\mathbf{A} \vee \mathbf{B})$, then **C** is assigned the tv $\mathfrak{t}$ if and only if at least one of **A** and **B** is assigned the tv $\mathfrak{t}$.
3. If **C** is $(\mathbf{A} \supset \mathbf{B})$, then **C** is assigned the tv $\mathfrak{f}$ if and only if **A** is assigned $\mathfrak{t}$ and **B** is assigned $\mathfrak{f}$.
4. If **C** is $(\mathbf{A} \equiv \mathbf{B})$, then **C** is assigned the tv $\mathfrak{t}$ if and only if **A** and **B** are assigned the same tv.
5. If **C** is $\sim\mathbf{A}$, then **C** is assigned the tv $\mathfrak{t}$ if and only if **A** is assigned $\mathfrak{f}$.

The assignment so far only takes care of the compound formulas of $F(P_1)$.

However, we must also make assignments to the propositional variables both for themselves and because the tv's assigned to compound formulas depend on it. Since the truth or falsity of a conjunction (disjunction, implication, etc.) depends on the truth or falsity of the conjuncts (disjuncts, etc.), if the conjuncts themselves correspond to propositional variables—e.g., (1) f is continuous; (2) it rains sometime during Feb. 12, 2000; (3) all men are mortal; (4) there exists a smallest positive integer; (5) for all x, if $x \in P$ then $x \in Q$;—then whether each is true or false depends on some sort of data or definition not of a logical nature. So, in the examples, the truth of (1) depends on the function f we are considering as well as the definition of continuous; that of (2) depends on what happens that day; that of (3) depends on the definitions of man and of mortal; that of (4) depends on the definitions of smallest and positive integer; and that of (5) depends on whether every member of P is a member of Q, which in turn depends on the definitions of the sets P and Q being considered.

From these considerations, we see that if $\mathbf{C}$ in $F(P_1)$ is, for example, $(\mathbf{p}_1 \supset \mathbf{p}_2)$, then the assignment of a tv to $\mathbf{C}$ depends on the assignments to $\mathbf{p}_1$ and to $\mathbf{p}_2$. As they are atomic, their truth or falsity depends on nonlogical considerations and hence, to those formulas of $F(P_1)$ which are propositional variables we must assign all possible truth values, i.e., both $\mathfrak{t}$ and $\mathfrak{f}$. Thus, if $\mathbf{p}_1$ and $\mathbf{p}_2$ are both assigned $\mathfrak{t}$, $\mathbf{C}$ is assigned $\mathfrak{t}$; if $\mathbf{p}_1$ is assigned $\mathfrak{t}$ and $\mathbf{p}_2$ is assigned $\mathfrak{f}$, then $\mathbf{C}$ is assigned $\mathfrak{f}$; if $\mathbf{p}_1$ is assigned $\mathfrak{f}$ and $\mathbf{p}_2$ is assigned $\mathfrak{t}$, then $\mathbf{C}$ is assigned $\mathfrak{t}$; if both $\mathbf{p}_1$ and $\mathbf{p}_2$ are assigned $\mathfrak{f}$, then $\mathbf{C}$ is assigned $\mathfrak{t}$. An easy way of writing all this is known as the *truth table* (tt). Thus, the tt for the tv's assigned to $(\mathbf{p}_1 \supset \mathbf{p}_2)$ as just discussed (briefly, "the tt for $\supset$") is

$\mathbf{p}_1$	$\mathbf{p}_2$	$(\mathbf{p}_1 \supset \mathbf{p}_2)$
$\mathfrak{t}$	$\mathfrak{t}$	$\mathfrak{t}$
$\mathfrak{t}$	$\mathfrak{f}$	$\mathfrak{f}$
$\mathfrak{f}$	$\mathfrak{t}$	$\mathfrak{t}$
$\mathfrak{f}$	$\mathfrak{f}$	$\mathfrak{t}$

Exercise 9.8 Make tt's for (a) $\sim \mathbf{p}_1$, (b) $(\mathbf{p}_1 \wedge \mathbf{p}_2)$, (c) $(\mathbf{p}_1 \vee \mathbf{p}_2)$, (d) $(\mathbf{p}_1 \vee \sim \mathbf{p}_1)$, (e) $(\mathbf{p}_1 \equiv \mathbf{p}_2)$, (f) $(\mathbf{p}_1 \wedge \sim \mathbf{p}_1)$, (g) $((\mathbf{p}_1 \supset \sim (\mathbf{p}_2 \wedge \mathbf{p}_3)) \vee \mathbf{p}_1)$.

9.5 TAUTOLOGIES AND THE CONSISTENCY OF P_0

If a formula $\mathbf{C}$ of $F(P_1)$ has the tv $\mathfrak{t}$ assigned to it for all tv assignments to its propositional variables, then we say that $\mathbf{C}$ is a *tautology*.

Exercise 9.9 (a) In exercise 9.8, are there any tautologies? (b) Is

$((\mathbf{p}_1 \supset \mathbf{p}_2) \supset \mathbf{p}_2)$ a tautology? (c) Are the axioms of P_0 tautologies? (d) If **A** is a tautology and $(\mathbf{A} \supset \mathbf{B})$ is a tautology, is **B** also a tautology?

Exercise 9.10 Show, at least intuitively, that the set of formulas of $\mathsf{F}(\mathsf{P}_0)$ that are tautologies is a recursive set.

Theorem 9.2 If $\vdash_{\mathsf{P}0} \mathbf{A}$, then **A** is a tautology.

Exercise 9.11 Prove the theorem.

Theorem 9.3 P_0 is consistent.

Exercise 9.12 Prove the theorem.

Exercise 9.13 In view of exercise 9.10 and Theorem 9.2, what single further fact would imply that P_0 has a solvable decision problem?

Exercise 9.14 In this exercise, we shall examine a different recursively axiomatizable propositional theory, call it Y, for which $\mathsf{L}(\mathsf{Y}) = \mathsf{L}(\mathsf{P}_0)$ and the axiom scheme of Y is $((\mathbf{A} \supset \mathbf{B}) \supset (\mathbf{A} \supset \mathbf{A}))$ and the rule of inference is *modus ponens*. The theorems of Y are to be just the formulas derivable from the axiom by means of the rule of inference. Answer the following:

(a) Is every theorem in Y a tautology?
(b) Prove: if $\vdash_{\mathsf{Y}} \mathbf{C}$, then **C** is of the form

$$((\mathbf{A} \supset \mathbf{B}) \supset (\mathbf{A} \supset \mathbf{A})) \quad \text{or} \quad ((\mathbf{A} \supset \mathbf{B}) \supset (\mathbf{A} \supset \mathbf{B})).$$

(c) $\vdash_{\mathsf{Y}}(\mathbf{A} \supset \mathbf{A})$?
(d) If $\mathbf{C} \in \mathsf{F}(\mathsf{Y})$ and **C** is a tautology, is **C** a theorem of Y?
(e) Are P_0 and Y equal?
(f) Does the deduction theorem hold in Y?
(g) Is Y consistent?
(h) Does Y have a solvable decision problem?

9.6 TRUTH-VALUE FUNCTIONS

Notice that if we substitute 1 for t and 0 for f, we can think of the tt for $(\mathbf{p}_1 \supset \mathbf{p}_2)$ as being a set of 3-tuples constituting the graph of a function from $\{0, 1\} \times \{0, 1\}$ to $\{0, 1\}$; the tt for $(\mathbf{p}_1 \vee \mathbf{p}_2)$ as another such (graph of a) function, etc. Of course, the table for $\sim\mathbf{p}_1$ is (the graph of a) function from $\{0, 1\}$ to $\{0, 1\}$. Further, these functions can be composed in obvious ways, e.g., (d), (f), and (g) of exercise 9.8, to form new functions from $\{0, 1\}^n$ to

$\{0, 1\}$ for various n. A function from $\{0, 1\}^n$ to $\{0, 1\}$ for some n is called a *truth-value function* (tvf). To each formula in $\mathsf{F}(\mathsf{P}_1)$ we have, in an obvious way, a tvf. In particular let $f_\wedge$ denote the function whose graph is the same as the tt for $(\mathbf{p}_1 \wedge \mathbf{p}_2)$ except that 1 replaces $\mathfrak{t}$ and 0 replaces $\mathfrak{f}$, and let $f_\vee$, $f_\supset$, $f_\equiv$, and $f_\sim$ be defined in a similar way. Then, inductively, for $(\mathbf{A} \wedge \mathbf{B})$ in $\mathsf{F}(\mathsf{P}_1)$, the corresponding tvf is defined by composition, $f_\wedge(f_A, f_B)$, where f_A and f_B are the tvf's corresponding to $\mathbf{A}$ and $\mathbf{B}$. We define $(\mathbf{A} \vee \mathbf{B})$, etc, in a similar way. Let $\mathscr{S}$ be the set of tvf's defined by: (1) $f_\vee$, $f_\wedge$, $f_\supset$, $f_\equiv$, $f_\sim$, and f_i (the identity function) are in $\mathscr{S}$, and (2) any function defined by a finite number of compositions of these is in $\mathscr{S}$. By the correspondence given above, we see that $\mathscr{S}$ is in a one–many correspondence with $\mathsf{F}(\mathsf{P}_1)$. (Notice that $(\mathbf{p}_1 \vee \mathbf{p}_2)$, $(\mathbf{p}_2 \vee \mathbf{p}_3)$, ... all correspond to $f_\vee$.) Further, the functions that are the constant 1 correspond to the tautologies. Let $\mathscr{B}$ be set of all functions from $\{0, 1\}^n$ to $\{0, 1\}$ for all $n \in N$. A reasonable question to ask is if $\mathscr{S} = \mathscr{B}$. If it is true that $\mathscr{S} = \mathscr{B}$, then an important result would be that all the (constantly) 1 functions in $\mathscr{B}$ correspond to tautologies in $\mathsf{F}(\mathsf{P}_1)$, so that everything that could conceivably be called "tautologous" is accounted for by $\mathsf{F}(\mathsf{P}_1)$. The theorem we prove is actually stronger and gives $\mathscr{S} = \mathscr{B}$ as a corollary.

Let $\mathsf{L}(\mathsf{P}_2)$ have symbols $\{\mathbf{p}_1, \mathbf{p}_2, \mathbf{p}_3, \ldots\} \cup \{(,)\} \cup \{\vee, \wedge, \sim\}$ and let the formulas $\mathsf{F}(\mathsf{P}_2)$ be defined in the obvious way. Define the corresponding set $\mathscr{S}^*$ of tvf's in a manner similar to that above, inductively from $f_\vee$, $f_\wedge$, $f_\sim$, and f_i.

Theorem 9.4 If f is a function from $\{0, 1\}^n$ to $\{0, 1\}$ for some n, then f is equal to a function defined by compositions of f_i, $f_\vee$, $f_\wedge$, and $f_\sim$. That is, $\mathscr{B} = \mathscr{S}^*$.

PROOF The proof is by induction on n. If $n = 1$, there are four possible tvf's:

(0, 0)	(0, 0)	(0, 1)	(0, 1)
(1, 0)	(1, 1)	(1, 0)	(1, 1)

The third is just the graph of $f_\sim$. The second is the graph of the identity function f_i. The fourth is the graph of $f_\vee(f_i, f_\sim)$, which corresponds to the tt for $(\mathbf{p}_1 \vee \sim\mathbf{p}_1)$. The first is the graph of $f_\wedge(f_i, f_\sim)$, which corresponds to the tt for $(\mathbf{p}_1 \wedge \sim \mathbf{p}_1)$.

Assume that the theorem is true for $n < k$ and show true for k: A k-ary tvf is a set of 2^k $(k + 1)$-tuples on 0 and 1 and we may assume that these $(k + 1)$-tuples are arranged in the usual way (like a tt), i.e., the first term of the first 2^{k-1} (half) of the tuples is 1, and is 0 for the second half; the second term is 1 for the first and third quarters, 0 for the second and fourth quarters, etc. If, in that arrangement, we take the first half of the tuples and the second half seperately and cross out the first term of each tuple, we have the graphs of two $(k-1)$-ary tvf's, call them f_1 and f_2. By the induction hypothesis, each is equivalent to some function in $\mathscr{S}^*$. But that is the same as saying that f_1 corresponds to a formula $\mathbf{A_1} \in \mathsf{F}(\mathsf{P}_2)$ and f_2 to a formula $\mathbf{A_2}$ of $\mathsf{F}(\mathsf{P}_2)$, where a total of $(k-1)$ propositional variables occur in $\mathbf{A_1}$ and $\mathbf{A_2}$; let those propositional variables be $\mathbf{p_2}, \mathbf{p_3}, \ldots, \mathbf{p_k}$. Consider those rows of the tt for the formula $\mathbf{C_1}$ which is $(\mathbf{p_1} \wedge \mathbf{A_1})$ in which $\mathbf{p_1}$ is assigned $\mathfrak{t}$: $\mathbf{C_1}$ has the same tv's as $\mathbf{A_1}$; in the rows in which $\mathbf{p_1}$ is assigned $\mathfrak{f}$, $\mathbf{C_1}$ is assigned $\mathfrak{f}$. In the same way, consider the tt for the formula $\mathbf{C_2}$ which is $(\sim\mathbf{p_1} \wedge \mathbf{A_2})$: for the rows in which $\mathbf{p_1}$ is assigned $\mathfrak{t}$, $\mathbf{C_2}$ is assigned $\mathfrak{f}$, and for the rows in which $\mathbf{p_1}$ is assigned $\mathfrak{f}$, $\mathbf{C_2}$ is assigned whatever $\mathbf{A_2}$ is assigned. Now, by considering the tt for $(\mathbf{C_1} \vee \mathbf{C_2})$, we see that (i) for the rows (the first half of the tt) in which $\mathbf{p_1}$ is assigned $\mathfrak{t}$, $\mathbf{C_2}$ is assigned $\mathfrak{f}$ and the assignment of $\mathbf{C_1}$ coincides with the assignment to $\mathbf{A_1}$ and therefore the assignment to $(\mathbf{C_1} \vee \mathbf{C_2})$ is the same as the assignment to $\mathbf{A_1}$; and (ii) for the rows (the second half) for which $\mathbf{p_1}$ is assigned $\mathfrak{f}$, $\mathbf{C_1}$ is assigned $\mathfrak{f}$ and the assignment to $\mathbf{C_2}$ coincides with the assignment to $\mathbf{A_2}$ and therefore the assignment to $(\mathbf{C_1} \vee \mathbf{C_2})$ is the same as the assignment to $\mathbf{A_2}$. But there is a tvf corresponding to $(\mathbf{C_1} \vee \mathbf{C_2})$ which is $f_\vee(f_{C_1}, f_{C_2})$, and by the construction, f_{C_1} is $f_\wedge(f_i, f_{A_1})$ and f_{C_2} is $f_\wedge(f_\sim, f_{A_2})$. By the induction hypothesis, f_{A_1} and $f_{A_2} \in \mathscr{S}^*$, and so by the definition of $\mathscr{S}^*$, the tvf corresponding to $(\mathbf{C_1} \vee \mathbf{C_2})$, i.e., $f_\vee(f_\wedge(f_i, f_{A_1}), f_\wedge(f_\sim, f_{A_2}))$, is in $\mathscr{S}^*$ and is equal to the given arbitrary k-ary truth-value function f. ▌

Corollary 1 For every tvf, there exists a formula of $\mathsf{F}(\mathsf{P}_1)$ that corresponds to it.

Corollary 2 For every tvf, there is a formula in $\mathsf{F}(\mathsf{P}_0)$ that corresponds to it.

Corollary 3 For every tvf which is the constant one function, there is a tautology in $\mathsf{F}(\mathsf{P}_0)$ that corresponds to it.

Two formulas $\mathbf{A}$ and $\mathbf{B}$ in $\mathsf{F}(\mathsf{P}_1)$ are defined to be *logically equivalent* if $(\mathbf{A} \equiv \mathbf{B})$ is a tautology. For example, $(\mathbf{A} \wedge \mathbf{B})$ is logically equivalent to $\sim(\mathbf{A} \supset \sim\mathbf{B})$.

Corollary 4 For every $\mathbf{A} \in \mathsf{F}(\mathsf{P}_0)$, there exists an $\mathbf{A}' \in \mathsf{F}(\mathsf{P}_2)$ such that $\mathbf{A}$ and $\mathbf{A}'$ are logically equivalent. For every $\mathbf{B} \in \mathsf{F}(\mathsf{P}_2)$, there exists a formula $\mathbf{B}' \in \mathsf{F}(\mathsf{P}_0)$ such that $\mathbf{B}$ and $\mathbf{B}'$ are logically equivalent.

Exercise 9.15 Prove Corollaries 1–4.

9.7 "COMPLETENESS" AND SOLVABILITY OF $\mathsf{P_0}$

Theorem 9.5 † If $\mathbf{A} \in \mathsf{F(P_0)}$ and $\mathbf{A}$ is a tautology, then $\vdash_{\mathsf{P0}} \mathbf{A}$.

As the proof of the theorem is a little lengthy, we first prepare the way with a lemma.

Lemma 9.2 Let $\mathbf{A} \in \mathsf{F(P_0)}$ and let $\mathbf{p_1}, \ldots, \mathbf{p_k}$ be the propositional variables occurring in $\mathbf{A}$. Consider each row of the truth table for $\mathbf{A}$ and for each $\mathbf{p_i}$ write the formula $\mathbf{B_i}$ as follows. If in that row, $\mathbf{p_i}$ has $\mathfrak{t}$ assigned to it, $\mathbf{B_i}$ is to be just $\mathbf{p_i}$, but if $\mathbf{p_i}$ has $\mathfrak{f}$ assigned to it, $\mathbf{B_i}$ is to be $\sim\mathbf{p_i}$. Define a formula $\mathbf{A'}$ in a similar way: If $\mathbf{A}$ is assigned $\mathfrak{t}$, $\mathbf{A'}$ is just $\mathbf{A}$; if $\mathbf{A}$ is assigned $\mathfrak{f}$, $\mathbf{A'}$ is to be $\sim\mathbf{A}$. Then, $\mathbf{B_1}, \ldots, \mathbf{B_k} \vdash_{\mathsf{P0}} \mathbf{A'}$.

PROOF The proof is by induction on the number n of propositional connectives in $\mathbf{A}$. If $n = 0$, then $\mathbf{A}$ must be a propositional variable $\mathbf{p_1}$. Thus, if $\mathbf{p_1}$ (and thus $\mathbf{A}$) is assigned $\mathfrak{t}$, we have $\mathbf{p_1} \vdash_{\mathsf{P0}} \mathbf{p_1}$, and if it is assigned $\mathfrak{f}$, we have $\sim\mathbf{p_1} \vdash_{\mathsf{P0}} \sim\mathbf{p_1}$. Assume that the lemma is true for $n < m$ and show that it is true for m.

1. Suppose $\mathbf{A}$ is of the form $\sim\mathbf{C}$. (a) If $\mathbf{A}$ is assigned $\mathfrak{t}$ and $\mathbf{A'}$ is $\mathbf{A}$, then $\mathbf{C}$ is assigned $\mathfrak{f}$ and so $\mathbf{C'}$ is $\sim\mathbf{C}$. As $\mathbf{C}$ has $m - 1$ propositional connectives, the induction hypothesis holds and hence $\mathbf{B_1}, \ldots, \mathbf{B_k} \vdash_{\mathsf{P0}} \mathbf{C'}$, but that is the same as $\mathbf{B_1}, \ldots, \mathbf{B_k} \vdash_{\mathsf{P0}} \mathbf{A}$. (b) If $\mathbf{A}$ is assigned $\mathfrak{f}$ and so $\mathbf{A'}$ is $\sim\mathbf{A}$ (which is $\sim\sim\mathbf{C}$), then $\mathbf{C}$ is assigned $\mathfrak{t}$, so $\mathbf{C'}$ is $\mathbf{C}$. By the induction hypothesis, $\mathbf{B_1}, \ldots, \mathbf{B_k} \vdash_{\mathsf{P0}} \mathbf{C}$. By Lemma 9.1, $\mathbf{B_1}, \ldots, \mathbf{B_k} \vdash_{\mathsf{P0}} (\mathbf{C} \supset \sim\sim\mathbf{C})$, so by *modus ponens*, $\mathbf{B_1}, \ldots, \mathbf{B_k} \vdash_{\mathsf{P0}} \sim\sim\mathbf{C}$.
2. Suppose $\mathbf{A}$ is of the form $(\mathbf{B} \supset \mathbf{C})$. (a) $\mathbf{A}$ is assigned $\mathfrak{f}$ only if $\mathbf{B}$ is assigned $\mathfrak{t}$ and $\mathbf{C}$ is assigned $\mathfrak{f}$. In that case, $\mathbf{A'}$ is $\sim\mathbf{A}$, i.e., is $\sim(\mathbf{B} \supset \mathbf{C})$; $\mathbf{B'}$ is $\mathbf{B}$ and $\mathbf{C'}$ is $\sim\mathbf{C}$. By the induction hypothesis, $\mathbf{B_1}, \ldots, \mathbf{B_k} \vdash_{\mathsf{P0}} \mathbf{B}$ and $\mathbf{B_1}, \ldots, \mathbf{B_k} \vdash_{\mathsf{P0}} \sim\mathbf{C}$. By Lemma 9.1, $\mathbf{B_1}, \ldots, \mathbf{B_k} \vdash_{\mathsf{P0}} (\mathbf{B} \supset (\sim\mathbf{C} \supset \sim(\mathbf{B} \supset \mathbf{C})))$, so, using *modus ponens* twice, $\mathbf{B_1}, \ldots, \mathbf{B_k} \vdash_{\mathsf{P0}} \sim(\mathbf{B} \supset \mathbf{C})$.

Exercise 9.16 Finish the proof of Lemma 9.1. That is, consider the other cases for when $\mathbf{A}$ is of the form $(\mathbf{B} \supset \mathbf{C})$; e.g., under what circumstances is $\mathbf{A}$ assigned the tv $\mathfrak{t}$? ▮

PROOF OF THEOREM 9.5 If $\mathbf{A}$ is a tautology, for every row of its truth table, it is assigned $\mathfrak{t}$, so $\mathbf{A'}$ (as defined in the lemma) is just $\mathbf{A}$. If $\mathbf{p_1}, \ldots, \mathbf{p_k}$ are the

† This is often called the completeness theorem for the propositional calculus. It has nothing to do with the word "complete" defined earlier.

propositional variables occurring in **A**, then the tt for **A** has 2^k rows. In half of them, $\mathbf{p_k}$ is assigned t, in which case, $\mathbf{B_k}$ is $\mathbf{p_k}$ and so by the lemma, $\mathbf{B_1}, \ldots, B_{k-1}, \mathbf{p_k} \vdash_{P0} \mathbf{A}$. In the other half of the rows, $\mathbf{p_k}$ is assigned f, and for those, we have, by the lemma, $\mathbf{B_1}, \ldots, \mathbf{B_{k-1}}, \sim\mathbf{p_k} \vdash_{P0} \mathbf{A}$. By using the deduction theorem for both cases, we obtain $\mathbf{B_1}, \ldots, \mathbf{B_{k-1}} \vdash_{P0} (\mathbf{p_k} \supset \mathbf{A})$ and $\mathbf{B_1}, \ldots, \mathbf{B_{k-1}} \vdash_{P0} (\sim\mathbf{p_k} \supset \mathbf{A})$. By Lemma 9.1, we have

$$\mathbf{B_1}, \ldots, \mathbf{B_{k-1}} \vdash_{P0} ((\mathbf{p_k} \supset \mathbf{A}) \supset ((\sim \mathbf{p_k} \supset \mathbf{A}) \supset \mathbf{A}))$$

and so, using *modus ponens* twice, we have $\mathbf{B_1}, \ldots, \mathbf{B_{k-1}} \vdash_{P0} \mathbf{A}$. We can repeat the same kind of argument $k - 1$ more times and finally conclude $\vdash_{P0} \mathbf{A}$.† ▮

Theorem 9.6 The decision problem for P_0 is solvable.

Exercise 9.17 Prove the theorem.

Exercise 9.18 Is P_0 complete in the sense of the definition given in section 9.2?

Section 9.2 was given the title, "A formulation of the propositional calculus." The term *the propositional calculus* is reserved for any propositional theory whose theorems coincide with the set of tautologies. We have now shown that it was correct to refer to P_0 as "a formulation of the propositional calculus." Obviously, a propositional theory is logically equivalent to P_0 if and only if it is (a formulation of) the propositional calculus. A propositional theory which is not the propositional calculus but is a subtheory (hence, consistent) is sometimes called a *fragment of the propositional calculus* or *a partial propositional calculus*. The theory Y of exercise 9.14 is an example of a fragment. A propositional theory (with *modus ponens* as rule of inference) whose set of theorems is closed under *modus ponens* and which includes more than just the tautologies is so unspeakable as to have no special name. The reason it is unspeakable is shown in the following.

Theorem 9.7 Let P be a propositional theory (based on the language $\mathsf{L}(\mathsf{P}_1)$) whose rules of inference include *modus ponens*. If the theorems of P include more than those formulas that are tautologies, then for all $\mathbf{A} \in \mathsf{F}(\mathsf{P}_1)$, $\vdash_P \mathbf{A}$.

Exercise 9.19 Prove the theorem.

† The student may wonder why we consider the possibility of **A** being assigned f (in the lemma) when the hypothesis of the theorem is that **A** is a tautology. It is true that we need only prove $\mathbf{B_1}, \ldots, \mathbf{B_k} \vdash \mathbf{A}$. However, that would be more difficult than the stronger statement proved above. See if you can see why. It is characteristic of induction proofs that it is often easier to prove a stronger statement than a similar but weaker one. This is not surprising since the more we can assume in the induction hypothesis, the easier the induction step should be.

Chapter X

INTRODUCTION TO FIRST-ORDER LANGUAGES AND RELATIONAL SYSTEMS

10.1 FIRST-ORDER LANGUAGE

We give here what we will call a general first-order language $\mathsf{L}(F)$, of which all other first-order languages considered will be subsets.† The symbols are:

1. The individual variables: $\mathbf{x}_1, \mathbf{x}_2, \mathbf{x}_3, \ldots$.
2. The predicate symbols: $\mathbf{P}_1^1, \mathbf{P}_2^1, \ldots, \mathbf{P}_1^2, \mathbf{P}_2^2, \ldots, \mathbf{P}_1^n, \mathbf{P}_2^n, \ldots$.
3. The function symbols: $\mathbf{f}_1^1, \mathbf{f}_2^1, \ldots, \mathbf{f}_1^2, \mathbf{f}_2^2, \ldots, \mathbf{f}_1^n, \mathbf{f}_2^n, \ldots$.
4. Names for constants: $\mathbf{a}_0, \mathbf{a}_1, \mathbf{a}_2, \mathbf{a}_3, \ldots$.
5. Propositional connectives: $\supset, \wedge, \vee, \equiv, \sim$.
6. Comma and parentheses: **, ( )**.
7. The universal quantifier: $\forall$.

There may also be the binary predicate constant $=$, in which case we say we have a *first-order language with equality.*

In order to define the formulas $\mathsf{F}(F)$, certain sets of words on the symbols must be defined first:

The *terms* of $\mathsf{L}(F)$: (1) Any individual variable is a term and any name for a constant is a term. (2) If $\mathbf{t}_1, \ldots, \mathbf{t}_n$ are terms and $\mathbf{f}_i^n$ is a function symbol, then $\mathbf{f}_i^n(\mathbf{t}_1, \ldots, \mathbf{t}_n)$ is a term.

The *atomic formulas* of $\mathsf{L}(F)$: If $\mathbf{t}_1, \ldots, \mathbf{t}_n$ are terms and $\mathbf{P}_i^n$ as a predicate symbol, then $\mathbf{P}_i^n(\mathbf{t}_1, \ldots, \mathbf{t}_n)$ is an atomic formula. When we have a first-order language with equality, if $\mathbf{t}_1$ and $\mathbf{t}_2$ are terms, then $\mathbf{t}_1 = \mathbf{t}_2$ is an atomic formula.

The *formulas* $\mathsf{F}(F)$ of $\mathsf{L}(F)$: (1) Every atomic formula is in $\mathsf{F}(F)$. (2) If $\mathbf{A}$ and $\mathbf{B}$ are in $\mathsf{F}(F)$, then $(\mathbf{A} \supset \mathbf{B})$, $(\mathbf{A} \wedge \mathbf{B})$, $(\mathbf{A} \vee \mathbf{B})$, $(\mathbf{A} \equiv \mathbf{B})$, and $\sim\mathbf{A}$ are

† It might be helpful for the reader to reread section 1.7 on first-order arithmetic.

in F(*F*). (3) If **A** is in F(*F*) and $\mathbf{x_i}$ is an individual variable, then $(\forall \mathbf{x_i})\mathbf{A}$ is in F(*F*).

A *first-order language* is any subset of L(*F*) which (1) has all the individual variables, (2) the propositional connectives $\supset$ and $\sim$ as well as the comma and the parentheses, (3) may have no, finitely many, or infinitely many names of constants, (4) may have no, finitely many, or infinitely many function symbols, (5) may have one, finitely many, or infinitely many predicate symbols not including $=$, (6) may have $=$, and no, finitely many, or infinitely many predicate symbols, (7) has the universal quantifier $\forall$ and, (8) has all the formulas formed from these according the definition stated above. When we wish to refer to a arbitrary first-order language, we will write L($*$).

When we wish to prove that the formulas of a first-order language have some particular property, we will usually do an "induction on the formation of formulas" by which we mean: (1) Prove that the property holds for terms —(a) show for individual variables and names of constants, and (b) assuming that it holds for terms $\mathbf{t_1}, \ldots, \mathbf{t_k}$ show that it holds for any term $\mathbf{f_i^k(t_1, \ldots, t_k)}$, where $\mathbf{f_i^k}$ is a functional symbol. (2) Prove that the property holds for atomic formulas; if it holds for terms $\mathbf{t_1}, \ldots, \mathbf{t_k}$, then it holds for $\mathbf{P_i^k(t_1, \ldots, t_k)}$ and for $\mathbf{t_i = t_j}$, where $\mathbf{P_i^k}$ and $=$ are symbols of the language. (3) Prove that the property holds for all formulas; if it holds for formulas **A** and **B**, then show that it holds for $(\mathbf{A} \supset \mathbf{B})$, $(\mathbf{A} \wedge \mathbf{B})$, $(\mathbf{A} \vee \mathbf{B})$, $(\mathbf{A} \equiv \mathbf{B})$, $\sim\mathbf{A}$, and $(\forall \mathbf{x_i})\mathbf{A}$.

Since, as we know from the previous chapter, $(\mathbf{A} \vee \mathbf{B})$ is logically equivalent to $(\sim\mathbf{A} \supset \mathbf{B})$, $(\mathbf{A} \wedge \mathbf{B})$ is logically equivalent to $\sim(\mathbf{A} \supset \sim\mathbf{B})$, and $(\mathbf{A} \equiv \mathbf{B})$ is logically equivalent to $((\mathbf{A} \supset \mathbf{B}) \wedge (\mathbf{B} \supset \mathbf{A}))$, we will use the connectives $\vee$, $\wedge$, $\equiv$ as abbreviations even if, as symbols, they are not specified as being in the language. Further, we abbreviate $\sim(\forall \mathbf{x_i}) \sim \mathbf{A}$ by $(\exists \mathbf{x_i})\mathbf{A}$. To make writing easier, we will often write **x**, **y**, **z** for individual variables as well as the proper symbols $\mathbf{x_1}$, $\mathbf{x_2}$, etc. Further, we shall use **P**, **Q**, **R**, etc. for predicate symbols as well as the proper ones, and omit subscripts and superscripts whenever this will not lead to any confusion. Some pairs of parentheses will also be omitted under the convention that they be restored by association to the left. For example, we may write $[\mathbf{A} \wedge \mathbf{B} \wedge \mathbf{C}]$ for $[[\mathbf{A} \wedge \mathbf{B}] \wedge \mathbf{C}]$.

Examples Formulas in F(*F*):

1. $(\exists \mathbf{x})(\forall \mathbf{y})(\mathbf{P}(\mathbf{x}, \mathbf{y}))$
2. $(\forall \mathbf{x})\mathbf{P}(\mathbf{x}) \equiv \sim(\exists \mathbf{x})(\sim\mathbf{P}(\mathbf{x}))$
3. $\mathbf{P}(\mathbf{a_1}) \supset (\forall \mathbf{x})\mathbf{P}(\mathbf{x})$
4. $(\forall \mathbf{x})\mathbf{P}(\mathbf{x}) \supset \mathbf{P}(\mathbf{a_1})$
5. $(\exists \mathbf{x})(\forall \mathbf{y})(\mathbf{P}(\mathbf{x}, \mathbf{y})) \supset (\forall \mathbf{x})(\exists \mathbf{y})(\mathbf{P}(\mathbf{x}, \mathbf{y}))$
6. $(\forall \mathbf{x})(\exists \mathbf{y})(\mathbf{P}(\mathbf{x}, \mathbf{y})) \supset (\exists \mathbf{x})(\forall \mathbf{y})(\mathbf{P}(\mathbf{x}, \mathbf{y}))$
7. $(\forall \mathbf{x})(\mathbf{P}(\mathbf{x}) \wedge \mathbf{Q}(\mathbf{x})) \equiv ((\forall \mathbf{x})\mathbf{P}(\mathbf{x}) \wedge (\forall \mathbf{x})\mathbf{Q}(\mathbf{x}))$

8. $(\forall x)E(x, x) \wedge (\forall x)(\forall y)(E(x, y) \equiv E(y, x)) \wedge (\forall x)(\forall y)(\forall z)(E(x, y) \supset (E(y, z) \supset E(x, z)))$
9. $\sim(\exists x)L(x, x) \wedge (\forall x)(\forall y)(L(x, y) \supset \sim L(y, x)) \wedge (\forall x)(\forall y)(\forall z)(L(x, y) \supset (L(y, z) \supset L(x, z)))$
10. $(\forall x)(\exists y)L(x, y) \wedge \sim(\exists x)(\forall y)L(y, x) \wedge (\forall x)L(a_1, x)$
11. $(\forall x)(\forall y)(\exists z)(L(x, z) \wedge L(z, y))$
12. $(\forall x)(\forall y)(L(x, y) \vee L(y, x))$
13. $(\forall x)(\exists y)(E(f^1(x), y)) \wedge \sim(\exists x)(E(f^1(x), a_1)) \wedge (\forall x)(\forall y)(E(f^1(x), f^1(y)) \supset E(x, y))$
14. $(\exists x)(\exists y)(\exists z)(\forall u)((\sim E(x, y) \wedge \sim E(y, z) \wedge \sim E(x, z)) \wedge (E(x, u) \vee E(y, u) \vee E(z, u)))$
15. $(\exists x)(\exists y)(\sim E(x, a) \wedge \sim E(y, a) \wedge E(f^2(x, y), a))$
16. $(\forall x)(\forall y)(E(f^2(f^1(x), y), f^1(f^2(x, y))))$
17. $(\forall x)P(x) \supset (\exists x)P(x)$
18. $(\exists x)P(x) \supset (\forall x)P(x)$
19. $(x = y) \supset (P(x) = P(y))$
20. $(x = y) \supset ((y = z) \supset (x = z))$
21. $(\forall x)(\exists y)P(x, y) \supset P(f^1(x), y)$

In designating $\mathsf{F}(F)$, we said if **A** is in $\mathsf{F}(F)$ and x_i is an individual variable, then $(\forall x_i)$**A** is in $\mathsf{F}(F)$. In this case, **A** is referred to as the *scope of* $(\forall x_i)$ and any occurrence of x_i in **A** is said to be *bound*. [Similarly for $(\exists x_i)$**A**.]

Examples

1. $(\exists x)(\forall y)P(x, y)$. The scope of $(\exists x)$ is $(\forall y)P(x, y)$ and the scope of $(\forall y)$ is $P(x, y)$. Both the individual variables **x** and **y** occur bound. We also say they are *bound variables*.

2. In $(\exists x)P(x, y) \supset (\forall y)Q(x, y)$, the scope of $(\exists x)$ is $P(x, y)$ and the scope of $(\forall y)$ is $Q(x, y)$. Both **x** and **y** have a bound occurrence and an unbound (free) occurrence.

3. In $(\exists x)(P(x, a) \supset (\forall y)Q(x, y))$, the scope of $(\exists x)$ is $(P(x, a) \supset (\forall y)Q(x, y))$, and the scope of $(\forall y)$ is $Q(x, y)$. In the formula, both **x** and **y** occur only as bound variables.

If x_i is a variable which occurs in a formula **A** but not in the scope of $(\forall x_i)$ or $(\exists x_i)$ (for that same i), then x_i has a *free occurrence* in **A**. In particular, if no quantifiers occur in **A** itself, then **A** is said to be *quantifier-free* and, of course, any individual variables which occur in **A** are free occurrences. Notice that [as in example 2 above] in a formula **A** which is not quantifier-free, an individual variable x_i can have both free occurrences and bound occurrences. If a formula **A** has k distinct individual variables which occur free, then we say **A** *has k free variables.*

Examples

1. In $(\exists x)P(x, y)$, the only occurrence of y is a free occurrence and x, of course, is bound. This formula has one free variable.
2. The formula $(P(x, y) \wedge Q(y, z))$ is a formula which is quantifier-free and has three free variables.
3. In the formula $((\forall x)P(x, y) \vee Q(y, z))$, x has both a free occurrence and a bound occurrence. The formula has three free variables.

If t_i is a term in which no individual variables occur, we say it is a *closed term* (or *variable free term*).

Examples

1. The following are closed terms: any name of a constant,

$$f_1{}^3(a_1, f^2(a_1, a_2), f^1(f_2{}^3(a_2, a_3, a_4))).$$

2. The following are not closed terms: any individual variable,

$$f^4(a_1, x_2, f^2(x_2, a_2), x^3).$$

If all the variables which occur in a formula are bound, then we say it is a *closed formula* or *sentence*.

Examples

1. The following are sentences:
(a) $(\forall x)(\exists y)P(x, y)$, (b) $(\exists x)P(f(a_1, a_2), x)$, (c) $(\forall x)(\exists y)P(f(x, a), y)$, (d) $(\forall x)(\forall y)(P(x) \vee Q(y))$, (e) $P(a_1, a_2)$. (f) $(\forall x)(\exists y)P(x, y) \supset (\exists z)Q(z)$.
2. The following are not sentences: (a) $(\forall x)P(x, y)$ has one free variable; (b) $P(f(a, x), x)$ has one free variable; (c) $(\forall x)(\exists y)P(x, y) \supset Q(z)$ has one free variable; (d) $(\forall x)P(x, y, z) \vee (\exists y)Q(x, y, z)$ has three free variables.

If A is a formula and the only variables which occur free in A are $x_1, \ldots, x_n$, then $(\forall x_1) \ldots (\forall x_n)A$ is the *universal closure of* A and $(\exists x_1) \ldots (\exists x_n)A$ is the *existential closure of* A. On many occasions, we will need to indicate some (or all) of the individual variables and/or names of constants occurring in a specific formula A. In that case, we write $A(a_{i_1}, \ldots, a_{i_n}, x_{j_1}, \ldots, x_{j_m})$. In the same way, we can refer to terms occurring in A by writing $A(t_1, \ldots, t_n)$. If $A(x_{j_1}, \ldots, x_{j_n})$ is a formula in which $x_{j_1}, \ldots, x_{j_n}$ are some or all of the free variables in A, then we denote by $A(t_{j_1}, \ldots, t_{j_n})$ (or, occasionally, by $A_{(x_1, \ldots, x_n)}[t_{j_1}, \ldots, t_{j_n}]$ the result of substituting the term t_{j_i}, $1 \leq i \leq n$, for *every free* occurrence of x_{j_i}, $1 \leq i \leq n$, in $A(x_{j_1}, \ldots, x_{j_n})$. For example, if $A(x, y, z)$ is $(\exists x)P(x, y, z) \vee (\forall y)Q(x, y, z)$ and there are terms t_1, t_2, t_3

which are $\mathbf{f(x, a)}$, $\mathbf{b}$, and $\mathbf{f^2(f^3(y, u, a), x)}$, respectively, then $\mathbf{A(t_1, t_2, t_3)}$ (or $\mathbf{A_{(x, y, z)}[t_1, t_2, t_3]}$) is

$$(\exists x)P(x, b, f^2(f^3(y, u, a), x)) \vee (\forall y)Q(f(x, a), y, f^2(f^3(y, u, a), x))$$

Suppose that $\mathbf{t}$ is a term in which the variable $\mathbf{x_j}$ occurs, then $\mathbf{t}$ *is free for* $\mathbf{x_i}$ *in* (formula) $\mathbf{A}$ if in $\mathbf{A}$ there is no free occurrence of $\mathbf{x_i}$ in the scope of $(\forall \mathbf{x_j})$ or $(\exists \mathbf{x_j})$. [One can remember this by saying that any place $\mathbf{x_i}$ in $\mathbf{A(x_i)}$ that is free should still be free in $\mathbf{A(t)}$ unless $\mathbf{t}$ is a closed term.]

Examples

1. Any closed term is free for any $\mathbf{x_i}$ in any formula $\mathbf{A}$.
2. Let $\mathbf{t}$ be the term $\mathbf{f(x, y, a_3)}$ and consider the formulas (a) $\mathbf{P(x, y, a_1)}$, (b) $(\forall \mathbf{y})\mathbf{P(x, y, f(x, a_1))}$, (c) $(\forall \mathbf{y})\mathbf{P(x, y)} \supset (\exists \mathbf{x})\mathbf{Q(x, y, z)}$. Then $\mathbf{t}$ is free for $\mathbf{x}$ and $\mathbf{y}$ in (a); $\mathbf{t}$ is not free for $\mathbf{x}$ and is free for $\mathbf{y}$ in (b); $\mathbf{t}$ is not free for $\mathbf{x}$, $\mathbf{y}$, or $\mathbf{z}$ in (c).

10.2 RELATIONAL SYSTEMS

Since a first-order language is merely a set of meaningless symbols unless we find a way to attach meanings to that set, the language is useless for mathematical purposes. If there were no possibility of making a correspondence between the symbols and mathematical reality (whatever that may be!) we would be just pushing symbols around. Hence, our goal now is to devise a way to give meaning to the terms and the formulas of a language. Usually when we do mathematics, we have some set of objects in mind (say the integers, the real numbers, the elements of a group, the set of polynomial functions over the rationals in one variable, etc.). As soon as we have such a set, call it the universe U (or domain of discourse), we automatically have, for every positive integer n, the set of all subsets of U^n, that is, for each n, we have all the n-ary relations on U. In order to do whatever it is we wish to do with this universe, we use certain of those relations. For example, we might be interested in the "less than" relation on the integers; or, since the polynomial functions over the rationals includes the rationals, we might be particularly interested in the binary relation which holds between an ordered pair of polynomials if the second (a rational) is a root of the first. Just as there are automatically many relations on a given universe, there are also many functions (or operations). For every n, there are many n-ary functions which associate to each n-tuple U^n a unique element of U. If S is a set, let $\bar{\bar{S}}$ denote the cardinality of S.

Exercise 10.1 (a) If $\bar{\bar{U}}$ is finite, for each n, what is the cardinality of the set of all subsets of U^n? For each n, what is the cardinality of the set of all n-ary functions from U^n to U? (b) Answer the same questions in the case that $\overline{\overline{U}}$ is not finite.

As in the case of relations, generally we are not really interested in all those functions, just certain ones. In the case of the polynomial functions, we may be interested in the unary operation which gives the derivative of a polynomial function. Or, in the case of a group, we are only interested in the binary function which is the group operation (call it $*$) and the unary function which is the inverse operation (call it the superscript -1). Thus, since U is a group, we know that if u_1 and u_2 are elements of U, then $u_1 * u_2$ and u_1^{-1} are elements of U and since U is the set of elements of a group, there is an element e such that $u_1 * e$ is the element u_1 and for each, $u_1 * u_1^{-1}$ is e.

With these ideas in mind, we define a *relational system* (*structure*) $\mathfrak{A}$ to be a nonempty set A with a certain number of functions from A^n into A (for various n), which we denote by f_i^n, and a certain number of n-ary relations on A, which we denote by R_i^n. As an abbreviation, we write

$$\mathfrak{A} = \langle A; f_1^{i_1}, \ldots, f_j^{i_j}; R_1^{k_1}, \ldots, R_t^{k_t} \rangle$$

We will call the f's the *functions* (or named functions) *of* $\mathfrak{A}$ and the R's the *relations* (or named relations) *of* $\mathfrak{A}$. [Since, as pointed out above, there are functions and relations on A besides those indicated by the f's and R's, it will sometimes be less confusing if we keep the distinction between the functions (relations) named by giving the relational system and those which are there automatically; hence, we sometimes use the term "named."] The set A is known as *the universe* (or *domain*) of the relational system. By the *cardinality of a relational system* $\mathfrak{A}$, we mean the cardinality of its universe A.

Examples Relational systems:

1. $\langle Z; ', +, \cdot; =, < \rangle$: Z is the set of all integers; $'$ is the successor function; the rest are obvious.
2. $\langle M; R^2 \rangle$: M is the set of all men throughout history and $(m_1, m_2) \in R^2$ if m_1 is the father of m_2.
3. $\langle N; ', Z, U_1^1, U_1^2, U_2^2, \ldots, U_j^n, \ldots; = \rangle$: N is the natural numbers, $'$ the successor function, Z the zero function, and U_j^n picks out the jth component of an n-tuple.
4. $\langle R_1; < \rangle$: R_1 is the closed interval of reals [0, 1].
5. $\langle Q; +, \cdot, -, \div; =, < \rangle$: Q is all the rationals.
6. $\langle Q; -, -1, +, \cdot; =, < \rangle$: $-$ is the unary function such that $q + q^- = 0$, and the superscript -1 is the unary function such that $q \cdot q^{-1} = 1$.
7. $\langle$all the stars; hotter than, bigger than$\rangle$.

8. $\langle P; D; Z_0, =\rangle$: P is the set of all polynomial functions in one variable with coefficients in Q; D is the unary function which for an element of P gives that element of P which is its derivative; Z_0 is the binary relation which holds between two elements of P if the second is a root of the first.
9. $\langle\{0, 1, 2\}; =; <\rangle$.
10. $\langle\{0, 1, 2, \ldots, n-1\}; \cdot_n; =\rangle$: $\cdot_n$ is the binary function of multiplication mod n.
11. $\langle N; \cdot; \equiv_n\rangle$: $\equiv_n$ is congruence mod n.

10.3 A RELATIONAL SYSTEM FOR A LANGUAGE

Suppose L(∗) is a first-order language and $\mathfrak{A}$ is a relational system. $\mathfrak{A}$ *is a relational system for* (*structure for, interpretation of*) L(∗) if the following hold:

1. (a) If L(∗) has =, then $\mathfrak{A}$ must include = among its (named) relations, and otherwise (b) $\mathfrak{A}$ has the same number of (named) n-ary relations as L(∗) has n-ary predicate symbols and there is a fixed correspondence between them. (Notice that even if L(∗) is not a language with equality, if it has a binary predicate symbol $\mathbf{P_i}^2$ and $\mathfrak{A}$ has =, $\mathfrak{A}$ may be a relational system for L(∗) by making these two correspond provided that everything else is as it should be.)

2. $\mathfrak{A}$ has the same number of (named) n-ary functions as L(∗) has n-place function symbols and there is a fixed correspondence between them.

3. A fixed correspondence is established between the names of constants in L(∗) and elements in the universe A of $\mathfrak{A}$.

Items 1, 2, and 3 are thus established by a correspondence (mapping) which we name Φ and we can say that $\mathfrak{A}$ is a relational system for L(∗) by means of Φ. Thus, if $\mathbf{P_i}^{\mathbf{n}}$ is a predicate symbol of L(∗), $\Phi\mathbf{P_i}^{\mathbf{n}}$ is the corresponding relation in $\mathfrak{A}$; if $\mathbf{f_i}^{\mathbf{n}}$ is a function symbol of L(∗), $\Phi\mathbf{f_i}^{\mathbf{n}}$ is the corresponding function in $\mathfrak{A}$ and if $\mathbf{a_i}$ is the name of a constant, $\Phi\mathbf{a_i}$ is the corresponding element of A. Notice that Φ is one–one onto betweeen the predicate symbols and relations and the function symbols and functions, but Φ need not be one–one or onto from the names of constants. Notice also that, given a language L(∗) and a relational system $\mathfrak{A}$, there may be more than one Φ which makes $\mathfrak{A}$ a relational system for L(∗). [E.g., if for some particular n there are, say k, n-ary predicate symbols and there are k (named) n-ary relations on $\mathfrak{A}$, then the different predicate symbols can be made to correspond to the different relations in $k!$ different ways.]

Example Let $\mathsf{L}(g)$ (first-order language for groups) be a first order language with one name of a constant $\mathbf{a}_0$, with one unary function symbol $\mathbf{f}_1{}^1$, with one binary function symbol $\mathbf{f}_1{}^2$, and equality. Let $\mathfrak{G} = \langle Z; {}^-, +; =\rangle$, where Z is the integers, ${}^-$ is the unary function giving the additive inverse, and $+$ is the usual addition. Let $\Phi\mathbf{a}_0$ be 0 in Z, let $\Phi\mathbf{f}_1{}^1$ be ${}^-$ and $\Phi\mathbf{f}_1{}^2$ be $+$. Thus, $\mathfrak{G}$ is a relational system for $\mathsf{L}(g)$ by means of Φ.

Exercise 10.2 Take some other group and show that it is a relational system for $\mathsf{L}(g)$.

Exercise 10.3 For each example of a relational system in section 10.2, make up a first-order language so that relational system is a relational system for that language.

Exercise 10.4 If $\mathsf{L}(f)$ is a first-order language that has some names of constants and some function symbols, show that there is a relational system for $\mathsf{L}(f)$ which has as its universe the set of all closed terms of $\mathsf{L}(f)$.

10.4 A FORMULA IS TRUE IN A RELATIONAL SYSTEM

Suppose we have a language $\mathsf{L}(*)$ and a relational system $\mathfrak{A}$ which is a relational system for $\mathsf{L}(*)$ by means of Φ. If we consider some formula of $\mathsf{L}(*)$ we already have a fairly clear intuitive idea of what that formula should "mean" in reference to the relational system. In the example about groups just above, if we innocently asked, what do the formulas

$$\text{(i)} \qquad (\forall \mathbf{x})(\mathbf{f}_1{}^2(\mathbf{x}, \mathbf{a}_0) = \mathbf{x})$$
$$\text{(ii)} \qquad (\forall \mathbf{x})(\forall \mathbf{y})(\mathbf{f}_1{}^2(\mathbf{x}, \mathbf{y}) = \mathbf{f}_1{}^2(\mathbf{y}, \mathbf{x}))$$

"mean" with reference to a group, we would probably answer with something like, "Oh, the first means that any element times its inverse is the identity element, and the second means that the elements commute." We have now come to the point where we must choose between (i) carrying on by means of our intuitive notions about the truth of a formula in a relational system (see section 1.7), even though it sometimes leaves one with a queasy feeling, and (ii) defining rigorously what it means for a formula to be true in a relational system, even though it is a fairly long and boring procedure. We choose the latter course, though not joyfully.

Let $\mathsf{L}(*)$ be a language and $\mathfrak{A}$ a relational system for $\mathsf{L}(*)$ by means of Φ. Consider the set of all mappings from the set of individual variables of $\mathsf{L}(*)$, $\{\mathbf{x}_1, \mathbf{x}_2, \ldots\}$, into A, the universe of $\mathfrak{A}$. We call such a map *an assignment* (of members of A to the individual variables of $\mathsf{L}(*)$). Denote maps which are assignments by φ, φ', ψ, ψ', etc.

Examples (1) If $\bar{\bar{A}}$ is $\aleph_0$, then there are $2^{\aleph_0}$ possible assignments. Some possible assignments are as follows:

(a) For each enumeration (ordering) of A, there is a one–one onto assignment φ such that $\varphi(\mathbf{x}_i) = a_i$, where a_i is the ith element of A under the particular ordering.

(b) There are $\aleph_0$-many "constant" assignments; that is, for each particular a of A, there is the assignment φ which is $\varphi(\mathbf{x_i}) = a$ for all i.

(c) For each assignment φ and each i, there is a set of $\aleph_0$-many assignments φ' such that for all j not equal to i, $\varphi'(\mathbf{x_j}) = \varphi(\mathbf{x_j})$ and $\varphi'(\mathbf{x_i}) \neq \varphi(\mathbf{x_i})$.

(d) There are assignments for which there is an i so that for all $j \geq i$, $\varphi(\mathbf{x_j}) = a$, for some particular $a \in A$, but for $j, k < i$, $\varphi(\mathbf{x_j})$ may not be equal to $\varphi(\mathbf{x_k})$.

(2) If $\bar{\bar{A}} > \aleph_0$, there are at least $2^{\aleph_0}$ many assignments, none of which is onto.

(3) If $\bar{\bar{A}} < \aleph_0$, there are $\aleph_0$-many assignments, none of which is one one.

For a given $\mathsf{L}(*)$, $\mathfrak{A}$, and Φ, we define a set of mappings, one for each assignment φ, from the terms and formulas of $\mathsf{L}(*)$ into A and the truth values $\{\mathrm{t}, \mathfrak{f}\}$, i.e., for each φ,

$$V_\varphi : \begin{matrix} \{\text{terms}\} \longrightarrow A \cup \\ \cup \{\text{formulas}\} \rightarrow \{\mathrm{t}, \mathfrak{f}\} \end{matrix}$$

The V_φ's are known as the *valuations* (or valuation mappings). The definition is given inductively on the formation of formulas:

1. *Terms:* (a) $V_\varphi(\mathbf{x_i})$ is defined to be $\varphi(\mathbf{x_i})$; (b) $V_\varphi(\mathbf{a_i})$ is defined to be $\Phi(\mathbf{a_i})$; (c) If $\mathbf{t_1}, \ldots, \mathbf{t_n}$ are terms and $\mathbf{f_i^n}$ is an n-ary function symbol, then $V_\varphi(\mathbf{f_i^n}(\mathbf{t_1}, \ldots, \mathbf{t_n}))$ is defined to be $\Phi\mathbf{f_i^n}(V_\varphi(\mathbf{t_1}), \ldots, V_\varphi(\mathbf{t_n}))$. (Notice that if $\mathbf{t}$ is a term, $V_\varphi(\mathbf{t}) \in A$.)

2. *Atomic formulas:* (a) If $\mathbf{t_1}, \ldots, \mathbf{t_n}$ are terms and $\mathbf{P_i^n}$ is an n-ary predicate symbol then

$$V_\varphi(\mathbf{P_i^n}(\mathbf{t_1}, \ldots, \mathbf{t_n})) = \begin{cases} \mathrm{t} & \text{if } (V_\varphi(\mathbf{t_1}), \ldots, V_\varphi(\mathbf{t_n})) \in \Phi\mathbf{P_i^n} \\ \mathfrak{f} & \text{if not} \end{cases}$$

(b) If $\mathbf{t_1}$ and $\mathbf{t_2}$ are terms,

$$V_\varphi(\mathbf{t_1} = \mathbf{t_2}) = \begin{cases} \mathrm{t} & \text{if } V_\varphi(\mathbf{t_1}) \text{ is the same element of } A \text{ as } V_\varphi(\mathbf{t_2}) \\ \mathfrak{f} & \text{otherwise} \end{cases}$$

3. *Formulas*: (a) If $\mathbf{A}$ and $\mathbf{B}$ are formulas, then

$$V_\varphi(\mathbf{A} \supset \mathbf{B}) = \begin{cases} \mathrm{t} & \text{otherwise} \\ \mathfrak{f} & \text{if } V_\varphi(\mathbf{A}) = \mathrm{t} \text{ and } V_\varphi(\mathbf{B}) = \mathfrak{f} \end{cases}$$

$$V_\varphi(\mathbf{A} \wedge \mathbf{B}) = \begin{cases} \mathfrak{t} & \text{if both } V_\varphi(\mathbf{A}) \text{ and } V_\varphi(\mathbf{B}) \text{ are } \mathfrak{t} \\ \mathfrak{f} & \text{otherwise} \end{cases}$$

$$V_\varphi(\mathbf{A} \vee \mathbf{B}) = \begin{cases} \mathfrak{t} & \text{otherwise} \\ \mathfrak{f} & \text{if both } V_\varphi(\mathbf{A}) \text{ and } V_\varphi(\mathbf{B}) \text{ are } \mathfrak{f} \end{cases}$$

$$V_\varphi(\mathbf{A} \equiv \mathbf{B}) = \begin{cases} \mathfrak{t} & \text{if } V_\varphi(\mathbf{A}) = V_\varphi(\mathbf{B}) \\ \mathfrak{f} & \text{otherwise} \end{cases}$$

$$V_\varphi(\sim \mathbf{A}) = \begin{cases} \mathfrak{t} & \text{if } V_\varphi(\mathbf{A}) = \mathfrak{f} \\ \mathfrak{f} & \text{otherwise} \end{cases}$$

(b) If **A** is a formula of the form $(\forall \mathbf{x_i})\mathbf{B}(\mathbf{x_i})$, then

$$V_\varphi((\forall \mathbf{x_i})\mathbf{B}(\mathbf{x_i})) = \begin{cases} \mathfrak{t} & \text{if for all } \varphi' \text{ in the set} \\ & \text{of all assignments such that} \\ & \varphi'(\mathbf{x_j}) = \varphi(\mathbf{x_j}) \text{ for all } j \neq i, \\ & V_{\varphi'}(\mathbf{B}(\mathbf{x_i})) = \mathfrak{t} \\ \mathfrak{f} & \text{otherwise} \end{cases}$$

(End of definition of V_φ.)

Lemma 10.1 If **A** is a formula of the form $(\exists \mathbf{x_i})\mathbf{B}(\mathbf{x_i})$, show that

$$V_\varphi((\exists \mathbf{x_i})\mathbf{B}(\mathbf{x_i})) = \begin{cases} \mathfrak{t} & \text{if there is an assignment } \varphi' \text{ such that} \\ & \text{for all } j \neq i, \ \varphi'(\mathbf{x_j}) = \varphi(\mathbf{x_j}) \text{ and} \\ & V_{\varphi'}(\mathbf{B}(\mathbf{x_i})) = \mathfrak{t} \\ \mathfrak{f} & \text{otherwise} \end{cases}$$

Exercise 10.5 Prove the lemma. (Hint: Use the definition of $(\exists \mathbf{x_i})$ with the definition of valuation.)

A set (finite or infinite) of formulas Γ (in $\mathsf{L}(*)$) is (simultaneously) *satisfiable* in a relational system $\mathfrak{A}$ if (a) $\mathfrak{A}$ is a relational system for $\mathsf{L}(*)$ and (b) there is an assignment φ such that for all $\mathbf{A} \in \Gamma$, $V_\varphi(\mathbf{A}) = \mathfrak{t}$.

Lemma 10.2 For a given $\mathsf{L}(*)$, $\mathfrak{A}$, and Φ, if **t** is a closed term, then for all assignments φ and ψ, $V_\varphi(\mathbf{t}) = V_\psi(\mathbf{t})$.

Exercise 10.6 Prove the lemma. (Hint: By induction of the formation of terms.)

Lemma 10.3 For a given $\mathsf{L}(*)$, $\mathfrak{A}$, and Φ, if **A** is a closed atomic formula, then for all assignments φ and ψ, $V_\varphi(\mathbf{A}) = V_\psi(\mathbf{A})$.

Exercise 10.7 Prove the lemma. (Notice that by the definition, a closed atomic formula is either $(\mathbf{t_1} = \mathbf{t_2})$, where $\mathbf{t_1}$ and $\mathbf{t_2}$ are closed terms, or $\mathbf{P_i^n(t_1, \ldots, t_n)}$, where the $\mathbf{t}$'s are closed terms.)

Lemma 10.4 Let $\mathfrak{A}$ be a relational system for L(*) by means of Φ. Let $\mathbf{A(x_{i_1}, \ldots, x_{i_n})}$ be a formula with n free variables, as indicated. If φ and ψ are assignments such that for all j, $1 \leq j \leq n$, $\varphi(\mathbf{x_{i_j}}) = \psi(\mathbf{x_{i_j}})$, then

$$V_\varphi(\mathbf{A(x_{i_1}, \ldots, x_{i_n})}) = V_\psi(\mathbf{A(x_{i_1}, \ldots, x_{i_n})})$$

PROOF By induction of the formation of formulas.

1. *Terms*: For the term $\mathbf{x_i}$, if $\varphi(\mathbf{x_i}) = \psi(\mathbf{x_i})$, then by the definition of valuation, $V_\varphi(\mathbf{x_i}) = V_\psi(\mathbf{x_i})$. For the term $\mathbf{a_i}$, since both $V_\varphi(\mathbf{a_i}) = \Phi\mathbf{a_i}$ and $V_\psi(\mathbf{a_i}) = \Phi\mathbf{a_i}$, the lemma holds. If $\mathbf{f_j^k(t_1, \ldots, t_k)}$ is a term with free variables $\mathbf{x_{i_1}}, \ldots, \mathbf{x_{i_n}}$, by the induction hypothesis, if φ and ψ agree on these $\mathbf{x_{i_j}}$, then $V_\varphi(\mathbf{t_i}) = V_\psi(\mathbf{t_i})$ for $1 \leq t \leq k$. Since, by the definition of valuations,

$$V_\varphi(\mathbf{f_j^k(t_1, \ldots, t_k)}) = \Phi\mathbf{f_j^k}(V_\varphi(\mathbf{t_1}), \ldots\ V_\varphi(\mathbf{t_k}))$$
$$V_\psi(\mathbf{f_j^k(t_1, \ldots, t_k)}) = \Phi\mathbf{f_j^k}(V_\psi(\mathbf{t_1}), \ldots, V_\psi(\mathbf{t_k}))$$

the lemma follows from the induction hypothesis.

2. *Atomic formulas*: (a) $V_\varphi(\mathbf{t_1} = \mathbf{t_2}) = \mathfrak{t}$ if and only if $V_\varphi(\mathbf{t_1})$ is the same object as $V_\varphi(\mathbf{t_2})$. Further, by the induction hypothesis, if $(\mathbf{t_1} = \mathbf{t_2})$ has free variables $\mathbf{x_{i_1}}, \ldots, \mathbf{x_{i_n}}$ and φ and ψ are assignments which agree on those $(V_\psi\mathbf{t_1})$ variables, then $V_\varphi(\mathbf{t_1}) = V_\psi(\mathbf{t_1})$ and $V_\varphi(\mathbf{t_2}) = V_\psi(\mathbf{t_2})$. Then it follows that $V_\psi(\mathbf{t_1})$ and $V_\psi(\mathbf{t_2})$ are the same object and hence $V_\psi(\mathbf{t_1} = \mathbf{t_2}) = \mathfrak{t}$. If $V_\varphi(\mathbf{t_1} = \mathbf{t_2}) = \mathfrak{f}$, then $V_\varphi(\mathbf{t_1})$ is not the same as $V_\varphi(\mathbf{t_2})$, so, by the induction hypothesis, $V_\psi(\mathbf{t_1})$ cannot be the same as $V_\psi(\mathbf{t_2})$, so $V_\psi(\mathbf{t_1} = \mathbf{t_2}) = \mathfrak{f}$.

(b) If $\mathbf{P_j^k(t_1, \ldots, t_k)}$ has free variables $\mathbf{x_{i_1}}, \ldots, \mathbf{x_{i_n}}$, then as before, for assignments φ and ψ which agree on these $\mathbf{x_{i_j}}$, the induction hypothesis holds for the $\mathbf{t_i}$. Thus

$$V_\varphi(\mathbf{P_j^k(t_1, \ldots, t_k)}) = \mathfrak{t} \quad \text{if and only if} \quad (V_\varphi(\mathbf{t_1}), \ldots, V_\varphi(\mathbf{t_k})) \in \Phi\mathbf{P_j^k}$$

By the induction hypothesis,

$$(V_\varphi(\mathbf{t_1}), \ldots, V_\varphi(\mathbf{t_k})) \in \Phi\mathbf{P_j^k}, \quad \text{if and only if} \quad (V_\psi(\mathbf{t_1}), \ldots, V_\psi(\mathbf{t_k})) \in \Phi\mathbf{P_j^k}$$

But then, by the definition of valuation

$$(V_\psi(\mathbf{t_1}), \ldots, V_\psi(\mathbf{t_k})) \in \Phi\mathbf{P_j^k} \quad \text{if and only if} \quad V_\psi(\mathbf{P_j^k(t_1, \ldots, t_k)}) = \mathfrak{t}.$$

3. *Formulas*: (a) If $\mathbf{A}$ is $\sim\mathbf{B}$ or is $(\mathbf{B} \supset \mathbf{C})$, etc., the lemma follows immediately from the induction hypothesis and the definition of valuation.

(b) Suppose that $\mathbf{A}$ is $(\forall \mathbf{x_{i_{n+1}}})\mathbf{B(x_{i_1}, \ldots, x_{i_n}, x_{i_{n+1}})}$ and suppose also that φ and ψ are assignments which agree on $\mathbf{x_{i_1}}, \ldots, \mathbf{x_{i_n}}$. We must show that $V_\varphi(\mathbf{A}) = V_\psi(\mathbf{A})$. Now, $V_\varphi((\forall \mathbf{x_{i_{n+1}}})\mathbf{B(x_{i_1}, \ldots, x_{i_n}, x_{i_{n+1}})}) = \mathfrak{t}$ if and only if for every φ' in the set S_φ of all assignments which agree with φ everywhere except possibly at $\mathbf{x}_{i_{n+1}}$, $V_{\varphi'}(\mathbf{B}(\mathbf{x}_{i_1}, \ldots, \mathbf{x}_{i_n}, \mathbf{x}_{i_{n+1}})) = \mathfrak{t}$. Similarly, $V_\psi((\forall \mathbf{x}_{i_{n+1}}) \mathbf{B}(\mathbf{x}_{i_1}, \ldots, \mathbf{x}_{i_n}, \mathbf{x}_{i_{n+1}})) = \mathfrak{t}$ if and only if for every assignment ψ' in the set S_ψ of all assignments which agree with ψ everywhere except possibly at $\mathbf{x}_{i_{n+1}}$, $V_{\psi'}(\mathbf{B}(\mathbf{x}_{i_1}, \ldots, \mathbf{x}_{i_{n+1}})) = \mathfrak{t}$. By the definitions of S_φ and S_ψ, all assignments in $S_\varphi \cup S_\psi$ agree on $\mathbf{x}_{i_1}, \ldots, \mathbf{x}_{i_n}$. Further, for every $\varphi' \in S_\varphi$ there is a $\psi' \in S_\psi$ such that φ' and ψ' also agree on $\mathbf{x}_{i_{n+1}}$; and conversely. Thus, φ' and ψ' agree on the free variables of $\mathbf{B}(\mathbf{x}_{i_1}, \ldots, \mathbf{x}_{i_n}, \mathbf{x}_{i_{n+1}})$ and so by induction hypothesis, $V_{\varphi'}(\mathbf{B}(\mathbf{x}_{i_1}, \ldots, \mathbf{x}_{i_n}, \mathbf{x}_{i_{n+1}})) = V_{\psi'}(\mathbf{B}(\mathbf{x}_{i_1}, \ldots, \mathbf{x}_{i_n}, \mathbf{x}_{i_{n+1}}))$. Either for all assignments these are true and so $V_\varphi(\mathbf{A}) = V_\psi(\mathbf{A}) = \mathfrak{t}$, or in some cases they are not, and so $V_\varphi(\mathbf{A}) = V_\psi(\mathbf{A}) = \mathfrak{f}$. ∎

Theorem 10.1 Suppose $\mathsf{L}(*)$, $\mathfrak{A}$, and Φ are given so that $\mathfrak{A}$ is a relational system for $\mathsf{L}(*)$. If $\mathbf{A}$ is a closed formula, then for all assignments φ and ψ, $V_\varphi(\mathbf{A}) = V_\psi(\mathbf{A})$.

Exercise 10.8 Prove the theorem.

In view of the theorem, when speaking of the valuation of closed formulas, we can drop the subscript and simply write $V(\mathbf{A})$.

Theorem 10.2 Let $\mathbf{A}$ be a formula and $\mathbf{A}'$ its universal closure. $V(\mathbf{A}') = \mathfrak{t}$ if and only if for all φ, $V_\varphi(\mathbf{A}) = \mathfrak{t}$.

PROOF If $\mathbf{A}$ is closed, the theorem follows from the previous theorem. Suppose $\mathbf{A}'$ is $(\forall \mathbf{x}_{i_1}) \ldots (\forall \mathbf{x}_{i_k})\mathbf{A}$. The proof is by induction on k, the number of quantifiers necessary to close $\mathbf{A}$, and is given for the case $k = 1$, i.e., $\mathbf{A}'$ is $(\forall \mathbf{x}_i)\mathbf{A}$. (1) Suppose that $V(\mathbf{A}') = \mathfrak{t}$ and that for some assignment φ, $V_\varphi(\mathbf{A}) = \mathfrak{f}$. Then by Lemma 10.4, for any assignment φ' which agrees with φ on $\mathbf{x}_i$, $V_{\varphi'}(\mathbf{A}) = \mathfrak{f}$. On the other hand, $V(\mathbf{A}') = \mathfrak{t}$ means that for all assignments ψ, $V_\psi(\mathbf{A}') = \mathfrak{t}$, which by the definition of valuations, is true if and only if for *every* ψ' which agrees with ψ except at $\mathbf{x}_i$, $V_{\psi'}(\mathbf{A}) = \mathfrak{t}$. But there is a ψ' such that $\psi'(\mathbf{x}_j) = \psi(\mathbf{x}_j)$ for all $j \neq i$ and $\psi'(\mathbf{x}_i) = \varphi(\mathbf{x}_i)$ and so by Lemma 10.4 and the assumption that $V_\varphi(\mathbf{A}) = \mathfrak{f}$, $V_{\psi'}(\mathbf{A}) = \mathfrak{f}$, hence a contradiction.
(2) Suppose $V(\mathbf{A}') = \mathfrak{f}$ and show that for some φ, $V_\varphi(\mathbf{A}) = \mathfrak{f}$. ∎

Exercise 10.9 Finish the proof of the theorem.

A *closed formula A is true* (valid) *in* $\mathfrak{A}$ *if* $V(\mathbf{A}) = \mathfrak{t}$. *A formula* $\mathbf{A}$ *is true in* $\mathfrak{A}$ *if* its universal closure is true in $\mathfrak{A}$. If Γ is a set of formulas such that every formula of Γ is true in $\mathfrak{A}$, then we say that $\mathfrak{A}$ *is a model of* Γ.

Exercise 10.10 Show that if a closed formula is satisfiable in a relational system then it is true there.

Exercise 10.11 Show that if **A** is not true in $\mathfrak{A}$, then $\sim$**A** is satisfiable in $\mathfrak{A}$.

Exercise 10.12 Using the example formulas of section 10.1 and the example relational systems of section 10.2, (a) find which relational systems are relational systems for the languages of the different formulas, and (b) determine, when appropriate, which formulas are satisfiable and/or true in which relational systems.

10.5 FURTHER DISCUSSION OF A LANGUAGE AND ITS RELATIONAL SYSTEMS

So as to better understand some of the concepts discussed above, we will take an example and try to look at it from the points of view indicated by those concepts and then introduce some new concepts.

Let $\mathsf{L}(r)$ be the first-order language with $=$ having two names of constants, $\mathbf{a}_0$ and $\mathbf{a}_1$, and two binary function symbols, $\mathbf{f}_1{}^2$ and $\mathbf{f}_2{}^2$. Let the relational system $\mathfrak{R}$ be $\langle R^+;\ +,\ \cdot\ ;\ =\rangle$, where R^+ denotes the nonnegative reals. Then, $\mathfrak{R}$ is a relational system for $\mathsf{L}(r)$ in the obvious way with $\mathbf{a}_0$ corresponding to 0 and $\mathbf{a}_1$ corresponding to 1, $\mathbf{f}_1{}^2$ to $+$, and $\mathbf{f}_2{}^2$ to $\cdot$.

1. We will consider the valuation in $\mathfrak{R}$ of certain formulas of $\mathsf{L}(r)$.

(a) Claim that for all assignments φ, $V_\varphi(\mathbf{f}_2{}^2(\mathbf{x}, \mathbf{y}) = \mathbf{f}_2{}^2(\mathbf{y}, \mathbf{x})) = \mathrm{t}$: By the definition of V_φ, this is the case if $\Phi\mathbf{f}_2{}^2(\varphi(\mathbf{x}), \varphi(\mathbf{y}))$ is the same real number as $\Phi\mathbf{f}_2{}^2(\varphi(\mathbf{y}), \varphi(\mathbf{x}))$, which is certainly true since multiplication is commutative in the reals.

Exercise 10.13 Show that for all φ, $V_\varphi(\mathbf{f}_2{}^2(\mathbf{x}, \mathbf{a}_0) = \mathbf{a}_0) = \mathrm{t}$.

(b) Claim that $(\forall\mathbf{x})(\exists\mathbf{y})((\mathbf{x} = \mathbf{a}_0) \vee \mathbf{f}_2{}^2(\mathbf{x}, \mathbf{y}) = \mathbf{a}_1)$ is true in $\mathfrak{R}$, i.e., every positive real number has a multiplicative inverse. By the definition of truth in a relational system, the claim holds if and only if for all assignments φ,

$$V_\varphi\big((\forall\mathbf{x})(\exists\mathbf{y})((\mathbf{x} = \mathbf{a}_0) \vee \mathbf{f}_2{}^2(\mathbf{x}, \mathbf{y}) = \mathbf{a}_1)\big) = \mathrm{t} \quad \text{in } \mathfrak{R} \qquad (*)$$

But $(*)$ holds if and only if for every assignment φ' which is equal to φ except perhaps at $\mathbf{x}$,

$$V_{\varphi'}\big((\exists\mathbf{y})((\mathbf{x} = \mathbf{a}_0) \vee \mathbf{f}_2{}^2(\mathbf{x}, \mathbf{y}) = \mathbf{a}_1)\big) = \mathrm{t} \quad \text{in } \mathfrak{R} \qquad (**)$$

For such a φ', $(**)$ holds if and only if there is some assignment φ'' differing from φ' at most on $\mathbf{y}$ such that

$$V_{\varphi''}\big((\mathbf{x} = \mathbf{a}_0) \vee \mathbf{f}_2{}^2(\mathbf{x}, \mathbf{y}) = a_1\big) = \mathrm{t} \quad \text{in } \mathfrak{R}$$

But this is the case if and only if

$$V_{\varphi''}(\mathbf{x} = \mathbf{a}_0) = \mathrm{t} \qquad \text{or} \qquad V_{\varphi''}(\mathbf{f}_2{}^2(\mathbf{x}, \mathbf{y}) = \mathbf{a}_1) = \mathrm{t} \qquad \text{in } \mathfrak{R}$$

Finally, the preceding statement holds if and only if

$$\left.\begin{array}{r} \text{either } \varphi''(\mathbf{x}) \text{ is } \varphi''(\mathbf{a}_0), \text{ namely } 0, \text{ or} \\ \Phi\mathbf{f}_2{}^2(\varphi''(\mathbf{x}), \varphi''(\mathbf{y})) = \Phi\mathbf{a}_1, \text{ i.e., } \varphi''(\mathbf{x}) \cdot \varphi''(\mathbf{y}) = 1 \end{array}\right\} \qquad (***)$$

Thus, for (*) to be true, there must be an assignment φ'' differing from φ' at most on **y** such that (***) is true. Notice that $\varphi''(\mathbf{x}) = 0$ if and only if $\varphi'(\mathbf{x}) = 0$. So, on the one hand, if $\varphi'(\mathbf{x}) = 0$, let $\varphi'' = \varphi'$. On the other hand, if $\varphi'(\mathbf{x}) \neq 0$, then there is a positive real r such that $r \cdot \varphi'(\mathbf{x}) = 1$, namely $r = 1/\varphi'(\mathbf{x})$. Let φ'' be the assignment equal to φ' except at **y** and let $\varphi''(\mathbf{y}) = 1/\varphi'(\mathbf{x})$; then $\varphi''(\mathbf{y}) = 1/\varphi''(\mathbf{x})$, and so $\varphi''(\mathbf{x}) \cdot \varphi''(\mathbf{y}) = 1$.

Exercise 10.14 Show that $(\forall\mathbf{x})(\forall\mathbf{y})(\sim(\mathbf{y} = \mathbf{a}_0) \supset (\exists\mathbf{z})(\mathbf{f}_2{}^2(\mathbf{x}, \mathbf{z}) = \mathbf{y}))$ is true in $\mathfrak{R}$, i.e., divisibility.

2. As already mentioned, there are many relations on the nonnegative reals besides the "equals" named in $\mathfrak{R}$. For example, "less than." It is natural to ask if there is a formula $\mathbf{A} \in \mathsf{F}(r)$ with two free variables such that for each assignment φ, $V_\varphi(\mathbf{A}(\mathbf{x}, \mathbf{y})) = \mathrm{t}$ in $\mathfrak{R}$ if and only if $\varphi(\mathbf{x})$ is less than $\varphi(\mathbf{y})$. In terms of the nonnegative reals, we know that s is less than t if and only if there is an r such that $r \neq 0$ and $s + r = t$. Using this idea, in $\mathsf{L}(r)$ we write

$$(\exists\mathbf{z})(\sim(\mathbf{z} = \mathbf{a}_0) \wedge \mathbf{f}_1{}^2(\mathbf{x}, \mathbf{z}) = \mathbf{y}) \qquad (*)$$

and check to see that for any assignment φ, V_φ of (*) is t if and only if $\varphi(\mathbf{x})$ is less than $\varphi(\mathbf{y})$. In this case, we say that "the less-than relation on the nonnegative reals is defined in $\mathsf{L}(r)$." In general, let $\mathfrak{A}$ be a relational system for a language $\mathsf{L}(*)$. If $M \subseteq U^n$, where U is the universe of $\mathfrak{A}$ and **A** is a formula of $\mathsf{L}(*)$ with n free variables, $\mathbf{x}_1, \ldots, \mathbf{x}_n$, such that for all assignments φ,

$$V_\varphi(\mathbf{A}) = \mathrm{t} \quad \text{in } \mathfrak{A} \quad \text{if and only if} \quad (\varphi(\mathbf{x}_1), \ldots, \varphi(\mathbf{x}_n)) \in M$$

then we say the *relation M of $\mathfrak{A}$ is defined in* $\mathsf{L}(*)$ by the formula **A**. Once "less than" is defined, other facts about the nonnegative reals can be expressed. For example, the fact that they are dense, i.e., for every pair of reals r and s which are not equal, there is another real between them.

Exercise 10.15 (a) Write a sentence of $\mathsf{F}(r)$ expressing that the relational system $\mathfrak{R}$ is dense. (b) Is $(\forall\mathbf{y})(\exists\mathbf{x})(\exists\mathbf{z})(\sim(\mathbf{z} = \mathbf{a}_0) \wedge \mathbf{f}_1{}^2(\mathbf{y}, \mathbf{z}) = \mathbf{x})$ true in $\mathfrak{R}$?

Exercise 10.16 What relation in $\mathfrak{R}$ does the formula of $\mathsf{L}(r)$, $((\mathbf{x} = \mathbf{x}) \wedge \sim(\mathbf{x} = \mathbf{x}))$, define? The formula $((\mathbf{x} = \mathbf{x}) \vee \sim(\mathbf{x} = \mathbf{x}))$?

Exercise 10.17 Consider the relational system with domain the closed interval [0, 1] on the reals, with "special" addition, multiplication, and equals. "Special" addition $\oplus$ is defined as follows: for any pair $r, s \in [0, 1]$,

$$r \oplus s = \begin{cases} 1 & \text{if } \; r + s \geq 1 \\ r + s & \text{otherwise} \end{cases}$$

(a) is L(r) a language for this relational system? (b) Does the formula (*) define "less than" in this relational system? (c) Write sentences in F(r) expressing the fact that 0 is the least element and 1 the greatest element in [0, 1].

One can also ask if similar things can be done for functions. For instance with reference to our example, we know that each nonnegative real has a square root. That is, there is a square root function. Is there some sort of formula in the language to indicate that? We say that a *function is defined in a language* if the relation which is its graph is defined in that language. Speaking ordinary mathematics, we know that the graph of the square root function is the relation $\{(x, y) | y \cdot y = x\}$ and to be sure that the relation is the graph of a function, we must add that if $y \cdot y = x$ and $z \cdot z = x$, then $y = z$. In L(r), we have

$$\mathbf{f}_2{}^2(\mathbf{y}, \mathbf{y}) = \mathbf{x} \wedge (\forall \mathbf{z})((\mathbf{f}_2{}^2(\mathbf{y}, \mathbf{y}) = \mathbf{x} \wedge \mathbf{f}_2{}^2(\mathbf{z}, \mathbf{z}) = \mathbf{x}) \supset (\mathbf{y} = \mathbf{z}))$$

and one can check to see if its works out as it should.

Exercise 10.18 Consider the 3-ary relation on R^+, which is

$$\{(r_1, r_2, r_3) | r_1 + \sqrt{r_2} = r_3\}$$

Write a formula **A** in L(r) with three free variables so that for all assignments φ, $V_\varphi(\mathbf{A}) = \mathrm{t}$ if and only if $(\varphi(\mathbf{x}), \varphi(\mathbf{y}), \varphi(\mathbf{z}))$ belongs to the above relation.

3. There are several questions one may ask about a formula **A** in a language L(*) with regard to any relational system $\mathfrak{A}$ which is a relational system for the language. So far, we have considered whether there is an assignment φ such that $V_\varphi(\mathbf{A}) = \mathrm{t}$ (i.e., **A** is satisfiable in $\mathfrak{A}$) and we have considered whether for all assignments φ, $V_\varphi(\mathbf{A}) = \mathrm{t}$, i.e., if **A** is true in $\mathfrak{A}$. Consider the following formula in reference to our example:

$$\begin{aligned} &(\forall \mathbf{x})((\mathbf{f}_1{}^2(\mathbf{a}_0, \mathbf{y}) = \mathbf{y}) \supset (\mathbf{f}_1{}^2(\mathbf{x}, \mathbf{a}_1) = \mathbf{x})) \\ &\quad \supset ((\mathbf{f}_1{}^2(\mathbf{a}_0, \mathbf{y}) = \mathbf{y}) \supset (\forall \mathbf{x})(\mathbf{f}_1{}^2(\mathbf{x}, \mathbf{a}_1) = \mathbf{x})) \end{aligned} \qquad (*)$$

Exercise 10.19 (a) Show that for any assignment φ, $V_\varphi(\mathbf{f}_1{}^2(\mathbf{a}_0, \mathbf{y}) = \mathbf{y}) = \mathrm{t}$ and $V_\varphi(\mathbf{f}_1{}^2(\mathbf{x}, \mathbf{a}_1) = \mathbf{x}) = \mathfrak{f}$. (b) Does there exist an assignment φ such that V_φ

of the given formula is t? (i.e., is it satisfiable in $\mathfrak{N}$?). (c) Is V_φ of it true for all assignments φ? (d) If $\mathfrak{A}$ were any relational system for $\mathsf{L}(r)$ other than $\mathfrak{N}$ would (∗) be true in $\mathfrak{A}$?

Suppose we consider all relational systems which are relational systems for $\mathsf{L}(r)$. A third question one can ask of a formula **A** of $\mathsf{F}(r)$, say (∗) above, is whether it is true in all relational systems for its language.

10.6 UNIVERSALLY VALID FORMULAS

A formula **A** in a language $\mathsf{L}(*)$ is *universally valid* if it is true in every relational system which is a relational system for $\mathsf{L}(*)$. Denote by $\mathscr{V}(*)$ the set of all universally valid formulas on a language $\mathsf{L}(*)$. Notice that if the next lemma is true, then (∗) in the previous section is universally valid.

Lemma 10.5 Any formula **A** of the form

$$(\forall \mathbf{x_i})(\mathbf{B} \supset \mathbf{C}(\mathbf{x_i})) \supset (\mathbf{B} \supset (\forall \mathbf{x_i})\mathbf{C}(\mathbf{x_i})),$$

where **B** contains no free occurrence of $\mathbf{x_i}$, is universally valid.

PROOF To show that **A** is universally valid, we assume that it is not and try to get a contradiction. In assuming that **A** is not universally valid, we are claiming that there is a relational system for $\mathsf{L}(*)$, $\mathfrak{A}$, and an assignment φ such that

$$V_\varphi\big((\forall \mathbf{x_i})(\mathbf{B} \supset \mathbf{C}(\mathbf{x_i})) \supset (\mathbf{B} \supset (\forall \mathbf{x_i})\mathbf{C}(\mathbf{x_i}))\big) = \mathfrak{f} \qquad \text{in} \quad \mathfrak{A}$$

That can be the case if and only if

$V_\varphi((\forall(\mathbf{x_i})(\mathbf{B} \supset \mathbf{C}(\mathbf{x_i})) = \mathrm{t}$ and $V_\varphi((\mathbf{B} \supset (\forall \mathbf{x_i})\mathbf{C}(\mathbf{x_i})) = \mathfrak{f}$

which can be the case if and only if for all assignments φ' such that for all $j \neq i$,

$$\varphi'(\mathbf{x_j}) = \varphi(\mathbf{x_j})$$

and

$$V_{\varphi'}(\mathbf{B} \supset \mathbf{C}(\mathbf{x_i})) = \mathrm{t}$$

which can be the case if and only if

$V_\varphi(\mathbf{B}) = \mathrm{t}$ and $V_\varphi((\forall \mathbf{x_n})\mathbf{C}(\mathbf{x_i})) = \mathfrak{f}$

since $\mathbf{x_i}$ does not occur free in **B**, $\varphi' = \varphi$ on all free variables in **B**, so by Lemma 10.4

$$V_{\varphi'}(\mathbf{B}) = V_\varphi(\mathbf{B}) = \mathrm{t}.$$

which is the case if and only if for *all* assignments φ'' such that for all $j \neq i$,

$$\varphi''(\mathbf{x_j}) = \varphi(\mathbf{x_j}).$$

if and only if

(i) $V_{\varphi'}(\mathbf{B}) = \mathfrak{t}$ and $V_{\varphi'}(\mathbf{C}(\mathbf{x_i})) = \mathfrak{t}$, or

(ii) $V_{\varphi'}(\mathbf{B}) = \mathfrak{f}$ and $V_{\varphi'}(\mathbf{C}(\mathbf{x_i})) = \mathfrak{t}$, or

(iii) $V_{\varphi'}(\mathbf{B}) = \mathfrak{f}$ and $V_{\varphi'}(\mathbf{C}(\mathbf{x_i})) = \mathfrak{f}$.

Therefore, cases (i) and (ii) in the left column cannot hold.

But that means that the set of assignments each of which is denoted by φ' is the same as the set each of which is denoted by φ''.
Thus, in the left column, we have

$$V_{\varphi'}(\mathbf{C}(\mathbf{x_i})) = \mathfrak{t}$$

and in this column we have

$$V_{\varphi''}(\mathbf{C}(\mathbf{x_i})) = \mathfrak{f}.$$

Contradiction. ∎

Notice that in showing the formula to be universally valid we never had to refer to the actual relational system or values of the assignments. Just as for a propositional language we have certain formulas (tautologies) which have the truth value $\mathfrak{t}$ no matter what truth values are assigned to its propositional variables, for first-order languages we have certain formulas (the universally valid formulas) which are true no matter what elements of what universe are assigned to its terms. That is, such formulas are true regardless of context. Naturally, we wish to distinguish those formulas which are true regardless of context from those whose truth depends on the context.

Exercise 10.20 (a) Contemplate the following statement, whose truth is obvious. Let S be a set, and let f, g_1, g_2, g_3, and h be functions on S^n, for the appropriate n, to S, where h is defined by

$$h(x_1, x_2, x_3, x_4) = f(g_1(x_1, x_2), g_2(x_1, x_2, x_3, x_4), g_3(x_4))$$

Then the image of h is a subset of the image of f. Thus, if all the elements of the image of f hold in some relation, then all the elements of the image of h must also hold in the same relation. Think of some real example. (b) Prove the following: Suppose $\mathfrak{A}$ is a relational system for $\mathsf{L}(*)$ by means of Φ. Let $\bar{\mathbf{t}}$ be any term and let $\mathbf{t}$ be a term containing $\mathbf{x_i}$. Let $\mathbf{t}'$ be the term which results by substituting $\bar{\mathbf{t}}$ for $\mathbf{x_i}$ in $\mathbf{t}$. Let $R = \{V_\varphi(\mathbf{t}) \mid \text{for all assignments } \varphi\}$ and let $S = \{V_\varphi(\mathbf{t}') \mid \text{for all assignments } \varphi\}$. Then $S \subseteq R$.

Lemma 10.6 Suppose $\mathsf{L}(*)$, $\mathfrak{A}$, and Φ given as usual. Let $\mathbf{A}(\mathbf{x_i})$ be a formula in which $\mathbf{x_i}$ occurs free. Let $\bar{\mathbf{t}}$ be a term whose individual variables are

$\mathbf{x_{i_1}}, \ldots, \mathbf{x_{i_n}}$ and which is free for $\mathbf{x_i}$ in $\mathbf{A}$. For any pair of assignments φ and ψ such that $\psi(\mathbf{x_i}) = V_\varphi(\bar{\mathbf{t}})$ and $\psi(\mathbf{x_j}) = \varphi(\mathbf{x_j})$ for all $j \neq i$,

$$V_\psi(\mathbf{A}(\mathbf{x_i})) = V_\varphi(\mathbf{A}(\bar{\mathbf{t}}))$$

Proof By induction on the formation of $\mathbf{A}(\mathbf{x_i})$.

1. *Terms*: If $\mathbf{x_i}$ is the term, then by definition, $V_\psi(\mathbf{x_i}) = V_\varphi(\bar{\mathbf{t}})$. If $\mathbf{t_1}, \ldots, \mathbf{t_k}$ are terms such that $\mathbf{x_i}$ occurs in some of them, let $\mathbf{t_1}', \ldots, \mathbf{t_k}'$ be the terms which result from replacing $\mathbf{x_i}$ by $\bar{\mathbf{t}}$. By definition,

$$V_\psi(\mathbf{f_j^k}(\mathbf{t_1}, \ldots, \mathbf{t_k})) = \Phi\mathbf{f_j^k}(V_\psi(\mathbf{t_1}), \ldots, V_\psi(\mathbf{t_k}))$$

By induction hypothesis, $V_\psi(\mathbf{t_j}) = V_\varphi(\mathbf{t_j}')$, $1 \le j \le k$, so the right side of the equation equals $\Phi\mathbf{f_j^k}(V_\varphi(\mathbf{t_1}'), \ldots, V_\varphi(\mathbf{t_k}'))$, which is $V_\varphi(\mathbf{f_j^k}(\mathbf{t_1}', \ldots, \mathbf{t_k}'))$.

2. *Atomic formulas*: If $\mathbf{A}(\mathbf{x_i})$ is $\mathbf{P_j^k}(\mathbf{t_1}, \ldots, \mathbf{t_k})$, then as above, $\mathbf{A}(\bar{\mathbf{t}})$ is $\mathbf{P_j^k}(\mathbf{t_1}', \ldots, \mathbf{t_k}')$ and the proof can easily be done by the reader, who can also supply the case that $\mathbf{A}(\mathbf{x_i})$ is $(\mathbf{t_1} = \mathbf{t_2})$.

3. *Formulas*: (a) For the cases that $\mathbf{A}(\mathbf{x_i})$ is $\sim\mathbf{B}(\mathbf{x_i})$ or $\mathbf{A}(\mathbf{x_i})$ is $(\mathbf{B}(\mathbf{x_i}) \supset \mathbf{C}(\mathbf{x_i}))$, the reader can carry out the proof from the induction hypothesis.

(b) Consider $V_\psi((\forall \mathbf{x_k})\mathbf{B}(\mathbf{x_i}))$. Since $\mathbf{x_i}$ has free occurrences in $\mathbf{A}(\mathbf{x_i})$, $k \neq i$, and since $\bar{\mathbf{t}}$ is free for $\mathbf{x_i}$ in $\mathbf{A}(\mathbf{x_i})$, $k \neq i_1, \ldots, i_n$. To determine whether this is $\mathfrak{t}$ or $\mathfrak{f}$, we must consider the set S_ψ of all assignments which agree with ψ except possibly at $\mathbf{x_k}$. Thus, for each element a of the universe A, there is a $\psi_a' \in S_\psi$ such that $\psi_a'(\mathbf{x_k}) = a$ and $\psi_a'(\mathbf{x_j}) = \psi(\mathbf{x_j})$ for all $j \neq k$. In particular, then, $\psi_a'(\mathbf{x_i}) = \psi(\mathbf{x_i}) = V_\varphi(\bar{\mathbf{t}})$. For each such ψ_a', let φ_a' be the assignment such that $\varphi_a'(\mathbf{x_k}) = \psi_a'(\mathbf{x_k})$ and $\varphi_a'(\mathbf{x_j}) = \varphi(\mathbf{x_j})$ for all $j \neq k$. Claim that $\psi_a'(\mathbf{x_i}) = V_{\varphi'a}(\bar{\mathbf{t}})$ and $\psi_a'(\mathbf{x_j}) = \varphi_a'(\mathbf{x_j}), j \neq i$: Since $k \neq i_1, \ldots, i_n$, the individual variables of $\bar{\mathbf{t}}$, $\varphi_a'(\mathbf{x_{i_j}}) = \varphi(\mathbf{x_{i_j}})$ and so by Lemma 10.4, $V_{\varphi'a}(\bar{\mathbf{t}}) = V_\varphi(\bar{\mathbf{t}})$. But $\psi_a'(\mathbf{x_i}) = \psi(\mathbf{x_i}) = V_\varphi(\bar{\mathbf{t}}) = V_{\varphi'a}(\bar{\mathbf{t}})$. Also, $\psi_a'(\mathbf{x_j}) = \psi(\mathbf{x_j}) = \varphi(\mathbf{x_j}) = \varphi_a'(\mathbf{x_j})$ for $j \neq i$. Thus, by the induction hypothesis, for all $a \in A$,

$$V_{\psi'a}(\mathbf{B}(\mathbf{x_i})) = V_{\varphi'a}(B(\bar{\mathbf{t}}))$$

Since ψ_a' agrees with ψ everywhere except possibly $\mathbf{x_k}$ and since φ_a' agrees with φ everywhere except possibly $\mathbf{x_k}$, $V_\psi((\forall \mathbf{x_k})\mathbf{B}(\mathbf{x_i})) = V_\varphi((\forall \mathbf{x_k})\mathbf{B}(\bar{\mathbf{t}}))$. ▌

Exercise 10.21 Let $\mathbf{A}$ be a quantifier-free formula with $\mathbf{x_{i_1}}, \ldots, \mathbf{x_{i_k}}$ as its individual variables. Let $\mathbf{B}$ be the formula obtained from $\mathbf{A}$ by substituting the terms $\mathbf{t_{i_1}}, \ldots, \mathbf{t_{i_k}}$ for the variables in $\mathbf{A}$, respectively. Let φ be any assignment and let ψ be an assignment such that $\psi(\mathbf{x_{i_j}}) = V_\varphi(\mathbf{t_{i_j}})$, $1 \le j \le k$. Show that $V_\psi(\mathbf{A}) = V_\varphi(\mathbf{B})$.

Theorem 10.3 Let $\mathbf{A}(\mathbf{x_i})$ be a formula in which $\mathbf{x_i}$ occurs free and $\bar{\mathbf{t}}$ be a term free for $\mathbf{x_i}$ in $\mathbf{A}(\mathbf{x_i})$. If $\mathbf{A}(\mathbf{x_i})$ is universally valid, then $\mathbf{A}(\bar{\mathbf{t}})$ is universally valid.

Exercise 10.22 Prove the theorem.

Exercise 10.23 Show that if $\mathbf{A(x_i)}$ is a formula in which $\mathbf{x_i}$ has free occurrences and $\bar{\mathbf{t}}$ is a term *not* free for $\mathbf{x_i}$ in $\mathbf{A(x_i)}$, then even if $\mathbf{A(x_i)}$ is true in a relational system $\mathfrak{A}$, $\mathbf{A(\bar{t})}$ need not be true in $\mathfrak{A}$. [Hint: Try to use an $\mathbf{A(x_i)}$ of the form $\mathbf{(\exists x_j)B(x_i)}$.]

Lemma 10.7 The following formulas are universally valid:

(1) $\mathbf{(\forall x_i)(A(x_i) \supset B(x_i)) \supset ((\forall x_i)A(x_i) \supset (\forall x_i)B(x_i))}$.
(2) $\mathbf{(\forall x_i)(A(x_i) \wedge B(x_i)) \equiv ((\forall x_i)A(x_i) \wedge (\forall x_i)B(x_i))}$.

PROOF Before giving a formal proof of (1), let us see what the natural set-theoretic interpretation of the two formulas is. The first one says that if A is a subset of B [i.e., $\mathbf{(\forall x_i)(A(x_i) \subset B(x_i))}$], then, if all x are in A, all x must be in B. Notice that to make the right side false, we must have an interpretation in which all x are in A but not all are in B; however, this means that A is not a subset of B, i.e., the left side is also false. The second formula says that every x is in $A \cap B$ [i.e. $\mathbf{(\forall x_i)(A(x_i) \wedge B(x_i))}$] if and only if every x is in A and every x is in B. We show that (1) is universally valid by assuming that it is not. Thus, for some relational system $\mathfrak{A}$ and valuation V_φ, we have that

$V_\varphi(\mathbf{(\forall x_i)(A(x_i) \supset B(x_i))}) = \mathrm{t}$ and $V_\varphi(\mathbf{(\forall x_i)A(x_i) \supset (\forall x_i)B(x_i)}) = \mathfrak{f}$

Consider the set S_φ of all assignments φ' which differ from φ at most on $\mathbf{x_i}$.
Then for all $\varphi' \in S_\varphi$,

$$V_{\varphi'}(\mathbf{A(x_i) \supset B(x_i)}) = \mathrm{t}$$

so in particular there is *no* $\varphi'' \in S_\varphi$ such that

$$V_{\varphi''}(\mathbf{A(x_i)}) = \mathrm{t}$$

and

$$V_{\varphi''}(\mathbf{B(x_i)}) = \mathfrak{f}.$$

This means that

$V_\varphi(\mathbf{(\forall x_i)A(x_i)}) = \mathrm{t}$ and $V_\varphi(\mathbf{(\forall x_i)B(x_i)}) = \mathfrak{f}$

Therefore, for every $\varphi' \in S_\varphi$,

$$V_{\varphi'}(\mathbf{A(x_i)}) = \mathrm{t}$$

and in particular,

$$V_{\varphi''}(\mathbf{A(x_i)}) = \mathrm{t}.$$

Therefore, there is some $\varphi'' \in S_\varphi$ for which

$$V_{\varphi''}(\mathbf{B(x_i)}) = \mathfrak{f}.$$

Contradiction. ▮

Exercise 10.24 Prove that the second formula of the lemma is universally valid.

Exercise 10.25 Show that

$$((\forall x_i)A(x_i) \supset (\forall x_i)B(x_i)) \supset (\forall x_i)(A(x_i) \supset B(x_i))$$

is not universally valid.

We extend the definition of logical equivalence given for propositional formulas so as to include first-order formulas. Two first-order formulas, **A** and **B** are defined to be *logically equivalent* if $(A \equiv B)$ is universally valid.

Exercise 10.26: (1) Show that if **A** and **B** are logically equivalent, then for every relational system $\mathfrak{A}$, **A** is true in $\mathfrak{A}$ if and only if **B** is true in $\mathfrak{A}$. (2) That the converse is false can be shown as follows: Suppose the language has two unary predicate symbols **P** and **Q**. Let **A** be the formula $(P(x) \wedge (\exists z) \sim P(z))$ and **B** be the formula $(Q(x) \wedge (\exists z) \sim Q(z))$. It must be shown that

$$\text{for all } \mathfrak{A}, \quad A \text{ is true in } \mathfrak{A} \quad \text{if and only if} \quad B \text{ is true in } \mathfrak{A} \qquad (*)$$

does not imply that

$$(A \equiv B) \quad \text{is true in all} \quad \mathfrak{A} \qquad (**)$$

(a) Show that neither **A** nor **B** is true in any relational system for the language. (b) Consider the relational system $\mathfrak{A}$ with three elements $\{a, b, c\}$ and two unary relations R_1 and R_2, where R_1 is $\{a\}$ and R_2 is $\{b\}$. Find an assignment φ so that $V_\varphi(A \equiv B) = \mathfrak{f}$ in $\mathfrak{A}$. (c) Show that (*) does not imply (**).

Lemma 10.8 The following pairs of formulas are logically equivalent:

1. $(\forall x_i)A(x_i) \supset B$ and $(\exists x_j)(A(x_j) \supset B)$, x_j not free in $A(x_i)$ or **B**.
2. $(\exists x_i)A(x_i) \supset B$ and $(\forall x_j)(A(x_j) \supset B)$, x_j not free in $A(x_i)$ or **B**.
3. $(A \supset (\forall x_i)B(x_i))$ and $(\forall x_j)(A \supset B(x_j))$, x_j not free in **A** or $B(x_i)$.
4. $(A \supset (\exists x_i)B(x_i))$ and $(\exists x_j)(A \supset B(x_j))$, x_j not free in **A** or $B(x_i)$.

PROOF 1. Suppose they are not logically equivalent. Then, for some relational system $\mathfrak{A}$ and valuation V_φ, V_φ of one is $\mathfrak{t}$ and V_φ of the other is $\mathfrak{f}$. Notice first that by virtue of the definition of the existential quantifier and the definition of logically equivalent for propositional formulas, $(\forall x_i) A(x_i) \supset B$ is logically equivalent to $(\exists x_i) \sim A(x_i) \vee B$ and $(\exists x_j)(A(x_j) \supset B)$ is logically equivalent to $(\exists x_j)(\sim A(x_j) \vee B)$. We will prove the lemma using the newly introduced pair of formulas.

(a) Suppose

$$V_\varphi((\exists x_i) \sim A(x_i) \vee B) = \mathfrak{f} \qquad \text{and} \qquad V_\varphi((\exists x_j)(\sim A(x_j) \vee B)) = \mathfrak{t}$$

This holds if and only if

$$V_\varphi((\exists x_i) \sim A(x_i)) = \mathfrak{f}$$

and

$$V_\varphi(B) = \mathfrak{f},$$

but this is the case if and only if there is an assignment φ' differing from φ at most at x_i such that

$$V_{\varphi'}(\sim A(x_i)) = \mathfrak{f}$$

Thus,

$$V_{\varphi'}(A(x_i)) = t \text{ and } V_\varphi(B) = \mathfrak{f}.$$

if and only if there is an assignment φ'' differing from φ at most at x_j such that

$$V_{\varphi''}(\sim A(x_j) \vee B) = t,$$

which is the case if and only if

$$V_{\varphi''}(\sim A(x_j)) = t \text{ or } V_{\varphi''}(B) = t.$$

Thus,

$$V_{\varphi''}(A(x_j)) = \mathfrak{f} \text{ or } V_{\varphi''}(B) = t.$$

Since x_j is not free in **B** and since φ'' agrees with φ on all individual variables except possibly x_j, then φ'' and φ agree on all the free variables of **B** and so by Lemma 10.4, $V_\varphi(B) = V_{\varphi''}(B)$. Consequently, for the analysis in both columns to be maintained, it must be the case that $V_{\varphi''}(A(x_j)) = \mathfrak{f}$. Notice that φ' and φ'' agree on all individual variables except possibly x_i and x_j. Further, $A(x_j)$ has no free occurrence of x_i [because by definition, x_j has replaced *all* free occurrences of x_i in $A(x_i)$] and $A(x_i)$ has no free occurrences of x_j by hypothesis of the lemma. Therefore, φ' and φ'' agree on the free variables of $A(x_j)$ and $A(x_i)$, so by Lemma 10.4, $V_{\varphi'}(A(x_i)) = V_{\varphi''}(A(x_j))$. But this contradicts $V_{\varphi'}(A(x_i)) = t$ in the left column and $V_{\varphi''}(A(x_j)) = \mathfrak{f}$ in the right.

(b) Suppose $V_\varphi((\exists x_j) \sim A(x_i) \vee B) = t$ and $V_\varphi((\exists x_j)(\sim A(x_j) \vee B)) = t$. Since the analysis of part (a) was all in terms of "if and only if," (b) follows immediately.

3. Suppose they are not logically equivalent. Then there, is some relational system $\mathfrak{A}$ and valuation V_φ for which V_φ of one formula is t and of the other $\mathfrak{f}$. Notice that $(A \supset (\forall x_i)B(x_i))$ is logically equivalent to $(\sim A \vee (\forall x_i)B(x_i))$ and $(\forall x_j)(A \supset B(x_j))$ is logically equivalent to $(\forall x_j)(\sim A \vee B(x_j))$.

(a) Suppose

$$V_\varphi(\sim A \vee (\forall x_i)B(x_i)) = \mathfrak{f} \quad \text{and} \quad V_\varphi((\forall x_j)(\sim A \vee B(x_j))) = t$$

if and only if

$$V_\varphi(\sim A) = \mathfrak{f} \text{ and } V_\varphi((\forall x_i)B(x_i)) = \mathfrak{f}$$

if and only if

$$V_\varphi(A) = t$$

if and only if for all φ'' in the set $S_\varphi^{\,j}$ of all assignments which differ from φ at most at x_j,

$$V_{\varphi''}(\sim A \vee B(x_j)) = t$$

and for all φ' in the set $S_\varphi{}^i$ of all assignments which differ from φ at most at $\mathbf{x_i}$,

$$V_{\varphi'}(\mathbf{B}(\mathbf{x_i})) = \mathfrak{f}.$$

Thus,

$$V_\varphi(\mathbf{A}) = \mathfrak{t} \text{ and } V_{\varphi'}(\mathbf{B}(\mathbf{x_i})) = \mathfrak{f}$$

for all φ'.

if and only if

$$V_{\varphi''}(\sim\mathbf{A}) = \mathfrak{t} \text{ or } V_{\varphi''}(\mathbf{B}(\mathbf{x_j})) = \mathfrak{t}.$$

Thus, for all φ'',

$$V_{\varphi''}(\mathbf{A}) = \mathfrak{f} \text{ or } V_{\varphi''}(\mathbf{B}(\mathbf{x_j})) = \mathfrak{t}.$$

Since $\mathbf{x_j}$ is not free in **A**, φ and φ'' agree on the free variables of **A** and so $V_\varphi(\mathbf{A}) = V_{\varphi''}(\mathbf{A})$. Therefore, to maintain the analysis in both columns, it must be that $V_{\varphi''}(\mathbf{B}(\mathbf{x_j})) = \mathfrak{t}$. Notice that any $\varphi \in S_\varphi{}^i$ and any $\varphi' \in S_\varphi{}^j$ differ at most on $\mathbf{x_i}$ and $\mathbf{x_j}$. But since $\mathbf{x_i}$ is not free in $B(\mathbf{x_j})$ and $\mathbf{x_j}$ is not free in $B(\mathbf{x_i})$, φ and φ' agree on the free variables of **B** and so $V_\varphi(\mathbf{B}) = V_{\varphi'}(\mathbf{B})$. Thus, it is impossible for the last statements in the two columns both to be true. ▮

Exercise 10.27 Prove the lemma for the second and fourth pairs of formulas.

10.7 DEFINITION OF AND INTRODUCTION TO FIRST-ORDER THEORIES

A *set* Γ *of first-order formulas is deductively closed* if the two following conditions are met: (1) if **A** and **(A ⊃ B)** are in Γ, then **B** is in Γ, and (2) if **A** is in Γ, then, for any individual variable $\mathbf{x_i}$, $(\forall\mathbf{x_i})\mathbf{A}$ is in Γ.

Theorem 10.4 The set of universally valid formulas on a given first-order language is deductively closed.

Exercise 10.28 Prove the theorem.

The *deductive closure* of a set of formulas Γ is the smallest deductively closed set of formulas containing Γ. Denote the deductive closure of Γ by D(Γ).

A *first-order* (logical) *theory* is any deductively closed set of formulas (the theorems of the theory) on a first-order language which includes the set of universally valid formulas on that language. A first-order theory is *consistent* (as before) if there is no formula **A** such that both **A** and **∼A** belong to the theory. A first-order theory is *complete* if for every pair of *closed* formulas **A** and **∼A**, at least one of them belongs to the theory. A first-order theory is *recursively axiomatizable* (as before) if there is a recursive set of formulas whose deductive closure is that theory. A first-order theory is *decidable* if the set (of Gödel numbers) of its theorems is a recursive set.

Theorem 10.5 Let $\mathfrak{A}$ be a relational system for $\mathsf{L}(*)$ and let Γ be the set of formulas of $\mathsf{L}(*)$ which are true in $\mathfrak{A}$. Then Γ is a consistent, complete first-order theory.

Exercise 10.29 Prove the theorem. (It is an easy consequence of the definitions.)

Theorem 10.6 If a deductively closed set of formulas has a model, then it is consistent.

Exercise 10.30 Prove the theorem. (Hint: By contradiction.)

Corollary If a first-order theory has a model, then it is consistent.

We say that a set of formulas Γ (on $\mathsf{L}(*)$) is consistent if there is no formula $\mathbf{A}$ such that both $\mathbf{A}$ and $\sim\mathbf{A}$ are in Γ. It should be clear from the foregoing that if T is a first-order theory and Γ is a set of formulas, then $\mathsf{D}(\mathsf{T} \cup \Gamma)$ is a first-order theory.

Exercise 10.31 Show that if a relational system $\mathfrak{A}$ is a model of Γ (on $\mathsf{L}(*)$), then (a) Γ is consistent; (b) $\mathfrak{A}$ is a model of $\mathsf{D}(\mathscr{V}(*) \cup \Gamma)$; and (c) if $\Gamma' \subset \Gamma$, then $\mathfrak{A}$ is a model of Γ'.

We will be considering several first-order theories and investigating the above properties in connection with them. Here, we mention briefly two cases.

1. The set of universally valid formulas $\mathscr{V}(*)$ on language $\mathsf{L}(*)$. The next chapter will be devoted to showing that $\mathscr{V}(*)$ is recursively axiomatizable and thus that the set of Gödel numbers of the theorems of $\mathscr{V}(*)$ is an r.e. set. By its very definition, it is consistent. Is it complete? $\mathscr{V}(*)$ is *not* decidable. That result will be shown in chapter XIV for only one particular language.

2. First-order number theory. Let $\mathsf{L}(N)$ be the first-order language with equality, the function symbols $\mathbf{f}_1{}^1$, $\mathbf{f}_1{}^2$, $\mathbf{f}_2{}^2$, and the name of a constant $\mathbf{a}_0$. Consider the relational system $\mathfrak{N} = \langle N; {}', +, \cdot\,; = \rangle$, where $'$ is the successor function and N is the natural numbers. Then $\mathfrak{N}$ is a rational system for $\mathsf{L}(N)$ in the obvious way, taking 0 for $\mathbf{a}_0$. The name "first-order number theory" (or "elementary number theory") is given to the set of sentences of $\mathsf{F}(N)$ that are true in $\mathfrak{N}$, and we shall denote that set by $\mathscr{N}$. In considering the possibility of finding a recursive axiomatization of $\mathscr{N}$, one feels fairly confident in trying as axioms, in addition to those which give a recursive axiomatization of $\mathscr{V}(N)$, a formalization of the Peano postulates. (This will be done in detail in a later chapter.) Let P be the set of formulas derivable from

these axioms. If one wishes this to be a recursive axiomatization of $\mathcal{N}$, then one must prove: "$\mathbf{A} \in \mathcal{N}$ if and only if $\vdash_{\mathsf{P}}\mathbf{A}$." If, on the other hand, one wishes that P not be a recursive axiomatization of $\mathcal{N}$, one approach to proving it would be to find a sentence $\mathbf{A} \in \mathcal{N}$ such that neither $\vdash_{\mathsf{P}}\mathbf{A}$ nor $\vdash_{\mathsf{P}} \sim\mathbf{A}$. (That is, to show that P is not complete.) Then, since if $\mathbf{A} \in \mathsf{F}(N)$, either $\mathbf{A}$ is true in $\mathfrak{N}$ or $\sim\mathbf{A}$ is true in $\mathfrak{N}$, we would have a sentence which is true in $\mathfrak{N}$ but not provable in P, i.e., P is not a recursive axiomatization of $\mathcal{N}$. One of the most famous theorems of the 20th century is that P is not complete and that, indeed, this is not due merely to a poor choice of axioms, but that *any* recursively axiomatizable theory which is sufficient to account for addition and multiplication and for which $\mathfrak{N}$ is a model is incomplete. (The theorem is, of course, the Gödel incompleteness theorem.) This result does not immediately imply that the set $\mathcal{N}$ is not r.e. nor that the set of theorems of P is not recursive. However, those two results do follow with a little effort. The incompleteness theorem, that P is undecidable, and that $\mathcal{N}$ is not r.e. will be studied in chapter XIV.

Chapter XI

FIRST-ORDER THEORIES WITHOUT EQUALITY

11.1 THE AXIOMS FOR V

Let L(∗) be a first-order language without equality. Our first concern is to give a set of axioms and rules of inference which will eventually be shown to be a recursive axiomatization of $\mathscr{V}(*)$.

The axioms are:

$(\mathbf{A} \supset (\mathbf{B} \supset \mathbf{A}))$
$((\mathbf{A} \supset (\mathbf{B} \supset \mathbf{C})) \supset ((\mathbf{A} \supset \mathbf{B}) \supset (\mathbf{A} \supset \mathbf{C})))$
$((\sim \mathbf{A} \supset \sim \mathbf{B}) \supset (\mathbf{B} \supset \mathbf{A}))$
$(\forall \mathbf{x}_i)\mathbf{A}(\mathbf{x}_i) \supset \mathbf{A}(\mathbf{t})$ if $\mathbf{t}$ is free for $\mathbf{x}_i$ in $\mathbf{A}$
$(\forall \mathbf{x}_i)(\mathbf{A} \supset \mathbf{B}) \supset (\mathbf{A} \supset (\forall \mathbf{x}_i)\mathbf{B})$ if $\mathbf{A}$ contains no free occurrence of $\mathbf{x}_i$

The rules of inference are:

If $\mathbf{A}$ and $\mathbf{A} \supset \mathbf{B}$, then $\mathbf{B}$ modus ponens (mp)
If $\mathbf{A}$, then $(\forall \mathbf{x}_i)\mathbf{A}$ universal generalization (ug)

[Remark that the above hold for all individual variables $\mathbf{x}_i$ and for all formulas **A**, **B**, and **C** of F(∗). As in P_0, the axioms are actually axiom schemes. For example, if L(∗) includes the predicate symbol $\mathbf{P}^2$ and the function symbol $\mathbf{f}^1$, then $(\forall \mathbf{x}_1)\mathbf{P}^2(\mathbf{x}_1, \mathbf{x}_2) \supset \mathbf{P}^2(\mathbf{x}_1, \mathbf{f}^1(\mathbf{x}_1))$ is an instance of the fourth scheme.]

Let V(∗) be the name of the logical theory whose axioms are those just given. Certainly it is a recursively axiomatizable theory. The set of theorems of V(∗) is to be just those formulas of F(∗) that are derivable from these axioms by means of the two rules of inference given, i.e., the deductive closure of the axioms. Notice that we do not yet know if the theory V(∗) is a first-order theory as defined in section 10.7. To say that the set of axioms just given is a recursive axiomatization of $\mathscr{V}(*)$ is equivalent to saying that the set of theorems derivable in V(∗) coincides with the set of formulas $\mathscr{V}(*)$. This is exactly what we wish to prove, and once it is proved we will have proved also that V(∗) is a first-order theory.

Since everything we will be doing will be based on V(∗), we define ⊢**A** to mean that **A** is a theorem of V(∗). The theory V(∗) is called *the predicate calculus* (for L(∗)). If the language L(∗) for V(∗) contains infinitely many predicate symbols but no function symbols or names of constants, then V(∗) is a *pure predicate calculus.* Let Γ be a set (finite or infinite) of formulas of L(∗) and define $\vdash_\Gamma$**A** to mean that **A** is derivable in the logical theory whose axioms are the formulas of Γ in addition to the axioms of V(∗). We will use "Γ" ambiguously, both to denote a set of formulas (not necessarily deductively closed) and to denote the theory whose theorems are D(V(∗)∪Γ) as just described. If the set of formulas Γ and the set of axioms of V(∗) are disjoint, then we refer to the set Γ as the *nonlogical axioms* of the theory and refer to the axioms of V(∗) as the *logical axioms.* If Γ is a finite set of formulas (actual formulas, not schemes), we say that the theory Γ is *finitely axiomatizable.* If Γ is a recursive set, then the theory Γ is, naturally, a recursively axiomatizable theory. We chose a temporary (and very nonstandard) name for theories whose axioms include the axioms of V(∗): let them be called "V(∗)-based theories." Once we have accomplished our task of showing that the set of formulas of $\mathscr{V}$(∗) equals the set of theorems of V(∗), we will have shown that all V(∗)-based theories are first-order theories and conversely. We will then immediately, and gladly, drop the name "V(∗)-based theory."

Exercise 11.1 Show that **(∼(A ⊃ B) ⊃ A)** and **(∼(A ⊃ B) ⊃ ∼B)** are provable in V(∗).

Exercise 11.2 Show that **A(t) ⊃ (∃**$\mathbf{x_i}$**)A(**$\mathbf{x_i}$**)**, where **t** is free for $\mathbf{x_i}$ is a theorem of V(∗).

Exercise 11.3 Is **(A(x) ⊃ (∀x)A(x))** a theorem of V(∗)?

Exercise 11.4 Let T be a V(∗)-based theory, **A** a formula, and **A′** the universal closure of **A**. Show that $\vdash_T$**A** if and only if $\vdash_T$**A′**.

Exercise 11.5 Prove that if V(∗) is not consistent, then the theorems of V(∗) do not coincide with the formulas of $\mathscr{V}$(∗). So what would be the wisest thing to try to prove next?

Theorem 11.1 V(∗) is consistent.

PROOF We wish to show that there is a mapping θ of F(∗) to F(P_0) such that θ(∼**A**) = ∼θ(**A**) and θ(**A ⊃ B**) = (θ(**A**) ⊃ θ(**B**)); then, if we could also show that $\vdash \mathbf{A}$ implies that $\theta(\mathbf{A})$ is a tautology, we would have shown that V(*) is consistent. For, if there was a formula **A** such that both $\vdash \mathbf{A}$ and $\vdash \sim\mathbf{A}$, then that would mean that both $\theta(\mathbf{A})$ and $\sim\theta(\mathbf{A})$ are tautologies, which is impossible. θ is a map which projects only a sort-of shadow of a formula **A**. Given **A**, $\theta(\mathbf{A})$ is to be the result of removing all terms, all quantifiers with their variables, and all the parentheses and commas associated with these. Thus, for example, if **A** is $(\forall \mathbf{x})(\exists \mathbf{y})(\mathbf{P}(\mathbf{x}, \mathbf{y}, \mathbf{a}) \supset \mathbf{Q}(\mathbf{x}))$, then $\theta(\mathbf{A})$ is $(\mathbf{P} \supset \mathbf{Q})$; or if **B** is $(\forall \mathbf{x}) \sim (\forall \mathbf{y})\mathbf{P}(\mathbf{x}, \mathbf{y}) \supset \mathbf{Q}(\mathbf{y}, \mathbf{a})$, then $\theta(\mathbf{B})$ is $\sim\mathbf{P} \supset \mathbf{Q}$. The demonstration that $\vdash \mathbf{A}$ implies $\theta(\mathbf{A})$ is a tautology is to be filled in by the reader. ∎

Theorem 11.2 If $\vdash \mathbf{A}$, then **A** is universally valid, i.e., $\mathbf{A} \in \mathscr{V}(*)$.

Exercise 11.6 Prove it. (Hint: The hard work was done in the previous chapter.)

Exercise 11.7 Is V(*) complete?

As we learned in our investigations of the propositional calculus, life is much easier if we have a deduction theorem.

Theorem 11.3 Let T be any V(*)-based theory, and let $\mathbf{A}_1, \ldots, \mathbf{A}_n$ be a set of closed formulas. If $\mathbf{A}_1, \ldots, \mathbf{A}_n \vdash_T \mathbf{B}$, then $\mathbf{A}_1, \ldots, \mathbf{A}_{n-1} \vdash_T (\mathbf{A}_n \supset \mathbf{B})$.

PROOF The proof is by induction on the number of steps in the proof from hypotheses, $\mathbf{A}_1, \ldots, \mathbf{A}_n \vdash_T \mathbf{B}$. In thinking about what is to be done, the reader should see that the proof is exactly like the proof of the deduction theorem for the propositional calculus, with one exception at the induction step. (That is, we assume that the theorem is true for all proofs from hypotheses of less than k steps and show that it is true for k steps.) The extra case that we must consider is if the kth step results from a previous step by a use of ug. In that case, we have

⋮

$\mathbf{A}_1, \ldots, \mathbf{A}_n \vdash_T \mathbf{C}$ — However, $\mathbf{A}_1, \ldots, \mathbf{A}_{n-1} \vdash_T (\mathbf{A}_n \supset \mathbf{C})$ induction hypothesis

$\mathbf{A}_1, \ldots, \mathbf{A}_{n-1} \vdash_T (\forall \mathbf{x})(\mathbf{A}_n \supset \mathbf{C}) \supset (\mathbf{A}_n \supset (\forall \mathbf{x})\mathbf{C})$ by axiom 5 since $\mathbf{A}_n$ is closed, **x** is not free in $\mathbf{A}_n$

⋮

$\mathbf{A}_1, \ldots, \mathbf{A}_n \vdash_T (\forall \mathbf{x})\mathbf{C}$ — $\mathbf{A}_1, \ldots, \mathbf{A}_{n-1} \vdash_T (\mathbf{A}_n \supset (\forall \mathbf{x})\mathbf{C})$ (mp) ∎

Corollary Let T be a V(*)-based theory and Γ a finite set of sentences (closed formulas), then $\Gamma \vdash_T \mathbf{B}$ if and only if $\vdash_{T \cup \Gamma} \mathbf{B}$.

Exercise 11.8 In the deduction theorem, is it necessary that the formulas that are the hypotheses be closed? Suppose **A** is not closed; is the following true: If $\mathbf{A} \vdash \mathbf{B}$, then $\vdash \mathbf{A} \supset \mathbf{B}$.

Lemma 11.1 If T is a consistent V(*)-based theory and **A** is a closed formula such that not $\vdash_T \mathbf{A}$, then the theory $T \cup \{\sim\mathbf{A}\}$ is consistent.

PROOF Suppose not. Then there are formulas **C** and $\sim\mathbf{C}$ such that $\vdash_{T \cup \{\sim A\}} \mathbf{C}$ and $\vdash_{T \cup \{\sim A\}} \sim\mathbf{C}$. By the corollary, this is equivalent to $\sim\mathbf{A} \vdash_T \mathbf{C}$ and $\sim\mathbf{A} \vdash_T \sim\mathbf{C}$. By the deduction theorem, $\vdash_T \sim\mathbf{A} \supset \mathbf{C}$ and $\vdash_T \sim\mathbf{A} \supset \sim\mathbf{C}$. Since it is a tautology, $\vdash_T (\sim\mathbf{A} \supset \mathbf{C}) \supset ((\sim\mathbf{A} \supset \sim\mathbf{C}) \supset \mathbf{A})$. Hence, by mp, $\vdash_T \mathbf{A}$, which contradicts the hypothesis of the lemma. ▮

Exercise 11.9 (a) Let T be a finitely axiomatizable V(*)-based theory with axioms $\mathbf{A}_1, \ldots, \mathbf{A}_k$ and let $\mathbf{A}_1', \ldots, \mathbf{A}_k'$ be the closures of each. Prove that $\vdash_T \mathbf{B}$ if and only if $\vdash \mathbf{A}_1' \supset (\mathbf{A}_2' \supset \ldots \supset (\mathbf{A}_k' \supset \mathbf{B}))$. (b) Let T_1 and T_2 be two V(*)-based theories on the same language. Show that if T_2 is undecidable and T_2 is a finite extension of T_1 (i.e., T_1 is a subtheory of T_2 and T_2 has only finitely many more axioms than T_1), then T_1 is undecidable.

Exercise 11.10 If T is a consistent, not complete V(*)-based theory, then there is a sentence **A** such that $T \cup \{\sim\mathbf{A}\}$ is a consistent V(*)-based theory and $T \cup \{\mathbf{A}\}$ is a consistent V(*)-based theory.

Exercise 11.11 If T is a decidable, consistent, not complete V(*)-based theory, then there is a sentence **A** such that $T \cup \{\mathbf{A}\}$ is a decidable, consistent V(*)-based theory.

We set out to show that V(*) is a recursive axiomatization of $\mathscr{V}(*)$ and so far we have shown that $V(*) \subseteq \mathscr{V}(*)$. That is, that provability in V(*) implies universal validity. So we must show that universal validity implies provability in V(*). But this is equivalent to showing that:

$$\text{if a formula is not provable in V(*),} \quad \text{then it is not universally valid} \qquad (*)$$

On the one hand, for a formula **A** not to be provable in V(*) is equivalent to $V(*) \cup \{\sim\mathbf{A}\}$ being consistent, since V(*) is consistent. On the other hand, for a formula **A** not to be universally valid is equivalent to $V(*) \cup \{\sim\mathbf{A}\}$ having a model. Thus, (*) is equivalent to: if $V(*) \cup \{\sim\mathbf{A}\}$ is consistent, then $V(*) \cup \{\sim\mathbf{A}\}$ has a model. The truth of that statement is certainly established if we show:

$$\text{for any formula } \mathbf{B}, \quad \text{if} \quad V(*) \cup \{\mathbf{B}\} \text{ is consistent,} \quad \text{then} \quad V(*) \cup \{\mathbf{B}\} \text{ has a model} \quad (**)$$

In view of the forgoing analysis, we change our previously established aim of proving (*) to proving (**) from which (*) follows immediately.

11.2 ADDING EXTRA NAMES OF CONSTANTS

Suppose $\mathsf{L}(*)$ is a first-order language and $\mathsf{L}(*') = \mathsf{L}(*) \cup \{\mathbf{c}_1, \mathbf{c}_2, \ldots\}$, where the $\mathbf{c}_i$ are names of constants which do not belong to $\mathsf{L}(*)$. Obviously, $\mathsf{F}(*) \subseteq \mathsf{F}(*')$. Suppose we have a $\mathsf{V}(*)$-based theory, $\mathsf{V}(*) \cup \Gamma$, and consider what is meant by the theory $\mathsf{V}(*') \cup \Gamma$ for the same set Γ. [Notice that $\Gamma \subseteq \mathsf{F}(*)$.] The axioms of $\mathsf{V}(*')$ are, of course, just the usual schemes, but the set of formulas which are instances of each scheme properly includes the set of formulas whose elements are all instances of each of the axiom schemes of $\mathsf{V}(*)$. Obviously, $\mathsf{D}(\mathsf{V}(*') \cup \Gamma)$ includes $\mathsf{D}(\mathsf{V}(*) \cup \Gamma)$. But one might wonder about the formulas in $\mathsf{F}(*') - \mathsf{F}(*)$ which belong to $\mathsf{D}(\mathsf{V}(*') \cup \Gamma)$.

Theorem 11.4 Let $\mathsf{L}(*') = \mathsf{L}(*) \cup \{\mathbf{c}_1, \mathbf{c}_2, \ldots\}$, where

$$\mathsf{L}(*) \cap \{\mathbf{c}_1, \mathbf{c}_2, \ldots\} = \varnothing.$$

If $\mathbf{A} \in \mathsf{F}(*)$ and has $\mathbf{x}_{i_1}, \ldots, \mathbf{x}_{i_n}$ as its free variables, then for all n-tuples $(\mathbf{c}_{j_1}, \ldots, \mathbf{c}_{j_n})$ from $\{\mathbf{c}_1, \mathbf{c}_2, \ldots\}$,

$$\vdash_{\mathsf{V}(*) \cup \Gamma} \mathbf{A}(\mathbf{x}_{i_1}, \ldots, \mathbf{x}_{i_n}) \qquad \text{if and only if} \qquad \vdash_{\mathsf{V}(*') \cup \Gamma} \mathbf{A}(\mathbf{c}_{j_1}, \ldots, \mathbf{c}_{j_n})$$

PROOF To make the writing easier, let $\mathsf{V}(*) \cup \Gamma$ be called T and let $\mathsf{V}(*') \cup \Gamma$ be called T'. (1) If $\vdash_{\mathsf{T}} \mathbf{A}(\mathbf{x}_{i_1}, \ldots, \mathbf{x}_{i_n})$, then $\vdash_{\mathsf{T}'} \mathbf{A}(\mathbf{x}_{i_1}, \ldots, \mathbf{x}_{i_n})$, and so by ug, $\vdash_{\mathsf{T}'} (\forall \mathbf{x}_{i_1}) \ldots (\forall \mathbf{x}_{i_n}) \mathbf{A}(\mathbf{x}_{i_1}, \ldots, \mathbf{x}_{i_n})$, and so by axiom 5, since $\mathbf{c}_j$ are free for everything,

$$\vdash_{\mathsf{T}'} (\forall \mathbf{x}_{i_1}) \ldots (\forall \mathbf{x}_{i_n}) \mathbf{A}(\mathbf{x}_{i_1}, \ldots, \mathbf{x}_{i_n}) \supset \mathbf{A}(\mathbf{c}_{j_1}, \ldots, \mathbf{c}_{j_n})$$

then by mp $\vdash_{\mathsf{T}'} \mathbf{A}(\mathbf{c}_{j_1}, \ldots, \mathbf{c}_{j_n})$.

(2) To say $\vdash_{\mathsf{T}'} \mathbf{A}(\mathbf{c}_{j_1}, \ldots, \mathbf{c}_{j_n})$ means that there is a sequence of formulas constituting a proof of $\mathbf{A}(\mathbf{c}_{j_1}, \ldots, \mathbf{c}_{j_n})$ in T'. Let $\mathbf{c}_{j_1}, \ldots, \mathbf{c}_{j_n}, \ldots, \mathbf{c}_t$ be a list of all the constants from $\{\mathbf{c}_1, \mathbf{c}_2, \ldots\}$ that appear in the proof. Let $\mathbf{x}_{m_1}, \ldots, \mathbf{x}_{m_n}, \ldots, \mathbf{x}_{m_t}, \mathbf{x}_{m_{t+1}}, \ldots, \mathbf{x}_{m_{t+n}}$ be a set of individual variables that do not occur in the proof and are not any of the $\mathbf{x}_{i_1}, \ldots, \mathbf{x}_{i_n}$. In each formula in the proof, replace each occurrence of the $\mathbf{c}_{j_k}$ by $\mathbf{x}_{m_k}$ for all k between 1 and t. At the same time, if $\mathbf{x}_{i_j} \in \{\mathbf{x}_{i_1}, \ldots, \mathbf{x}_{i_n}\}$ appears in the proof, replace it by $\mathbf{x}_{m_{t+j}}$. Because the axioms of T' are exactly those of T but written on $\mathsf{L}(*)$, the resulting sequence of formulas is a proof in T. (See the next exercise.) Thus, from $\vdash_{\mathsf{T}'} \mathbf{A}(\mathbf{c}_{i_1}, \ldots, \mathbf{c}_{i_n})$, we have $\vdash_{\mathsf{T}} \mathbf{A}(\mathbf{x}_{m_1}, \ldots, \mathbf{x}_{m_n})$. By axiom 4

$$\vdash_{\mathsf{T}} (\forall \mathbf{x}_{m_1}) \ldots (\forall \mathbf{x}_{m_n}) \mathbf{A}(\mathbf{x}_{m_1}, \ldots, \mathbf{x}_{m_n}) \supset \mathbf{A}(\mathbf{x}_{i_1}, \ldots, \mathbf{x}_{i_n})$$

since, by construction, each $\mathbf{x}_{i_j}$ is free for everything. Therefore, by ug and mp, $\vdash_{\mathsf{T}} \mathbf{A}(\mathbf{x}_{i_1}, \ldots, \mathbf{x}_{i_n})$. ▮

Exercise 11.12 Let $\{\mathbf{A}_1, \ldots, \mathbf{A}_k\}$ be a proof of $\mathbf{A}(\mathbf{c}_{i_1}, \ldots, \mathbf{c}_{i_n})$ in T' (where the languages and theories are defined as in the theorem). Certain formulas of $\{\mathbf{A}_1, \ldots, \mathbf{A}_k\}$ are axioms of T'; let $\{\mathbf{A}_1, \ldots, \mathbf{A}_s\}$ be that set. Let $\mathbf{x}_{m_1}, \ldots, \mathbf{x}_{m_{t+n}}$ (as in the above proof) be a set of individual variables such that none occurs (free or bound) in any $\mathbf{A}_i$ of the proof of $\mathbf{A}(\mathbf{c}_{i_1}, \ldots, \mathbf{c}_{i_n})$. Let $\mathbf{c}_{i_1}, \ldots, \mathbf{c}_{i_n}, \ldots, \mathbf{c}_t$ be a list of the constants occurring in the $\mathbf{A}_i$. If $\mathbf{B} \in \mathsf{F}(*')$, let $\theta\mathbf{B}$ be the formula of $\mathsf{F}(*)$ that results by replacing each occurrence of $\mathbf{c}_{j_k}$ by $\mathbf{x}_{m_k}$ and each occurrence of $\mathbf{x}_{i_j}$ by $\mathbf{x}_{m_{t+j}}$ (also as in the proof above). Show that there is a proof in T of $\theta\mathbf{A}(\mathbf{c}_{i_1}, \ldots, \mathbf{c}_{i_n})$ using only the axioms $\theta\mathbf{A}_1, \ldots, \theta\mathbf{A}_s$. [Hint: The main difficulty is understanding the problem. After that, the proof is by induction on k, the number of steps in a proof of $\mathbf{A}(\mathbf{c}_{i_1}, \ldots, \mathbf{c}_{i_n})$.]

Corollary 1 Given the conditions of the theorem, the theory $\mathsf{V}(*) \cup \Gamma$ is consistent if and only the theory $\mathsf{V}(*') \cup \Gamma$ is consistent.

Corollary 2 Given the conditions of the theorem, the theory $\mathsf{V}(*) \cup \Gamma$ is decidable if and only if the theory $\mathsf{V}(*') \cup \Gamma$ is decidable.

Exercise 11.13 Prove the corollaries.

Exercise 11.14 (a) Let T be a $\mathsf{V}(*)$-based theory. Show that $\vdash_{\mathsf{T}} (\exists\mathbf{x})\mathbf{B}(\mathbf{x})$ if and only if $\vdash_{\mathsf{T}} (\exists\mathbf{y})\mathbf{B}(\mathbf{y})$ if $\mathbf{x}$ does not occur in $\mathbf{B}(\mathbf{y})$ and $\mathbf{y}$ does not occur in $\mathbf{B}(\mathbf{x})$.

(b) One of the following is true, one false; which is which, and why? If $\vdash_{\mathsf{T}} \mathbf{B}(\mathbf{a})$, then $\vdash_{\mathsf{T}} \mathbf{B}(\mathbf{y})$ if $\mathbf{y}$ is an individual variable not occurring in $\mathbf{B}(\mathbf{a})$. If $\vdash_{\mathsf{T}} \mathbf{B}(\mathbf{y})$, then $\vdash_{\mathsf{T}} \mathbf{B}(\mathbf{a})$ if $\mathbf{y}$ is an individual variable not occurring in $\mathbf{B}(\mathbf{a})$. (Think about why the situation here is different from that of the theorem.)

11.3 $\mathsf{V} = \mathscr{V}$: THE COMPLETENESS THEOREM

If T is a $\mathsf{V}(*)$-based theory and T' is also a $\mathsf{V}(*)$-based theory and if the theorems of T' include the theorems of T, then we say that T' is an extension of T.

Theorem 11.5 (Lindenbaum) Let $\mathsf{L}(*)$ be a first-order language and T a consistent $\mathsf{V}(*)$-based theory, then T has a complete, consistent extension T_{E}.

PROOF In order to have a complete $\mathsf{V}(*)$-based theory, it must be the case that for every closed formula $\mathbf{C}$, either $\mathbf{C}$ or $\sim\mathbf{C}$ is a theorem in the theory. So as to guarantee that this is the case, we will build up the extension by adding one closed formula at a time. Let $\mathbf{C}_1, \mathbf{C}_2, \ldots$ be an enumeration of all the closed formulas of $\mathsf{F}(*)$. We will define T_{E} inductively as follows. Let $\mathsf{T}_0 = \mathsf{T}$, and let T_1 be $\mathsf{T}_0 \cup \{\mathbf{C}_1\}$ if it is *not* the case that $\vdash_{\mathsf{T}_0} \sim\mathbf{C}_1$; otherwise, let $T_1 = T_0$. At the nth stage, define T_n to be $T_{n-1} \cup \{\mathbf{C_n}\}$ if it is not the case that $\vdash_{T_{n-1}} \sim\mathbf{C_n}$; otherwise, $T_n = T_{n-1}$. Notice that by a slight variation of Lemma 11.1, each T_i is consistent. Let $T_E = \bigcup_{i \in N} T_i$. We wish to show that T_E is (1) an extension of T, (2) is consistent, and (3) is complete. Since $T \subset T_E$, certainly $\vdash_T \mathbf{A}$ implies $\vdash_{T_E} \mathbf{A}$, i.e., T_E is an extension of T. To show that T_E is consistent, we suppose that it is not. Then for some **A** (which we assume to be closed), $\vdash_{T_E} \mathbf{A}$ and $\vdash_{T_E} \sim\mathbf{A}$. To write $\vdash_{T_E} \mathbf{A}$ means there is a *finite* sequence of formulas which constitutes a proof. Therefore, there is an m such that all those formulas are in T_m. Similarly, to say that $\vdash_{T_E} \sim\mathbf{A}$ is to say that there is another finite sequence of formulas constituting a proof of $\sim\mathbf{A}$ and so there is a k such that all those formulas are in T_k. By the construction, either $T_m \subset T_k$ or $T_k \subset T_m$; so suppose the latter. Then, both $\vdash_{T_m} \mathbf{A}$ and $\vdash_{T_m} \sim\mathbf{A}$, which implies that T_m is not consistent. But that contradicts the construction, which was such as to ensure that each T_i is consistent. To show that T_E is complete, we must show that for every closed formula, either it or its negation is a theorem in T_E. But the closed formulas are just the $\mathbf{C_i}$ originally listed. So for any $\mathbf{C_i}$, if not $\vdash_{T_{i-1}} \sim\mathbf{C_i}$, then $\vdash_{T_i} \mathbf{C_i}$, so $\vdash_{T_E} \mathbf{C_i}$; and if $\vdash_{T_{i-1}} \sim\mathbf{C_i}$. then $\vdash_{T_E} \sim \mathbf{C_i}$. Hence, T_E is complete. ∎

Exercise 11.15 Suppose T is a consistent V(∗)-based theory and we define an infinite sequence of extensions (each of which we assume to be consistent) by $T_0 = T$, $T_1 = T_0 \cup \{\mathbf{A_1}\}, \ldots, T_n = T_{n-1} \cup \{\mathbf{A_n}\}, \ldots$. Then, define $T^* = \bigcup_{i \in N} T_i$. Prove that T^* is a consistent V(∗)-based theory.

Exercise 11.16 If T is a consistent, decidable first-order theory [or V(∗)-based theory], then T has a consistent, decidable, complete extension.

Exercise 11.17 Show that the following formulas are provable in V(∗):

(a) $(\forall y)(A(y) \supset C) \equiv ((\exists y)A(y) \supset C)$, y not free in **C**.
(b) $(\exists y)(C \supset A(y)) \equiv (C \supset (\exists y)A(y))$, y not free in **C**.
(c) $(A \supset B) \supset ((A \equiv C) \supset (C \supset B))$.
(d) $(\exists x)B(x) \supset (\exists y)B(y)$.

Theorem 11.6 † (Gödel) Every consistent V(∗)-based theory has a model.

PROOF (1) Let L(∗) be a first-order language without equality and let T be a consistent V(∗)-based theory with axioms $V(*) \cup \Gamma$. (Notice that Γ need not be a recursive or even an r.e. set.) Let $\{\mathbf{s_1}, \mathbf{s_2}, \ldots\}$ be a countably

† This theorem was first proved in Gödel (1930). Most present-day proofs, including the one given here, are developed from Henkin (1949).

infinite set of names of constants which are distinct from any belonging to $\mathsf{L}(*)$. Define $\mathsf{L}(*') = \mathsf{L}(*) \cup \{\mathbf{s}_1, \mathbf{s}_2, \ldots\}$. Define T' to be the $\mathsf{V}(*')$-based theory with axioms $\mathsf{V}(*') \cup \Gamma$ for the same Γ. By the first corollary to Theorem 11.4, since T is assumed to be consistent, T' is consistent.

(2) Consider the set of sentences of $\mathsf{F}(*')$ that begin with an existential quantifier and put them into some enumeration:

$$(\exists \mathbf{x}_{i_1})\mathbf{B}_1(\mathbf{x}_{i_1}), \quad (\exists \mathbf{x}_{i_2})\mathbf{B}_2(\mathbf{x}_{i_2}), \ldots, (\exists \mathbf{x}_{i_j})\mathbf{B}_j(\mathbf{x}_{i_j}), \ldots$$

Of course, only finitely many $\mathbf{s}_i$ can occur in any $\mathbf{B}_j$. With this in mind, form the following infinite sequence of formulas: $\mathbf{H}_1$ is $(\exists \mathbf{x}_{i_1})\mathbf{B}_1(\mathbf{x}_{i_1}) \supset \mathbf{B}_1(\mathbf{s}_{i_1})$, where $\mathbf{s}_{i_1}$ is the first in the list $\{\mathbf{s}_1, \mathbf{s}_2, \ldots\}$ that does not occur in $\mathbf{B}_1(\mathbf{x}_{i_1})$. Inductively, $\mathbf{H}_j$ is $(\exists \mathbf{x}_{i_j})\mathbf{B}_j(\mathbf{x}_{i_j}) \supset \mathbf{B}_j(\mathbf{s}_{i_j})$, where $\mathbf{s}_{i_j}$ is the first in the list $\{\mathbf{s}_1, \mathbf{s}_2, \ldots\}$ that does not occur in $\mathbf{B}_j(\mathbf{x}_{i_j})$ or in $\mathbf{H}_1, \ldots, \mathbf{H}_{j-1}$.

(3) We shall use these $\mathbf{H}$'s to make an infinite sequence of extensions of T'. Let $\mathsf{T}_0 = \mathsf{T}'$, let $\mathsf{T}_1 = \mathsf{T}_0 \cup \{\mathbf{H}_1\}$, and, inductively, let $\mathsf{T}_n = \mathsf{T}_{n-1} \cup \{\mathbf{H}_n\}$. Finally, define T^* to be $\bigcup_{i \in N} \mathsf{T}_i$. By exercise 11.15, T^* is consistent *if* each T_i is consistent.

(4) We shall show that each T_i is consistent by induction on i. For $i = 0$, we have T_0, i.e., T', which is already known to be consistent. So, for the induction hypothesis, we assume that T_i is consistent for all $i < k$ and show that T_k is consistent. We do this by assuming that T_k is not consistent and deriving a contradiction in T_{k-1}. If T_k is not consistent, then there is some closed formula $\mathbf{A}$ such that both

$$\vdash_{\mathsf{T}_k} \mathbf{A} \quad \text{and} \quad \vdash_{\mathsf{T}_k} \sim\mathbf{A}$$

$$\mathbf{H}_k \vdash_{\mathsf{T}_{k-1}} \mathbf{A} \quad \text{and} \quad \mathbf{H}_k \vdash_{\mathsf{T}_{k-1}} \sim\mathbf{A}$$

$$\vdash_{\mathsf{T}_{k-1}} \mathbf{H}_k \supset \mathbf{A} \quad \text{and} \quad \vdash_{\mathsf{T}_{k-1}} \mathbf{H}_k \supset \sim\mathbf{A}$$

Since the $\mathbf{s}_{i_k}$ in $\mathbf{H}_k$ was chosen so that it occurs in no $\mathbf{H}_j$, $j < k$, we can take the proofs of $(\mathbf{H}_k \supset \mathbf{A})$ and $(\mathbf{H}_k \supset \sim\mathbf{A})$ and replace $\mathbf{s}_{i_k}$ everywhere by an individual variable $\mathbf{y}$ that does not occur in either proof. Thus,

$$\vdash_{\mathsf{T}_{k-1}} \big((\exists \mathbf{x}_{i_k})\mathbf{B}(\mathbf{x}_{i_k}) \supset \mathbf{B}_k(\mathbf{y})\big) \supset \mathbf{A} \quad \text{and} \quad \vdash_{\mathsf{T}_{k-1}} \big((\exists \mathbf{x}_{i_k})\mathbf{B}(\mathbf{x}_{i_k}) \supset \mathbf{B}_k(\mathbf{y})\big) \supset \sim\mathbf{A}$$

In the left column

$\vdash_{\mathsf{T}_{k-1}} (\forall \mathbf{y})\big(((\exists \mathbf{x}_{i_k})\mathbf{B}_k(\mathbf{x}_{i_k}) \supset \mathbf{B}_k(\mathbf{y})) \supset \mathbf{A}\big)$	by ug
$\vdash_{\mathsf{T}_{k-1}} (\forall \mathbf{y})\big(((\exists \mathbf{x}_{i_k})\mathbf{B}_k(\mathbf{x}_{j_k}) \supset \mathbf{B}_k(\mathbf{y})) \supset \mathbf{A}\big)$	by exercise 11.17 (a)
$\equiv (\exists \mathbf{y})\big((\exists \mathbf{x}_{i_k})\mathbf{B}_k(\mathbf{x}_{i_k}) \supset \mathbf{B}_k(\mathbf{y})\big) \supset \mathbf{A}$	$\mathbf{y}$ not free in $\mathbf{A}$ because $\mathbf{A}$ is closed
$\vdash_{\mathsf{T}_{k-1}} (\exists \mathbf{y})\big((\exists \mathbf{x}_{i_k})\mathbf{B}_k(\mathbf{x}_{i_k}) \supset \mathbf{B}_k(\mathbf{y})\big) \supset \mathbf{A}$	by mp
$\vdash_{\mathsf{T}_{k-1}} (\exists \mathbf{y})\big((\exists \mathbf{x}_{i_k})\mathbf{B}_k(\mathbf{x}_{i_k}) \supset \mathbf{B}_k(\mathbf{y})\big)$	by exercise 11.17 (b),
$\equiv \big((\exists \mathbf{x}_{i_k})\mathbf{B}_k(\mathbf{x}_{i_k}) \supset (\exists \mathbf{y})\mathbf{B}_k(\mathbf{y})\big)$	$\mathbf{y}$ not free in $(\exists \mathbf{x}_{i_k})\mathbf{B}_k(\mathbf{x}_{i_k})$
$\vdash_{\mathsf{T}_{k-1}} \big((\exists \mathbf{x}_{i_k})\mathbf{B}_k(\mathbf{x}_{i_k}) \supset (\exists \mathbf{y})\mathbf{B}_k(\mathbf{y})\big) \supset \mathbf{A}$	by exercise 11.17(c) and mp
$\vdash_{\mathsf{T}_{k-1}} \mathbf{A}$	by exercise 11.17(d), the previous step, and mp

One can carry through the same sort of derivation from the right column to obtain $\vdash_{T_{k-1}} \sim \mathbf{A}$. But that contradicts the induction hypothesis saying that T_{k-1} is consistent.

(5) Since T^* is a consistent $V(*')$-based theory, by Theorem 11.5, it has a complete, consistent extension. Call it T^{**}.

(6) Define the relational system $\mathfrak{H}$ as follows: The universe H is the set of all closed terms of $L(*')$. Henceforth, when we refer to a closed term as an element of $L(*')$, we will write $\mathbf{h_i}$. The same closed term considered as an element of H will be written $[\mathbf{h_i}]$. Notice that the cardinality of H is $\aleph_0$. For each function symbol $\mathbf{f_j^n}$ in $L(*')$, there is to be a function f_j^n in $\mathfrak{H}$ such that $f_j^n([\mathbf{h_1}], \ldots, [\mathbf{h_n}])$ has the value $[\mathbf{f_j^n(h_1, \ldots, h_n)}]$ in H. For each predicate symbol $\mathbf{P_j^n}$ in $L(*')$, there is to be a corresponding relation P_j^n in $\mathfrak{H}$ defined by

$$P_j^n = \{([\mathbf{h_1}], \ldots, [\mathbf{h_n}]) \mid \vdash_{T^{**}} \mathbf{P(h_1, \ldots, h_n)}\}$$

that is, $([\mathbf{h_1}], \ldots, [\mathbf{h_n}]) \in P_j^n$ if and only if $\vdash_{T^{**}} \mathbf{P(h_1, \ldots, h_n)}$. (As we will see in the next chapter, it is just at this step that we are using the fact that $L(*')$ is without equality.)

(7) If $\mathfrak{H}$ is to be a model of T^{**}, we must first be sure that it is a relational system for $L(*')$. That is obvious from the definition of $\mathfrak{H}$. Formally, we have: If $\mathbf{h}$ is the name of a constant in $L(*')$, then $\Phi(\mathbf{h}) = [\mathbf{h}]$. If $\mathbf{f_j^n}$ is a function symbol of $L(*')$, then $\Phi\mathbf{f_j^n}$ is f_j^n as defined above. Similarly, if $\mathbf{P_j^n}$ is a predicate symbol, $\Phi\mathbf{P_j^n}$ is the relation P_j^n defined above.

(8) We claim that $\mathfrak{H}$ is a model of T^{**} and so we must prove that if $\mathbf{A}$ is a closed formula of $F(*')$ such that $\vdash_{T^{**}} \mathbf{A}$, then $\mathbf{A}$ is true in $\mathfrak{H}$. However, what we will actually prove is stronger: Let $\mathbf{A}$ be a sentence; $\vdash_{T^{**}} \mathbf{A}$ if and only if $\mathbf{A}$ is true in $\mathfrak{H}$. Naturally, this will be done by induction on the formation of $\mathbf{A}$. Since all $\mathbf{A}$ are closed formulas, as a result of Theorem 10.1, we may consider valuations V without reference to particular assignments.

(8.1) *Atomic formulas.* If $\mathbf{A}$ is a closed atomic formula, then it is of the form $\mathbf{P_j^n(h_1, \ldots, h_n)}$. By the definition of $\mathfrak{H}$ and valuations, $V(\mathbf{h_i}) = [\mathbf{h_i}]$, and $V(\mathbf{P_j^n(h_1, \ldots, h_n)}) = \mathfrak{t}$ if and only if $(V(\mathbf{h_1}), \ldots, (V(\mathbf{h_n})) \in P_j^n$, i.e., $([\mathbf{h_1}], \ldots, [\mathbf{h_n}]) \in P_j^n$. But by the definition of $\mathfrak{H}$, $\vdash_{T^{**}} \mathbf{P_j^n(h_1, \ldots, h_n)}$ if and only if $([\mathbf{h_1}], \ldots, [\mathbf{h_n}]) \in P_j^n$.

(8.2.) $\mathbf{A}$ is of the form $\sim\mathbf{B}$. Then, $V(\mathbf{A}) = \mathfrak{t}$ in $\mathfrak{H}$ if and only if $V(\mathbf{B}) = \mathfrak{f}$ in $\mathfrak{H}$. Since $\mathbf{B}$ satisfies the induction hypothesis, $V(\mathbf{B}) = \mathfrak{f}$ in $\mathfrak{H}$ if and only if not $\vdash_{T^{**}} \mathbf{B}$. But since T^{**} is complete and consistent, not $\vdash_{T^{**}} \mathbf{B}$ if and only if $\vdash_{T^{**}} \sim \mathbf{B}$, i.e., $\vdash_{T^{**}} \mathbf{A}$.

(8.3.) $\mathbf{A}$ is of the form $(\mathbf{B} \supset \mathbf{C})$. We wish to show that $\vdash_{T^{**}} \mathbf{A}$ if and only if $V(\mathbf{A}) = \mathfrak{t}$ in $\mathfrak{H}$, but that is equivalent to showing that not $\vdash_{T^{**}} \mathbf{A}$ if and only if $V(\mathbf{A}) = \mathfrak{f}$ in $\mathfrak{H}$. But, further, by the completeness and consistency of T^{**}, that is equivalent to showing $\vdash_{T^{**}} \sim\mathbf{A}$ if and only if $V(\mathbf{A}) = \mathfrak{f}$ in $\mathfrak{H}$. It is this which we shall prove. $V(\mathbf{A}) = \mathfrak{f}$ if and only if $V(\mathbf{B}) = \mathfrak{t}$ and $V(\mathbf{C}) = \mathfrak{f}$. But, by the induction hypothesis, $V(\mathbf{B}) = \mathfrak{t}$ and $V(\mathbf{C}) = \mathfrak{f}$ if and only if $\vdash_{\mathsf{T}^{**}} \mathbf{B}$ and not $\vdash_{\mathsf{T}^{**}} \mathbf{C}$. By the completeness and consistency of T^{**}, not $\vdash_{\mathsf{T}^{**}} \mathbf{C}$ if and only if $\vdash_{\mathsf{T}^{**}} \sim\mathbf{C}$. Since it is a tautology,

$\vdash_{\mathsf{T}^{**}}(\mathbf{B} \supset (\sim\mathbf{C} \supset \sim(\mathbf{B} \supset \mathbf{C}))$,	so if
$\vdash_{\mathsf{T}^{**}} \mathbf{B}$	and
$\vdash_{\mathsf{T}^{**}} \sim\mathbf{C}$	then
$\vdash_{\mathsf{T}^{**}} \sim(\mathbf{B} \supset \mathbf{C})$,	i.e.,
$\vdash_{\mathsf{T}^{**}} \sim\mathbf{A}$.	Conversely, if
$\vdash_{\mathsf{T}^{**}} \sim(\mathbf{B} \supset \mathbf{C})$,	then, since
$\vdash_{\mathsf{T}^{**}}(\sim(\mathbf{B} \supset \mathbf{C}) \supset \mathbf{B})$	and
$\vdash_{\mathsf{T}^{**}}(\sim(\mathbf{B} \supset \mathbf{C}) \supset \sim\mathbf{C})$,	using mp, we have
$\vdash_{\mathsf{T}^{**}} \mathbf{B}$ and $\vdash_{\mathsf{T}^{**}} \sim\mathbf{C}$.	

Then, tracing back through the equivalences just above leads to $V(\mathbf{A}) = \mathfrak{f}$ in $\mathfrak{H}$.

(8.4.) $\mathbf{A}$ is of the form $(\exists \mathbf{x}_{\mathbf{i_j}})\mathbf{B_j}(\mathbf{x}_{\mathbf{i_j}})$. Since $\mathbf{A}$ is closed, all valuations of it are equal.

(a) If $\vdash_{\mathsf{T}^{**}}(\exists \mathbf{x}_{\mathbf{i_j}})\mathbf{B_j}(\mathbf{x}_{\mathbf{i_j}})$, then, since by the construction of T^{**},

$$\vdash_{\mathsf{T}^{**}}(\exists \mathbf{x}_{\mathbf{i_j}})\mathbf{B_j}(\mathbf{x}_{\mathbf{i_j}}) \supset \mathbf{B_j}(\mathbf{s}_{\mathbf{i_j}}), \text{ by mp, } \vdash_{\mathsf{T}^{**}} \mathbf{B_j}(\mathbf{s}_{\mathbf{i_j}}).$$

So, by the induction hypothesis,

$$V(\mathbf{B_j}(\mathbf{s}_{\mathbf{i_j}})) = \mathfrak{t}.$$

For any assignment φ, let φ' be an assignment such that $\varphi'(\mathbf{x}_{\mathbf{i_j}}) = [\mathbf{s}_{\mathbf{i_j}}]$ and φ' agrees with φ elsewhere. Then $V_\varphi(\mathbf{s}_{\mathbf{i_j}}) = [\mathbf{s}_{\mathbf{i_j}}] = \varphi'(\mathbf{x}_{\mathbf{i_j}})$ and so, by Lemma 10.6, $V_\varphi(\mathbf{B_j}(\mathbf{x}_{\mathbf{i_j}})) = V_{\varphi'}(\mathbf{B_j}(\mathbf{s}_{\mathbf{i_j}})) = \mathfrak{t}$. Therefore, for every φ, $V_\varphi((\exists \mathbf{x}_{\mathbf{i_j}})\mathbf{B_j}(\mathbf{x}_{\mathbf{i_j}})) = \mathfrak{t}$.

(b) Suppose $V((\exists \mathbf{x}_{\mathbf{i_j}})\mathbf{B_j}(\mathbf{x}_{\mathbf{i_j}})) = \mathfrak{t}$. Thus, for each assignment φ there is an assignment φ', agreeing with φ except perhaps at $\mathbf{x}_{\mathbf{i_j}}$, such that $V_{\varphi'}(\mathbf{B_j}(\mathbf{x}_{\mathbf{i_j}})) = \mathfrak{t}$. If $[\mathbf{h}] = \varphi'(\mathbf{x}_{\mathbf{i_j}})$ then, since $V_\varphi(\mathbf{h}) = [\mathbf{h}] = \varphi'(\mathbf{x}_{\mathbf{i_j}})$, by Lemma 10.6, $V_\varphi(\mathbf{B_j}(\mathbf{h})) = V_{\varphi'}(\mathbf{B}(\mathbf{x}_{\mathbf{i_j}})) = \mathfrak{t}$. Since $\mathbf{B_j}(\mathbf{h})$ is closed, $V(\mathbf{B}(\mathbf{h})) = \mathfrak{t}$ and so by the induction hypothesis, $\vdash_{\mathsf{T}^{**}} \mathbf{B_j}(\mathbf{h})$. By exercise 11.2, $\vdash_{\mathsf{T}^{**}} \mathbf{B_j}(\mathbf{h}) \supset (\exists \mathbf{x}_{\mathbf{i_j}})\mathbf{B_j}(\mathbf{x}_{\mathbf{i_j}})$, and so, by mp, $\vdash_{\mathsf{T}^{**}}(\exists \mathbf{x}_{\mathbf{i_j}})\mathbf{B_j}(\mathbf{x}_{\mathbf{i_j}})$.

(9) We have now shown that $\mathfrak{H}$ is a model of T^{**} and since $\mathsf{T} \subset \mathsf{T}^{**}$, $\mathfrak{H}$ is also a model of T, so the theorem is proved. ▮

Theorem 11.7 (Gödel Completeness) For a closed formula $\mathbf{A}$, $\vdash \mathbf{A}$ if and only if $\mathbf{A}$ is universally valid.

PROOF One direction was proved in proving Theorem 11.2. The other direction was indicated in the discussion at the end of section 11.1. However, we give it here also. We assume that $\mathbf{A}$ (closed) is universally valid and show that $\vdash \mathbf{A}$. Suppose not $\vdash \mathbf{A}$; then, $\mathsf{V}(*) \cup \{\sim\mathbf{A}\}$ is a consistent $\mathsf{V}(*)$-based theory by Lemma 11.1, and so by Theorem 11.6, it has a model $\mathfrak{A}$ in which, of course, $\sim\mathbf{A}$ is true. Also, as **A** is universally valid, it must be true in all relational systems, including $\mathfrak{A}$. But it is impossible that a formula and its negation both be true in the same relational system. ▮

We now banish the term "V(∗)-based" from our vocabulary and henceforth use only "first-order theory."

11.4 SOME IMMEDIATE CONSEQUENCES OF THE COMPLETENESS THEOREM

Theorem 11.8 (Löwenheim–Skolem) If a first-order theory has a model, it has countably infinite model.

PROOF T has a model, therefore T is consistent. But by the construction in the proof of Theorem 11.6, any consistent first-order theory has a countably infinite model. ▮

Exercise 11.18 Let T be a consistent first-order theory and **A** a sentence. Prove that $\vdash_{\mathsf{T}} \mathbf{A}$ if and only if **A** is true in every countably infinite model of T.

Exercise 11.19 Consider the first-order language without =, $\mathsf{L}(E)$, which has no function symbols, no names of constants, and only one binary predicate symbol, **E**. Let $\mathsf{T}(E)$ be a first-order theory whose axioms are those of $\mathsf{V}(E)$ union

$$(\forall x)\mathbf{E}(x, x) \wedge (\forall x)(\forall y)(\mathbf{E}(x, y) \equiv \mathbf{E}(y, x)) \\ \wedge (\forall x)(\forall y)(\forall z)(\mathbf{E}(x, y) \supset (\mathbf{E}(y, z) \supset \mathbf{E}(x, z)))$$

and

$$(\exists x)(\exists y)(\exists z)(\forall u)((\sim\mathbf{E}(x, y) \wedge \sim\mathbf{E}(y, z) \wedge \sim\mathbf{E}(x, z)) \\ \wedge (\mathbf{E}(x, u) \vee \mathbf{E}(y, u) \vee \mathbf{E}(z, u)))$$

(a) Is $\langle\{0, 1, 2\}; =\rangle$ a model of $\mathsf{T}(E)$? (b) Is $\langle\{5, 7, 9, 11\}; =\rangle$ a model of $T(E)$? (c) What does the Löwehheim Skolem theorem say about $\mathsf{T}(E)$? (d) Is there a first-order theory which has only finite models? (e) Can we represent "equals" in its most decisive meaning in a first-order theory?

Theorem 11.9 If α and β are two cardinal numbers such that $\beta > \alpha$ and if a first-order theory T has a model of cardinality α, then it has a model of cardinality β.

PROOF The idea of the proof is to take the model $\mathfrak{A}$ which has a universe A of cardinality α and add enough distinct new elements to have a set B of cardinality β and then to force these new elements to behave (in the sense of what happens to them in relations and in the domains of the functions) exactly like some elements in A. That way, although we have a larger universe, the new elements do not contribute anything new or interesting beyond their mere existence—they just shadow some old elements. The construction of $\mathfrak{B}$ from $\mathfrak{A}$ is as follows. The universe of $\mathfrak{B}$ is A plus a set whose cardinality is $\beta - \alpha$ (i.e., a set of cardinality γ so that $\alpha + \gamma = \beta$). Take some fixed $a_0 \in A$ and define the n-ary functions f_i^n of $\mathfrak{B}$ to extend those g_i^n in $\mathfrak{A}$ in such a way that

$$f_i^n(b_{j_1}, \ldots, b_{j_n}) = g_i^n(\cdots)$$

where a_0 replaces each $b_{i_j} \in B - A$. In the same manner, a given n-tuple $(b_{j_1}, \ldots, b_{j_n})$ belongs to the n-ary relation P_i^n of $\mathfrak{B}$ if and only if the n-tuple which is obtained from it by replacing each $b_{i_j} \in B - A$ by a_0 belongs to R_j^n in $\mathfrak{A}$. By using the obvious correspondence between function symbols and functions, names of constants and elements of B (actually just A) and predicate symbols and relations, it is just a little exercise to show that $\mathfrak{B}$ is a model of T. ∎

Exercise 11.20 Show that $\mathfrak{B}$ is a model of T.

Corollary For every infinite cardinal α, any consistent first-order theory has a model of cardinality α.

Exercise 11.21 Prove the corollary.

Exercise 11.22 Let L(0) be a first-order language that has no function symbols, no names of constants, and does not have equality. (a) Assume that a sentence $\mathbf{A} \in$ F(0) is of the form $(\forall \mathbf{x}_1) \cdots (\forall \mathbf{x}_n)(\exists \mathbf{y}_1) \cdots (\exists \mathbf{y}_m)\mathbf{B}$ (i.e., all universal quantifiers precede all existential quantifiers) and $\mathbf{B}$ is quantifier-free. Show that $\mathbf{A}$ is universally valid if and only if $\mathbf{A}$ is valid in every relational system (for L(0)) of cardinality less than or equal to the number of universal quantifiers in $\mathbf{A}$. (b) Show that there is an effective procedure for deciding of an arbitrary sentence of the form in part (a) whether or not it is universally valid.

Exercise 11.23 Let L(E') be a first-order language without equality with one function symbol $\mathbf{f}_1^1$, one name of constant $\mathbf{a}_0$, and one binary predicate symbol $\mathbf{E}$. Let T(E') be a first-order theory with the axioms of V(E') and

$$(\forall x)E(x, x) \wedge (\forall x)(\forall y)(E(x, y) \equiv E(y, x))$$
$$\wedge (\forall x)(\forall y)(\forall z)(E(x, y) \supset (E(y, z) \supset E(x, z)))$$

and

$$(\forall x)(\exists y)(E(f_1{}^1(x), y) \wedge \sim(\exists x)(E(f_1{}^1(x), a_0)))$$
$$\wedge (\forall x)(\forall y)(E(f_1{}^1(x), f_1{}^1(y)) \supset E(x, y))$$

(a) Find a relational system which is a model of $\mathsf{T}(E')$.
(b) Is there a relational system with finite universe which is a model of $\mathsf{T}(E')$?

A summary of the cardinality results: (1) If a first-order theory T has a finite model of cardinality α, then for all cardinalites $\beta > \alpha$, T has a model of cardinality β. (2) If a first-order theory T has a model of any infinite cardinality, it has a model for each infinite cardinality.

Chapter XII

FIRST-ORDER THEORIES WITH EQUALITY

In chapter XI, we proved that for a first-order language without equality $\mathsf{L}(*)$, $\mathscr{V}(*)$, the set of universally valid formulas on $\mathsf{L}(*)$, is recursively axiomatizable by means of the set of axioms $V(*)$. We now consider a first-order language with equality, $\mathsf{L}(*^{=})$, and will show that the set of universally valid formulas $\mathscr{V}(*^{=})$ is also recursively axiomatizable. Let $\mathsf{V}(*^{=})$ denote the theory whose axioms are those of $\mathsf{V}(*)$, but written on the language $\mathsf{L}(*^{=})$, plus †

$$(\forall \mathbf{x})(\mathbf{x} = \mathbf{x})$$
$$(\mathbf{x} = \mathbf{y}) \supset (\mathbf{A}(\mathbf{x}) \equiv \mathbf{A}(\mathbf{y})) \qquad \text{for each} \quad \mathbf{A} \in \mathsf{F}(*^{=})$$

We refer to these two axiom schemes as "the axioms for equality." Temporarily, we call any logical theory whose axioms include the axioms of $\mathsf{V}(*^{=})$ a "$V(*^{=})$-based theory." We want to show that any $\mathsf{V}(*^{=})$-based theory is a first-order theory with equality [i.e., is a deductively closed set which includes $\mathscr{V}(*^{=})$]. We abbreviate $\sim(\mathbf{x} = \mathbf{y})$ by $(\mathbf{x} \neq \mathbf{y})$.

Exercise 12.1 Show that the three following are theorems in any $\mathsf{V}(*^{=})$-based theory: (a) $(\mathbf{t} = \mathbf{t})$ for any term $\mathbf{t}$, (b) $(\mathbf{x} = \mathbf{y}) \supset (\mathbf{y} = \mathbf{x})$, (c) $((\mathbf{x} = \mathbf{y}) \supset ((\mathbf{y} = \mathbf{z}) \supset (\mathbf{x} = \mathbf{z})))$.

Lemma 12.1 In any $\mathsf{V}(*^{=})$-based theory, the following are provable:

(i) $\quad ((\mathbf{t}_1 = \mathbf{t}_1') \wedge \cdots \wedge (\mathbf{t}_n = \mathbf{t}_n')) \supset (\mathbf{A}(\mathbf{t}_1, \ldots, \mathbf{t}_n) \equiv \mathbf{A}(\mathbf{t}_1', \ldots, \mathbf{t}_n'))$

if $\mathbf{A}(\mathbf{x}_1, \ldots, \mathbf{x}_n)$ is a formula with at least n free variables and in which each $\mathbf{x}_i$ is free for each $\mathbf{t}_j$ and $\mathbf{t}_j'$; and

† Here, $\mathbf{A}(\mathbf{y})$ is obtained from $\mathbf{A}(\mathbf{x})$ by replacing *some* or all occurrences of $\mathbf{x}$ by $\mathbf{y}$, and $\mathbf{x}$ and $\mathbf{y}$ are free for each other in $\mathbf{A}$.

(ii) $((\mathbf{t}_1 = \mathbf{t}_1') \wedge \cdots \wedge (\mathbf{t}_n = \mathbf{t}_n')) \supset (\mathbf{f}^n(\mathbf{t}_1, \ldots, \mathbf{t}_n) = \mathbf{f}^n(\mathbf{t}_1', \ldots, \mathbf{t}_n'))$

for any terms $\mathbf{t}_j$, $\mathbf{t}_j'$, $1 \leq j \leq n$ and n-ary function symbol $\mathbf{f}^n$.

PROOF (i) (By induction on n) For $n = 1$, suppose $\mathbf{A}$ is a formula with at least one free variable, i.e., $\mathbf{A}(\mathbf{x})$. By the second axiom of equality,

$$(\mathbf{x} = \mathbf{y}) \supset (\mathbf{A}(\mathbf{x}) \equiv \mathbf{A}(\mathbf{y}))$$

By the fifth axiom,

$$(\forall \mathbf{x})((\mathbf{x} = \mathbf{y}) \supset (\mathbf{A}(\mathbf{x}) \equiv \mathbf{A}(\mathbf{y}))) \supset ((\mathbf{t} = \mathbf{y}) \supset (\mathbf{A}(\mathbf{t}) \equiv \mathbf{A}(\mathbf{y})))$$

Generalizing on the first and using *modus ponens*, we have

$$(\mathbf{t} = \mathbf{y}) \supset (\mathbf{A}(\mathbf{t}) \equiv \mathbf{A}(\mathbf{y}))$$

Repeating the same steps, in terms of $\mathbf{y}$ and $\mathbf{t}'$, we have

$$(\mathbf{t} = \mathbf{t}') \supset (\mathbf{A}(\mathbf{t}) \equiv \mathbf{A}(\mathbf{t}'))$$

For the induction step, suppose $\mathbf{A}$ is a formula with at least n free variables, i.e., $\mathbf{A}(\mathbf{x}_1, \ldots, \mathbf{x}_{n-1}, \mathbf{x}_n)$. By the induction hypothesis,

$$(\mathbf{t}_1 = \mathbf{t}_1') \wedge \cdots \wedge (\mathbf{t}_{n-1} = \mathbf{t}_{n-1}') \supset (\mathbf{A}(\mathbf{t}_1, \ldots, \mathbf{t}_{n-1}, \mathbf{x}_n) \equiv \mathbf{A}(\mathbf{t}_1', \ldots, \mathbf{t}_{n-1}', \mathbf{x}_n))$$

By the axiom for equality, we have both

$$(\mathbf{x}_n = \mathbf{y}) \supset (\mathbf{A}(\mathbf{t}_1, \ldots, \mathbf{t}_{n-1}, \mathbf{x}_n) \equiv \mathbf{A}(\mathbf{t}_1, \ldots, \mathbf{t}_{n-1}, \mathbf{y}))$$
$$(\mathbf{x}_n = \mathbf{y}) \supset (\mathbf{A}(\mathbf{t}_1', \ldots, \mathbf{t}_{n-1}', \mathbf{x}_n) \equiv \mathbf{A}(\mathbf{t}_1', \ldots, \mathbf{t}_{n-1}', \mathbf{y}))$$

The two following are tautologies and hence provable:

$$(\mathbf{C} \supset \mathbf{E}_1) \supset ((\mathbf{C} \supset \mathbf{E}_2) \supset (\mathbf{C} \supset (\mathbf{E}_1 \wedge \mathbf{E}_2)))$$
$$(\mathbf{B} \supset (\mathbf{D}_1 \equiv \mathbf{D}_2)) \supset ((\mathbf{C} \supset ((\mathbf{D}_1 \equiv \mathbf{D}_3) \wedge (\mathbf{D}_2 \equiv \mathbf{D}_4))) \supset ((\mathbf{B} \wedge \mathbf{C}) \supset (\mathbf{D}_1 \equiv \mathbf{D}_4)))$$

Using the last five steps and *modus ponens* a few times, we get

$$(\mathbf{t}_1 = \mathbf{t}_1') \wedge \cdots \wedge (\mathbf{t}_{n-1} = \mathbf{t}_{n-1}') \wedge (\mathbf{x}_n = \mathbf{y})$$
$$\supset (\mathbf{A}(\mathbf{t}_1, \ldots, \mathbf{t}_{n-1}, \mathbf{x}_n) \equiv \mathbf{A}(\mathbf{t}_1', \ldots, \mathbf{t}_{n-1}', \mathbf{y}))$$

The proof is now completed exactly as in the case for $n = 1$.

(ii) Let $\mathbf{A}(\mathbf{x}_1, \ldots, \mathbf{x}_n)$ be $(\mathbf{f}^n(\mathbf{x}_1, \ldots, \mathbf{x}_n) = \mathbf{f}^n(\mathbf{x}_1, \ldots, \mathbf{x}_n))$; then, according to the convention of replacing some occurrences of the $\mathbf{x}_i$, $\mathbf{A}(\mathbf{y}_1, \ldots, \mathbf{y}_n)$ can be $(\mathbf{f}^n(\mathbf{x}_1, \ldots, \mathbf{x}_n) = \mathbf{f}^n(\mathbf{y}_1, \ldots, \mathbf{y}_n))$. Similarly, then, $\mathbf{A}(\mathbf{t}_1, \ldots, \mathbf{t}_n)$ can be $(\mathbf{f}^n(\mathbf{t}_1, \ldots, \mathbf{t}_n) = \mathbf{f}^n(\mathbf{t}_1, \ldots, \mathbf{t}_n))$ and $\mathbf{A}(\mathbf{t}_1', \ldots, \mathbf{t}_n')$ can be $(\mathbf{f}^n(\mathbf{t}_1, \ldots, \mathbf{t}_n) = \mathbf{f}^n(\mathbf{t}_1', \ldots, \mathbf{t}_n'))$. So, using these in part (i), we have

$$(\mathbf{t}_1 = \mathbf{t}_1') \wedge \cdots \wedge (\mathbf{t}_n = \mathbf{t}_n') \supset ((\mathbf{f}^n(\mathbf{t}_1, \ldots, \mathbf{t}_n) = \mathbf{f}^n(\mathbf{t}_1, \ldots, \mathbf{t}_n))$$
$$\equiv (\mathbf{f}^n(\mathbf{t}_1, \ldots, \mathbf{t}_n) = \mathbf{f}^n(\mathbf{t}_1', \ldots, \mathbf{t}_n')))$$

Now, $\mathbf{f}^n(\mathbf{t}_1, \ldots, \mathbf{t}_n) = \mathbf{f}^n(\mathbf{t}_1, \ldots, \mathbf{t}_n)$ is provable by exercise 12.1, and $\mathbf{B} \supset ((\mathbf{A} \supset (\mathbf{B} \equiv \mathbf{C})) \supset (\mathbf{A} \supset \mathbf{C}))$ is provable because it is a tautology. Combining the last three steps and using *modus ponens* a few times, we have $(\mathbf{t}_1 = \mathbf{t}_1') \wedge \cdots \wedge (\mathbf{t}_n = \mathbf{t}_n') \supset (\mathbf{f}^n(\mathbf{t}_1, \ldots, \mathbf{t}_n) = \mathbf{f}^n(\mathbf{t}_1', \ldots, \mathbf{t}_n'))$ ∎

Exercise 12.2 Consider the $\mathsf{V}(_*{}^=)$-based theory which has for its axioms those of $\mathsf{V}(_*{}^=)$ plus

$$(\exists \mathbf{x})(\exists \mathbf{y})(\exists \mathbf{z})(\forall \mathbf{u})((\mathbf{x} \neq \mathbf{y} \wedge \mathbf{y} \neq \mathbf{z} \wedge \mathbf{x} \neq \mathbf{y}) \wedge (\mathbf{x} = \mathbf{u} \vee \mathbf{y} = \mathbf{u} \vee \mathbf{z} = \mathbf{u}))$$

Is $\langle\{0, 1, 2\}; =\rangle$ a model of this theory? Does it have any models of cardinality other than three?

Apparently a Löwenheim–Skolem theorem for $\mathsf{V}(_*{}^=)$-based theories must be different from the Löwenheim–Skolem theorem for first-order theories without equality. Since the proof of Theorem 11.7 depended on the proof of Theorem 11.6, we shall now look at the analog of Theorem 11.6 for theories with equality to see how it is proved. Given a consistent $\mathsf{V}(_*{}^=)$-based theory, the problems involved in constructing a model for it include those that we had before, but further, our model must be such that whenever $(\mathbf{t}_1 = \mathbf{t}_2)$ is a theorem of the theory, in the model, the same element of the universe must correspond to both $\mathbf{t}_1$ and $\mathbf{t}_2$.

Theorem 12.1 Every consistent $\mathsf{V}(_*{}^=)$-based theory has a model.

SKETCH OF PROOF Follow the proof of Theorem 11.6 up through step 4, i.e., just up to beginning to construct the model. The new step 5 will be the following.

$5^=$. As before, let $\mathbf{h}_i$ denote a closed term of the language. This time, we define an equivalence relation on the set of closed terms. We define $\mathbf{h}_i$ to be equivalent to $\mathbf{h}_j$ if and only if $\vdash_{\mathsf{T}**}(\mathbf{h}_i = \mathbf{h}_j)$, and write $[\mathbf{h}_i] = [\mathbf{h}_j]$. That is, $[\mathbf{h}_i]$ denotes an equivalence class. (The reader should check that the relation just defined really is an equivalence relation.) Let $H^=$, the universe of the relational system $\mathfrak{H}^=$, be the set of these equivalence classes, i.e., $H^= = \{[\mathbf{h}_i] \mid \mathbf{h}_i$ is a closed term of $(\mathsf{L}_*{}^=)\}$. If $\mathsf{L}(_*{}^=)$ has function symbols, for each n-place function symbol $\mathbf{f}_j^n$, there is to be, in $\mathfrak{H}^=$, an n-ary function f_j^n such that for each n-tuple of elements in $H^=$, $([\mathbf{h}_1], \ldots, [\mathbf{h}_n])$, the value of $f_j^n([\mathbf{h}_1], \ldots, [\mathbf{h}_n])$ is to be $[\mathbf{f}_j^n(\mathbf{h}_1, \ldots, \mathbf{h}_n)]$. It must be verified that the function f_j^n is well defined, that is: if $[\mathbf{h}_i] = [\mathbf{h}_i']$, for $1 \leq i \leq n$, is $[\mathbf{f}_j^n(\mathbf{h}_1, \ldots, \mathbf{h}_n)] = [\mathbf{f}_j^n(\mathbf{h}_1', \ldots, \mathbf{h}_n')]$? That is the same as verifying that if for all i, $1 \leq i \leq n$, $\vdash_{\mathsf{T}**}(\mathbf{h}_i = \mathbf{h}_i')$, then $\vdash_{\mathsf{T}**}(\mathbf{f}_j^n(\mathbf{h}_1, \ldots, \mathbf{h}_n) = \mathbf{f}_j^n(\mathbf{h}_1', \ldots, \mathbf{h}_n'))$. But that is just what Lemma 12.1 says. For each n-ary predicate symbol $\mathbf{P}_j^n$, there is to be an n-ary relation P_j^n in $\mathfrak{H}^=$ such that

$$P_j^n = \{([\mathbf{h_1}], \ldots, [\mathbf{h_n}]) \mid \vdash_{\mathsf{T}^{**}} \mathbf{P}(\mathbf{h_1}, \ldots, \mathbf{h_n})\}$$

Again we must verify that the relation is well defined, that is, if $[\mathbf{h_i}] = [\mathbf{h_i}']$, for $1 \le i \le n$, then $\vdash_{\mathsf{T}^{**}} \mathbf{P_j^n}(\mathbf{h_1}, \ldots, \mathbf{h_n})$ if and only if $\vdash_{\mathsf{T}^{**}} \mathbf{P_j^n}(\mathbf{h_1}', \ldots, \mathbf{h_n}')$. However, since $[\mathbf{h_i}] = [\mathbf{h_i}']$ just in case $\vdash_{\mathsf{T}^{**}} (\mathbf{h_i} = \mathbf{h_i}')$, and since by Lemma 12.1,

$$\vdash_{\mathsf{T}^{**}} (\mathbf{h_1} = \mathbf{h_1}') \wedge \cdots \wedge (\mathbf{h_n} = \mathbf{h_n}') \supset (\mathbf{A}(\mathbf{h_1}, \ldots, \mathbf{h_n}) \equiv \mathbf{A}(\mathbf{h_1}', \ldots, \mathbf{h_n}'))$$

for the proper conditions, it follows that our requirement is met.

$6^=$ and $7^=$. To verify that $\mathfrak{H}^=$ is a model of T** is similar to the proof of Theorem 11.6 and is left to the reader. ▌

Corollary (Löwenheim–Skolem for first-order theories with equality) Every consistent first-order theory with equality has either a finite model or a countably infinite model.

PROOF We need only observe that in the construction of $\mathfrak{H}^=$ in the proof of the theorem, the cardinality of $H^=$ is at most $\aleph_0$. ▌

Corollary (Gödel completeness for first order theories with equality) Let **A** be a closed formula of a first order language with equality. **A** is a theorem of $\mathsf{V}(_*{}^=)$ if and only if **A** is universally valid.

Exercise 12.3 Suppose that a consistent first-order theory with equality has no infinite models. Show that if **A** is true in all (finite) models of the theory, then $\vdash_{\mathsf{T}} \mathbf{A}$.

Exercise 12.4 Let $\mathbf{E_n}$ be the formula

$$(\exists \mathbf{x_1}) \cdots (\exists \mathbf{x_n})(\mathbf{x_1} \ne \mathbf{x_2} \wedge \mathbf{x_1} \ne \mathbf{x_3} \wedge \cdots \wedge \mathbf{x_1} \ne \mathbf{x_n} \wedge \mathbf{x_2} \ne \mathbf{x_3} \wedge \cdots \wedge \mathbf{x_2} \ne \mathbf{x_n} \wedge \cdots \wedge \mathbf{x_{n-1}} \ne \mathbf{x_n})$$

and let $E^* = \bigcup_{n \in N} \mathbf{E_n}$. Designate by S_i a finite subset of E^*. Let $\mathsf{T_i}$ be the first-order theory with equality which has as its axioms the axioms of $\mathsf{V}(_*{}^=)$ plus the formulas of S_i. Let T* be the first-order theory with equality which has as its axioms those of $\mathsf{V}(_*{}^=)$ plus the set E^*. (a) Does every $\mathsf{T_i}$ have a model? (b) Does T* have a model?

Exercise 12.5 Show that if a first-order theory with equality has a model for infinitely many finite cardinals, then it has an infinite model.

Exercise 12.6 Show that if a first-order theory with equality has an infinite model, then it has a model of cardinality $\aleph_0$.

Although first-order theories with equality allow us to be somewhat more precise with respect to the meaning of "equals" than in the case of first-order theories without equality, we still cannot be as precise as we might like.

Consider the first-order language with equality $\mathsf{L}(_f{}^=)$, which has **0, 1** as names of constants and $+, \cdot$ as two-place function symbols. Define the following first-order theories with equality written on this language. T_1 has for its axioms those of $\mathsf{V}(_f{}^=)$ and

$$(\mathbf{x} + \mathbf{y}) + \mathbf{z} = \mathbf{x} + (\mathbf{y} + \mathbf{z})$$
$$\mathbf{x} + \mathbf{0} = \mathbf{x}$$
$$(\forall \mathbf{x})(\exists \mathbf{y})(\mathbf{x} + \mathbf{y} = \mathbf{0})$$
$$\mathbf{x} + \mathbf{y} = \mathbf{y} + \mathbf{x}$$
$$(\mathbf{x} \cdot \mathbf{y}) \cdot \mathbf{z} = \mathbf{x} \cdot (\mathbf{y} \cdot \mathbf{z})$$
$$\mathbf{x} \cdot \mathbf{1} = \mathbf{x}$$
$$\mathbf{x} \cdot \mathbf{y} = \mathbf{y} \cdot \mathbf{x}$$
$$\mathbf{x} \cdot (\mathbf{y} + \mathbf{z}) = (\mathbf{x} \cdot \mathbf{y}) + (\mathbf{x} \cdot \mathbf{z})$$
$$\mathbf{0} \neq \mathbf{1}$$

Theory $\mathsf{T}_2 = \mathsf{T}_1 \cup \{(\mathbf{x} \neq \mathbf{0}) \supset (\exists \mathbf{y})(\mathbf{x} \cdot \mathbf{y} = \mathbf{1})\}$

Theory $\mathsf{T}_3 = \mathsf{T}_2 \cup \{(\forall \mathbf{x})(\exists \mathbf{y})(\mathbf{y} \cdot \mathbf{y} = \mathbf{x})\}$

Consider also the following relational systems:

$\mathfrak{Z} = \langle Z; +, \cdot ; = \rangle$	Z is the integers
$\mathfrak{Q} = \langle Q; +, \cdot ; = \rangle$	Q is the rationals
$\mathfrak{F}_p = \langle 0, 1, \ldots, p-1; +_p, \cdot_p ; = \rangle$	for each prime p, $+_p$ and $\cdot_p$ are addition and multiplication mod p
$\mathfrak{R} = \langle R; +, \cdot ; = \rangle$	R is the reals

Of course, the cardinality of $\mathfrak{F}_p$ is the prime p, the cardinalities of Z and Q are $\aleph_0$, whereas that of $\mathfrak{R}$ is $2^{\aleph_0}$. All the relational systems are models of T_1; for each p, $\mathfrak{F}_p$, and $\mathfrak{Q}$, and $\mathfrak{R}$ are models of T_2 and $\mathfrak{R}$ is a model of T_3. But we also know that T_3 has a countably infinite model, though it is obviously not $\mathfrak{Q}$.

Theorem 12.2 (The Compactness Theorem) Let Γ be a set of sentences in a first-order language. Γ has a model if and only if each finite subset of Γ has a model.

Exercise 12.7 Prove the theorem.

This theorem is a very powerful tool employed frequently in logic. The following is an example. Consider once more the language $\mathsf{L}({}_f{}^{=})$. Let Fd be the theory T_2 just above. Any relational system which is a model of Fd is a *field*, and Fd is known as *elementary field theory* (or *first order field theory*). For each prime p, define $\mathbf{C_p}$ to be the sentence

$$((\underbrace{\mathbf{1} + \mathbf{1} + \cdots + \mathbf{1}}_{p}) = \mathbf{0})$$

For each prime p, define the theory $\mathsf{Fd}(p)$ by $\mathsf{Fd}(p) = \mathsf{Fd} \cup \{\mathbf{C_p}\}$. A model of $\mathsf{Fd}(p)$ is known as a *field of characteristic p;* for example, $\mathfrak{F}_p$ just above. Also, define

$$\mathsf{Fd}(0) = \mathsf{Fd} \cup \bigcup_{p=2,3,5,\ldots} \sim \mathbf{C_p}$$

This gives an axiomatization of the *fields of characteristic 0*, of which the rationals, the reals, and the complex numbers are examples.

Exercise 12.8 Prove the following:

(a) If $\mathbf{A}$ is a formula such that for infinitely many primes p, $\vdash_{\mathsf{Fd}(p)} \mathbf{A}$, then there is a model of $\mathsf{Fd}(0)$ in which $\mathbf{A}$ is true.

(b) If $\mathbf{B}$ is a formula such that $\vdash_{\mathsf{Fd}(0)} \mathbf{B}$, then there is a prime p^* such that for all primes $p > p^*$, $\vdash_{\mathsf{Fd}(p)} \mathbf{B}$.

(c) There can be no theory T which has all finite (and no infinite) fields as its models.

To complete the discussion, we state without proof the following.

Theorem (Tarski)† If a first-order theory with equality, T, has a model of infinite cardinality α, then for all cardinalities $\beta > \alpha$, T has a model of cardinality β.

To summarize: (1) If a first-order theory with equality has a finite model, it may, but need not, have a countably infinite model. (2) If a first-order theory has a model of any infinite cardinality, then it has a model for each infinite cardinal.

The most important implications are that one cannot define finiteness in the sense that one cannot have a first-order theory with or without equality which has models for all finite cardinals and no infinite cardinals, and second that any attempt to find a first-order theory (with or without equality) which has only the integers, or only the rationals, or only the reals for its model, is doomed to failure.

† For a proof, the reader can consult Shoenfield (1967, p. 78).

Chapter XIII

HERBRAND'S THEOREM

The goal of this chapter will be to prove the Herbrand theorem, which provides the basic idea of mechanical theorem-proving. Because of the relationships established by the theorem, if one wants to know whether or not a certain formula **A** is provable in a specified finitely axiomatizable theory, one can set up a fairly simple, effective (but possibly infinite) search for a certain kind of formula which is a contradiction, i.e., in its truth table, its only truth value is $\mathfrak{f}$. If the search comes across such a formula, then the formula **A** is provable. The search procedure does not give a solution to the decision problem for the theory, of course, because at no time can one conclude that the proper kind of contradictory formula is *not* going to be found eventually. A fair amount of work must be done before the Herbrand theorem can be proved, so the next few sections will be devoted to the necessary preliminaries.

13.1 PRENEX NORMAL FORM

Our aim in this section is to show that for every formula **A** in a first-order language (with or without equality), there is a closed formula **B** whose form is such that all the quantifiers come first and which is logically equivalent to **A**. For example, if **A** is $((\exists x)P(x, a) \supset (\forall y)Q(x, y))$, then **B** is

$$(\forall z)(\forall x)(\forall y)(P(x, a) \supset Q(z, y)).$$

The reader should now reread section 10.6, paying special attention to Lemma 10.8.

What is meant by a subformula should be clear. ***B** is a subformula of a formula **A*** if **B** is a formula and **B** occurs in **A**. Inductively: If **A** is atomic, **B** is **A**; if **A** is $(C \supset D)$, then **B** is a subformula of **C** or of **D**; if **A** is $\sim C$, then **B** is a subformula of **C**, and if **A** is $(\forall x)C$, then **B** is a subformula of **C**.

Exercise 13.1 Show that **A** and **B** are logically equivalent, where

$$\mathbf{A} \text{ is } (\exists x_1)(P(x_1, a_1) \supset (\forall x_2)((Q(x_2, a_1) \supset (\exists x_3)P(x_2, x_3)) \vee ((\forall x_4)P(x_2, x_4) \supset \sim Q(x_2, x_1))))$$

$$\mathbf{B} \text{ is } \sim(\forall x_1)(P(x_1, a_1) \wedge (\exists x_2)((\forall x_3)(Q(x_2, a_1) \wedge \sim P(x_2, x_3)) \wedge (\forall x_4)(P(x_2, x_4) \wedge Q(x_2, x_1))))$$

Theorem 13.1 (The Equivalence Theorem) Let $\{\mathbf{B}_1, \ldots, \mathbf{B}_n\}$ and $\{\mathbf{B}_1', \ldots, \mathbf{B}_n'\}$ be sets of formulas such that $\mathbf{B}_i$ and $\mathbf{B}_i'$ are logically equivalent for $1 \leq i \leq n$. Let **A** be a formula in which the $\mathbf{B}_i$ occur as subformulas and let **A′** be the formula obtained from **A** by replacing one or more occurrences of $\mathbf{B}_i$ by $\mathbf{B}_i'$. Then, **A** and **A′** are logically equivalent.

Exercise 13.2 Prove the theorem, by induction on the formation of **A**.

A formula is in *prenex normal form* if it is of the form $(Q_1 x_1) \cdots (Q_n x_n)\mathbf{B}$, where each Q_i is either $\forall$ or $\exists$ and **B** is quantifier-free.

Theorem 13.2 Given any formula **A**, there is a formula $\mathbf{A}_p$ in prenex normal form such that $\mathbf{A}_p$ is logically equivalent to **A**.

PROOF (Induction on the formation of **A**):

1. If **A** is atomic, then it is in prenex normal form.
2. If **A** is $\sim\mathbf{B}$, then, by the induction hypothesis, there is a formula $\mathbf{B}_p$ which is in prenex normal form and logically equivalent to **B**. Hence, by Theorem 13.1, **A** is logically equivalent to $\sim\mathbf{B}_p$. If $\mathbf{B}_p$ is of the form $(\cdots(\forall x)\cdots(\exists x)\cdots C(\cdots x\cdots y\cdots))$, then by the definition of the existential quantifier, $\sim\mathbf{B}_p$ is logically equivalent to $(\cdots(\exists x)\cdots(\forall y)\cdots\sim C(\cdots x\cdots y\cdots))$, which is in prenex normal form and, being logically equivalent to **A**, is $\mathbf{A}_p$. For $\mathbf{B}_p$ with different arrangements of quantifiers, the proofs are similar.
3. If **A** is $(\mathbf{B} \supset \mathbf{C})$, then by the induction hypothesis, there are formulas $\mathbf{B}_p$ and $\mathbf{C}_p$ in prenex normal form which are logically equivalent to **B** and **C**, respectively. So, by Theorem 13.1, **A** is logically equivalent to $(\mathbf{B}_p \supset \mathbf{C}_p)$. Suppose that $(\mathbf{B}_p \supset \mathbf{C}_p)$ is of the form

$$((\cdots(\forall x)\cdots(\exists y)\cdots \mathbf{B}_p'(\cdots x\cdots y\cdots)) \supset (\cdots(\exists z)\cdots(\forall w)\cdots \mathbf{C}_p'(\cdots z\cdots w\cdots))$$

then, by Lemma 10.8, it is logically equivalent to

$$(\cdots(\exists x')\cdots(\forall y')\cdots(\exists z')\cdots(\forall w')\cdots)(\mathbf{B}_p'(\cdots x'\cdots y'\cdots) \supset \mathbf{C}_p'(\cdots z'\cdots w'\cdots))$$

which is in prenex normal form and logically equivalent to **A**; thus, it is $\mathbf{A}_p$.

For $(\mathbf{B_p} \supset \mathbf{C_p})$ with different arrangements of quantifiers, the proofs are similar.

4. If **A** is $(\forall x)\mathbf{B}$, then by induction hypothesis, there is a formula $\mathbf{B_p}$ in prenex normal form and logically equivalent to **B**. But then, $(\forall x)\mathbf{B_p}$ is also in prenex normal form and is logically equivalent to $(\forall x)\mathbf{B}$ and hence to **A**. ▌

13.2 CONJUNCTIVE AND DISJUNCTIVE NORMAL FORMS

A quantifier-free formula **A** is in *conjunctive normal form* if it is a conjunction and each conjunct is either an atomic formula, the negation of an atomic formula, or a disjunction of atomic formulas and negations of atomic formulas.

Exercise 13.3 Find a formula in conjunctive normal form which is logically equivalent to **A** (usually called "the conjunctive normal form of **A**"), where **A** is

$$((\mathbf{B} \vee (\mathbf{C} \supset \sim\mathbf{D})) \supset \sim(\mathbf{B} \supset (\mathbf{D} \wedge \mathbf{E}))) \supset (\mathbf{C} \vee \mathbf{B})$$

and **B**, **C**, **D**, and **E** are assumed to be atomic formulas.

A quantifier-free formula **A** is in *disjunctive normal form* if it is a disjunction and each of its disjuncts is either an atomic formula, the negation of an atomic formula, or a conjunction of atomic formulas and negations of atomic formulas.

Exercise 13.4 Write down some horrible formula not in disjunctive normal form and then find its disjunctive normal form.

Exercise 13.5 Show that $((\mathbf{A} \wedge \mathbf{B}) \vee \mathbf{C})$ is logically equivalent to $(\mathbf{A} \vee \mathbf{C}) \wedge (\mathbf{B} \vee \mathbf{C})$ and that $((\mathbf{A} \vee \mathbf{B}) \wedge \mathbf{C})$ is logically equivalent to $(\mathbf{A} \wedge \mathbf{C}) \vee (\mathbf{B} \wedge \mathbf{C})$.

Exercise 13.6 (a) Suppose **B** is in conjunctive normal form and $\mathbf{A_1}, \ldots, \mathbf{A_n}$ is a list of the atomic formulas occurring in **B**. Show that $\sim\mathbf{B}$ is logically equivalent to a formula **C** in disjunctive normal form in which the list of atomic formulas is the same $\mathbf{A_1}, \ldots, \mathbf{A_n}$. (b) Similarly, suppose $\mathbf{B'}$ is in disjunctive normal form and $\mathbf{A_1}, \ldots, \mathbf{A_n}$ is a list of the atomic formulas occurring in $\mathbf{B'}$. Show that $\sim\mathbf{B'}$ is logically equivalent to a formula $\mathbf{C'}$ in conjunctive normal form in which the list of atomic formulas is the same $\mathbf{A_1}, \ldots, \mathbf{A_n}$. [Hint: Do parts (a) and (b) simultaneously by induction on the number of conjuncts for (a) and disjuncts for (b). If you have difficulty, read the proof of the next theorem and then try again.]

Theorem 13.3 For every quantifier-free formula **A**, there is a formula $\mathbf{A_c}$ in conjunctive normal form and a formula $\mathbf{A_d}$ in disjunctive normal form such that **A** is logically equivalent to $\mathbf{A_c}$ and to $\mathbf{A_d}$.

PROOF (By induction on the formation of **A**): The trick is to prove both the conjunctive and disjunctive forms simultaneously.

1. If **A** is an atomic formula, then it is both a conjunct (unique) and a disjunct (unique).
2. Suppose **A** is $\sim\mathbf{B}$. By the induction hypothesis, **B** has a conjunctive normal form $\mathbf{B_c}$, so **A** is logically equivalent to $\sim\mathbf{B_c}$. But then, by exercise 13.6, this is logically equivalent to a formula in disjunctive normal form. To get the conjunctive normal form is similar.
3. Suppose **A** is $(\mathbf{B} \supset \mathbf{C})$. (a) To find the conjunctive normal form, notice that $(\mathbf{B} \supset \mathbf{C})$ is logically equivalent to $(\sim\mathbf{B} \vee \mathbf{C})$ and **B** and **C** satisfy the induction hypothesis, so **A** is equivalent to $(\sim\mathbf{B_d} \vee \mathbf{C_c})$. By exercise 13.6, $\sim\mathbf{B_d}$ is equivalent to a formula in conjunctive normal form $(\mathbf{B_1} \wedge \cdots \wedge \mathbf{B_k})$, where the $\mathbf{B_i}$ are disjunctions. Thus, **A** is logically equivalent to $((\mathbf{B_1} \wedge \cdots \wedge \mathbf{B_k}) \vee \mathbf{C_c})$. By using exercise 13.5, we find that this is logically equivalent to $((\mathbf{B_1} \vee \mathbf{C_c}) \wedge \cdots \wedge (\mathbf{B_k} \vee \mathbf{C_c}))$. But $\mathbf{C_c}$ is, of course, a conjunction $(\mathbf{C_1} \wedge \cdots \wedge \mathbf{C_t})$, so we do the same thing again, i.e., we have $((\mathbf{B_1} \vee (\mathbf{C_1} \wedge \cdots \wedge \mathbf{C_t})) \cdots (\mathbf{B_k} \vee (\mathbf{C_1} \wedge \cdots \wedge \mathbf{C_t}))$; and so, once more by exercise 13.5, we have that this is logically equivalent to $(\mathbf{B_1} \vee \mathbf{C_1}) \wedge (\mathbf{B_1} \vee \mathbf{C_2}) \wedge \cdots \wedge (\mathbf{B_1} \vee \mathbf{C_m}) \wedge \cdots \wedge (\mathbf{B_k} \vee \mathbf{C_m})$, which is in conjunctive normal form since all $\mathbf{B_i}$ and $\mathbf{C_j}$ are disjunctions.

(b) To find the disjunctive normal form is easier and left as an exercise for the reader. ▌

13.3 SKOLEM FUNCTIONS

When we say that an n-ary function f is defined on some domain D, we mean exactly that *for every* n-tuple of elements of D, $(d_1, \ldots, d_n)$, *there is* an element $d' \in D$ associated to it by that function f, and we usually indicate d' by $f(d_1, \ldots, d_n)$. Using this idea, we notice that the formula **A** which is $(\forall \mathbf{x_1}) \cdots (\forall \mathbf{x_n})(\exists \mathbf{x_{n+1}})\mathbf{P^{n+1}}(\mathbf{x_1}, \ldots, \mathbf{x_n}, \mathbf{x_{n+1}})$ can very naturally be interpreted in a similar way. If we consider **A** and $\mathbf{A^S}$, which is

$$\mathbf{P^{n+1}}(\mathbf{x_1}, \ldots, \mathbf{x_n}, \mathbf{f^n}(\mathbf{x_1}, \ldots, \mathbf{x_n})),$$

we notice, among other things, that for each relational system $\mathfrak{A}$ in which **A** is true, there is a relational system $\mathfrak{A}'$ in which $\mathbf{A^S}$ is true. If we take the formula $(\exists \mathbf{x_1})\mathbf{P^1}(\mathbf{x_1})$, we can associate with it, in a similar manner, the formula $\mathbf{P}^1(\mathbf{a}_1)$ for some constant $\mathbf{a}_1$. Consider a more complicated example: **A** is $(\forall \mathbf{x}_1)(\forall \mathbf{x}_2)(\exists \mathbf{x}_3)(\forall \mathbf{x}_4)(\exists \mathbf{x}_5)(\exists \mathbf{x}_6)\mathbf{P}^6(\mathbf{x}_1, \mathbf{x}_2, \mathbf{x}_3, \mathbf{x}_4, \mathbf{x}_5, \mathbf{x}_6)$ and $\mathbf{A}^{\mathbf{S}}$ is

$$\mathbf{P}^6(\mathbf{x}_1, \mathbf{x}_2, \mathbf{f}_1{}^2(\mathbf{x}_1, \mathbf{x}_2), \mathbf{x}_4, \mathbf{f}_1{}^3(\mathbf{x}_1, \mathbf{x}_2, \mathbf{x}_4), \mathbf{f}_2{}^3(\mathbf{x}_1, \mathbf{x}_2, \mathbf{x}_4)).$$

Exercise 13.7 For each of the three examples, show that if there is a relational system $\mathfrak{A}$ in which **A** is true, then there is a relational system $\mathfrak{A}'$ in which $\mathbf{A}^{\mathbf{S}}$ is true.

If **A** is a closed formula in prenex normal form (in a first-order language with or without equality, L(∗)), of the form

$$(\exists \mathbf{x}_{0,1}) \cdots (\exists \mathbf{x}_{0,n_0})(\forall \mathbf{x}_{1,1}) \cdots (\forall \mathbf{x}_{1,n_1})(\exists \mathbf{x}_{2,1}) \cdots (\exists \mathbf{x}_{2,n_2})$$
$$(\forall \mathbf{x}_{3,1}) \cdots (\exists \mathbf{x}_{m,1}) \cdots (\exists \mathbf{x}_{m,n_m})(\forall \mathbf{x}_{m+1,1}) \cdots (\forall \mathbf{x}_{m+1,n_{m+1}})$$
$$\mathbf{B}(\mathbf{x}_{0,1}, \ldots, \mathbf{x}_{0,n_0}, \mathbf{x}_{1,1}, \ldots, \mathbf{x}_{1,n_1}, \mathbf{x}_{2,1}, \ldots, \mathbf{x}_{2,n_2}, \mathbf{x}_{3,1}, \ldots, \mathbf{x}_{m,1}, \ldots, \mathbf{x}_{m,n_m},$$
$$\mathbf{x}_{m+1,1}, \ldots, \mathbf{x}_{m+1,n_{m+1}})$$

then $\mathbf{A}^{\mathbf{S}}$ (the "Skolem function form of **A**") is defined to be

$$\mathbf{B}(\dot{\mathbf{a}}_{0,1}, \ldots, \dot{\mathbf{a}}_{0,n_0}, \mathbf{x}_{1,1}, \ldots, \mathbf{x}_{1,n_1}, \dot{\mathbf{f}}_1{}^{n_1}(\mathbf{x}_{1,1}, \ldots, \mathbf{x}_{1,n_1}), \ldots,$$
$$\dot{\mathbf{f}}_{n_2}^{n_1}(\mathbf{x}_{1,1}, \ldots, \mathbf{x}_{1,n_1}), \mathbf{x}_{3,1}, \ldots, \mathbf{x}_{3,m_3}, \ldots,$$
$$\dot{\mathbf{f}}_1^{(n_1+n_3+\cdots+n_{m-1})}(\mathbf{x}_{1,1}, \ldots, \mathbf{x}_{1,n_1}, \mathbf{x}_{3,1}, \ldots, \mathbf{x}_{3,n_3}, \ldots, \mathbf{x}_{m-1}, \ldots,$$
$$\mathbf{x}_{m-1,n_{m-1}}), \ldots,$$
$$\dot{\mathbf{f}}_{n_m}^{(n_1+n_3+\cdots+n_{m-1})}(\mathbf{x}_{1,1}, \ldots, \mathbf{x}_{1,n_1}, \mathbf{x}_{3,1}, \ldots, \mathbf{x}_{3,n_3}, \ldots, \mathbf{x}_{m-1,1}, \ldots, \mathbf{x}_{m-1,n_{m-1}}),$$
$$\mathbf{x}_{m+1,1}, \ldots, \mathbf{x}_{m+1,n_{m+1}})$$

where the $\dot{\mathbf{a}}_{i,j}$ and the $\dot{\mathbf{f}}_j{}^k$ are names of constants and function symbols which do not occur in L(∗). The convention is often made that a constant is a 0-ary function. The functions and constants introduced in obtaining $\mathbf{A}^{\mathbf{S}}$ from **A** are known as *Skolem functions*. We will say that "$\dot{\mathbf{a}}_{0,i}$ is associated with the *i*th existential quantifier" and that, in general, "$\dot{\mathbf{f}}_j^{n_1+\cdots+n_{i-1}}$ is associated with the $(n_1 + \cdots + n_{i-1} + j)$th existential quantifier." If we have a set of formulas Γ in prenex normal form, then Γ^S can be obtained from Γ by going through the above procedure for each member of Γ, taking new Skolem functions for each. If Γ is written on L(∗) then, of course, Γ^S is written on $\mathsf{L}(*^S) = \mathsf{L}(*) \cup \{\dot{\mathbf{a}}_{0,1}, \ldots, \dot{\mathbf{f}}_j{}^k, \ldots\}$, where $\mathsf{L}(*) \cap \{\dot{\mathbf{a}}_{0,1}, \ldots, \dot{\mathbf{f}}_i{}^k, \ldots\} = \varnothing$.

Exercise 13.8 If **A** is $(\exists \mathbf{x}_1)(\forall \mathbf{x}_2)(\forall \mathbf{x}_3)(\exists \mathbf{x}_4)(\exists \mathbf{x}_5)\mathbf{P}^6(\mathbf{a}_1, \mathbf{x}_1, \mathbf{x}_2, \mathbf{x}_3, \mathbf{x}_4, \mathbf{x}_5)$ and **B** is $(\forall \mathbf{x})(\forall \mathbf{y})(\exists \mathbf{z})(\mathbf{P}^3(\mathbf{f}^2(\mathbf{x}, \mathbf{y}), \mathbf{a}_1, \mathbf{z}) \supset \mathbf{P}^2(\mathbf{x}, \mathbf{a}_2))$, write $\mathbf{A}^{\mathbf{S}}$ and $\mathbf{B}^{\mathbf{S}}$.

Exercise 13.9 (a) Prove the following. Let $\mathfrak{A}$ be a relational system with universe *U* and let **A** be a closed formula in prenex normal form

$$(\forall \mathbf{y}_1)(\forall \mathbf{y}_2)(\exists \mathbf{x}_1)(\exists \mathbf{x}_2)(\forall \mathbf{y}_3)(\exists \mathbf{x}_3)\mathbf{B}$$

There are two binary and one 3-ary functions, F, G, and H, respectively, which are defined everywhere in U and have the following property: if φ is an assignment for which $V_\varphi(\mathbf{A}) = \mathrm{t}$ and ψ is any assignment such that

$$\psi(\mathbf{x}_1) = F(\psi(\mathbf{y}_1), \psi(\mathbf{y}_2))$$
$$\psi(\mathbf{x}_2) = G(\psi(\mathbf{y}_1), \psi(\mathbf{y}_2))$$
$$\psi(\mathbf{x}_3) = H(\psi(\mathbf{y}_1), \psi(\mathbf{y}_2), \psi(\mathbf{y}_3))$$

and $\psi = \varphi$ everywhere except at $\mathbf{y}_1$, $\mathbf{y}_2$, $\mathbf{y}_3$, $\mathbf{x}_1$, $\mathbf{x}_2$, $\mathbf{x}_3$, then $V_\psi(\mathbf{B}) = \mathrm{t}$. (b) State and prove the general form for which part (a) is a special case. (Warning: the statement of the theorem is very messy; hint: the proof is by induction on the number of quantifiers.)

Suppose T is a first-order theory and T^S is the first-order theory whose axioms are the Skolem function forms of the axioms of T. How do T and T^S compare? Of particular interest for proving the Herbrand theorem is the question of whether T is consistent if and only if T^S is. Notice that for a formula **A**, if $\vdash_{\mathsf{TS}} \mathbf{A}^\mathbf{S}$, then $\vdash_{\mathsf{TS}} \mathbf{A}$, by using exercise 11.2 as many times as necessary.

Exercise 13.10 Prove that if $\vdash_{\mathsf{TS}} \mathbf{B}(\mathbf{x}_1, \mathbf{x}_2, \mathbf{f}_1^2(\mathbf{x}_1, \mathbf{x}_2), \mathbf{f}_2^2(\mathbf{x}_1, \mathbf{x}_2), \mathbf{x}_5)$, then $\vdash_{\mathsf{TS}} (\forall \mathbf{x}_1)(\forall \mathbf{x}_2)(\exists \mathbf{x}_3)(\exists \mathbf{x}_4)(\forall \mathbf{x}_5)\mathbf{B}(\mathbf{x}_1, \mathbf{x}_2, \mathbf{x}_3, \mathbf{x}_4, \mathbf{x}_5)$.

Exercise 13.11 One of the following statements can be proved in one or two lines: (a) If T is consistent, then T^S is consistent. (b) If T^S is consistent, then T is consistent. Which is the easy one? Give the proof. What do you need to know about the provability in T^S of formulas in the language of T in order to prove the more difficult of these two statements?

If L(1) and L(2) are two first-order languages (with or without equality) such that $\mathsf{L}(1) \subset \mathsf{L}(2)$, and $\mathfrak{A}_2$ is a relational system for L(2), then by $\mathfrak{A}_2|_{\mathsf{L}(1)}$ we mean the relational system whose universe is the same as that of $\mathfrak{A}_2$, whose set of functions is the subset of functions of $\mathfrak{A}_2$ which correspond to the set of function symbols of L(1), and whose set of relations is the subset of the relations of L(2) which correspond to the set of predicate symbols (and possibly $=$) of L(1).

Exercise 13.12 Suppose that L(1) and L(2) are two first-order languages such that $\mathsf{L}(1) \subset \mathsf{L}(2)$, and that $\mathfrak{A}_2$ is a structure for L(2). Show that if $\mathbf{A} \in \mathsf{F}(1)$ is true in $\mathfrak{A}_2$, then **A** is true in $\mathfrak{A}_2|_{\mathsf{L}(1)}$. (Hint: Obvious.)

Exercise 13.13 Let $\mathsf{L}(G)$ be a first-order language with equality with no constants, no predicate symbols (other than $=$), and just one binary function symbol $+$. Let $\mathsf{T}(G)$ be the first-order theory on $\mathsf{L}(G)$ whose axioms are those of $\mathsf{V}(*^{=})$ plus:

$$(\forall x)(\forall y)(\forall z)((x + (y + z)) = ((x + y) + z))$$
$$(\exists y)(\forall x)((x + y) = x)$$
$$(\exists z)(\forall x)(\exists y)((x + y) = z)$$
$$(\forall x)(\forall y)(\forall z)((x = y) \supset (((x + z) = (y + z)) \wedge ((z + x) = (z + y))))$$

(a) Find a model $\mathfrak{A}$ for $\mathsf{T}(G)$. (b) Write the axioms for $\mathsf{T}^{\mathsf{S}}(G)$. (c) Construct a model $\mathfrak{A}'$ for $\mathsf{T}^{\mathsf{S}}(G)$ such that $\mathfrak{A}'|_{\mathsf{L}(G)} = \mathfrak{A}$.

Theorem 13.4 (Skolem's Theorem) Let $\mathsf{L}(*)$ be a first-order language, with or without equality. Let T be a first-order theory written on $\mathsf{L}(*)$ with nonlogical axioms $\{\mathbf{A_1}, \mathbf{A_2}, \ldots\}$, and let T^{S} be the first-order theory with nonlogical axioms $\{\mathbf{A_1}^{\mathbf{S}}, \mathbf{A_2}^{\mathbf{S}}, \ldots\}$, where $\mathbf{A_i}^{\mathbf{S}}$ is the Skolem function form of $\mathbf{A_i}$. For $\mathbf{A} \in \mathsf{F}(*)$, $\vdash_{\mathsf{T}} \mathbf{A}$ if and only if $\vdash_{\mathsf{T}^{\mathsf{S}}} \mathbf{A}$.

PROOF That $\vdash_{\mathsf{T}} \mathbf{A}$ implies $\vdash_{\mathsf{T}^{\mathsf{S}}} \mathbf{A}$ is an exercise. If T is not consistent, then anything is provable in T, so the theorem is proved. If T is consistent, then by the Löwenheim–Skolem theorems, it has countable (finite and/or infinite) models. For each countable model $\mathfrak{A}$, we will construct a model $\mathfrak{A}'$ such that (i) $\mathfrak{A}'$ is a model of T^{S} and (ii) $\mathfrak{A}'|_{\mathsf{L}(*)} = \mathfrak{A}$. Once this is done, the proof of the theorem is completed by remarking that $\vdash_{\mathsf{T}^{\mathsf{S}}} \mathbf{A}$ implies that $\mathbf{A}$ is true in $\mathfrak{A}'$ and therefore by exercise 13.12, $\mathbf{A}$ is true in $\mathfrak{A}$. Thus, $\mathbf{A}$ is true in all countable models of T and so $\vdash_{\mathsf{T}} \mathbf{A}$. We can assume that each axiom $\mathbf{A_i}$ is in closed prenex normal form. Let the Skolem functions in $\{\mathbf{A_1}^{\mathbf{S}}, \mathbf{A_2}^{\mathbf{S}}, \ldots\}$ be $\mathbf{f_{i,j}}$ where i, j indicates that $\mathbf{f_{i,j}}$ corresponds to the jth existential quantifier in $\mathbf{A_i}$. For each countable model $\mathfrak{A}$ of T, the universe U of $\mathfrak{A}$ is to be the universe of the relational system $\mathfrak{A}'$. $\mathfrak{A}'$ is to have all the functions and relations of $\mathfrak{A}$, but it must also have functions corresponding to the Skolem functions. Thus, the function Φ which associates the language $\mathsf{L}(*)$ with $\mathfrak{A}$ must be extended to Φ' so that for each $\mathbf{f_{i,j}}$ there is a $\Phi'\mathbf{f_{i,j}}$. For $\Phi'\mathbf{f_{i,j}}$ to be defined, we must be able to say what element of U $(\Phi'\mathbf{f_{i,j}})(u_1, \ldots, u_r)$ is for every r-tuple of elements in U, assuming that $\mathbf{f_{i,j}}$ is an r-ary function. For each $\mathbf{A_i}$, we will define all the $\Phi'\mathbf{f_{i,j}}$ simultaneously. Suppose $\mathbf{A_i}$ has m existential quantifiers. Let k_j be the number of universal quantifiers to the left of the jth existential quantifier. Write $\mathbf{A_i}$ as $(\cdots)(\exists \mathbf{x_m})\mathbf{B}$, where $\mathbf{B}$ is quantifier-free or in prenex normal form with only universal quantifiers. We make the convention that the individual variables quantified universally are $\mathbf{y_i}$ [i.e., $(\forall \mathbf{y_i})$] and those quantified existentially are $\mathbf{x_j}$ [i.e., $(\exists \mathbf{x_j})$]. It follows from exercise 13.9 that there are m (not named in $\mathfrak{A}$) functions $F_1, \ldots, F_m$ defined on U such that if ψ is an assignment with the property that

$$\psi(\mathbf{x_1}) = F_1(\psi(\mathbf{y_1}), \ldots, \psi(\mathbf{y_{k_1}}))$$
$$\vdots$$
$$\psi(\mathbf{x_m}) = F_m(\psi(\mathbf{y_1}), \ldots, \psi(\mathbf{y_{k_m}}))$$

then $V_\psi(\mathbf{B}) = \mathrm{t}$ in $\mathfrak{A}$. So we simply define the value of $(\Phi'\mathbf{f_{i,j}})(u_1, \ldots, u_{k_j})$ to be $F_j(u_1, \ldots, u_{k_j})$. That is, F_j is a (named) function in $\mathfrak{A}'$ and $\Phi'\mathbf{f_{i,j}}$ is F_j. For any assignment φ, we claim that $V_\varphi(\mathbf{A_i^S}) = \mathrm{t}$ in $\mathfrak{A}'$. For,

$V_\varphi(\mathbf{f_{i,j}}(\mathbf{y_1}, \ldots \mathbf{y_{k_j}}))$
$= (\Phi'\mathbf{f_{i,j}})(\varphi(\mathbf{y_1}), \ldots, \varphi(\mathbf{y}_{k_j}))$ by the definition of valuation
$= F_j(\varphi(\mathbf{y_1}), \ldots, \varphi(\mathbf{y_{k_j}}))$ by the definition of $\Phi'\mathbf{f_{i,j}}$.

In terms of $\mathfrak{A}$ let φ' be an assignment such that $\varphi' = \varphi$ except at $\mathbf{x_1}, \ldots, \mathbf{x_m}$ and $\varphi'(\mathbf{x_j}) = F_j(\varphi'(\mathbf{y_1}), \ldots, \varphi'(\mathbf{y_{k_j}}))$, $1 \le j \le m$. Since $V_{\varphi'}(\mathbf{B}) = \mathrm{t}$ in $\mathfrak{A}$, $V_\varphi(\mathbf{A_i^S})$ must be t in $\mathfrak{A}'$. It is clear from the construction that $\mathfrak{A}'|_{\mathsf{L}(*)} = \mathfrak{A}$. ∎

Corollary T is consistent if and only if $\mathsf{T^S}$ is consistent.

13.4 TREES AND KÖNIG'S LEMMA

The idea of a tree and the basic theorem about trees, König's lemma, are very useful in mathematics generally and will be used in particular in the next section. An example of a tree is given in figure 13.1. One may imagine other

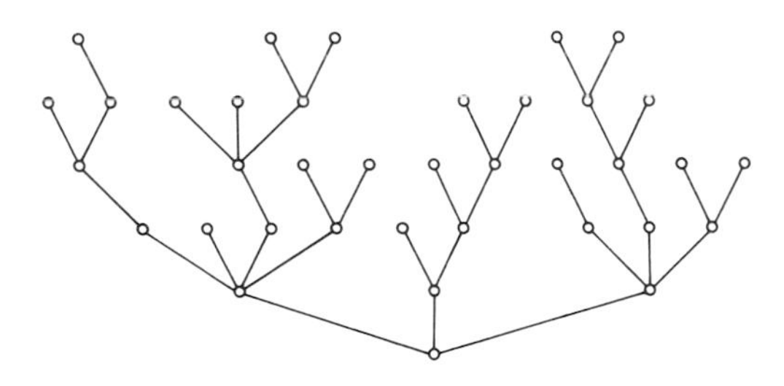

FIGURE 13.1

examples, some with infinitely long branches, some with infinitely many branches branching out from a single point, etc. A *tree* $\mathscr{T}$ is defined by a set (finite or infinite) of sets of points $\{V_0, V_1, \ldots, V_i, \ldots\}$, where for all $i \neq j$, $V_i \cap V_j = \varnothing$, and a set of directed line segments, called *arcs*, which satisfy the following properties:

1. $\overline{\overline{V}}_0 = 1$.

2. For every $v \in V_i$, for all $i > 0$, there is a unique $v' \in V_{i-1}$ such that there is an arc directed from v' to v [write arc(v', v)].
3. There are no arcs other than those required in 2.

We call the set V_i the ith *level vertices*. A *branch* with base v_{i_0} is a sequence of vertices $v_{i_0}, v_{i_1}, \ldots$ (finite or infinite) and arcs such that if $v_{i_0} \in V_{i_0}$, then $v_{i_1} \in V_{i_0+1}, \ldots, v_{i_j} \in V_{i_0+j}, \ldots$, and arc($v_{i_0}$, v_{i_1}), arc(v_{i_1}, v_{i_2}), ... We say that each vertex v_{i_j} of the sequence is a vertex belonging to that branch. The *branch has length n* if $n+1$ vertices (and therefore n arcs) belong to it. The branch has infinite length if for each n, the length of the branch is greater than that n. If for all i, the cardinality of V_i is finite, then the *tree is finitely branched.* The *tree is infinite* if the cardinality of the set of all its vertices is infinite.

Lemma 13.1 (König's Lemma) Let $\mathcal{T}$ be a finitely branched tree. If for every n, there is a branch of length greater than n, then there is a branch of infinite length.

PROOF If v is a vertex of $\mathcal{T}$, define the set of vertices D_v as follows: $v' \in D_v$ if v' is a vertex of a branch based at v. Let v_0 denote the unique vertex in V_0. Obviously, $\mathcal{T}$ is infinite if and only if the cardinality of D_{v_0} is infinite. $\mathcal{T}$ is infinite because if it were finite with, say, m vertices, then it could have no branch of length greater than m, in contradiction to the second hypothesis of the lemma. Since $\mathcal{T}$ is finitely branched, there must be a $v' \in V_1$ such that arc(v_0, v') and $D_{v'}$ has infinite cardinality. In the same way, there must be a $v'' \in V_2$ such that arc(v', v'') and $D_{v''}$ has infinite cardinality. The process can be continued without bound so as to obtain

$$\text{arc}(v_0, v'), \text{arc}(v', v''), \ldots, \text{arc}(v^{(n-1)}, v^{(n)}), \ldots$$

Thus, the sequence $v_0, v', v'', \ldots, v^{(n-1)}, v^{(n)}, \ldots$ and corresponding arcs is a branch with infinite length. ▮

13.5 HERBRAND'S THEOREM

Suppose we have a finitely axiomatizable theory with equality.† Let $\mathbf{F_1}, \ldots, \mathbf{F_n}$ be the nonlogical axioms, i.e., the axioms other than those of $\mathsf{V}(_*{}^=)$. We would like to decide, of an arbitrary formula $\mathbf{B}$ whether or not $\vdash_{\{\mathbf{F}, \ldots, \mathbf{F_n}\}} \mathbf{B}$. Generally, this cannot be done since there are many first-order theories which

† Herbrand's theorem for finitely axiomatizable theories without equality is somewhat easier to prove and is given as an exercise, exercise 13.15.

are not decidable. However, when the theory is recursively axiomatizable, as it is in this case, one can search for **B** by recursively enumerating the theorems and as each is enumerated checking to see if it is **B**. But, of course, such a procedure never allows one to conclude that **B** is *not* a theorem. (Such procedures are sometimes called "semi-decisions procedures.") Naturally, the Herbrand theorem cannot help us out of this basic difficulty. What is does do is allow us to search in a set of formulas much easier to generate than the set of theorems. Rather than consider all the theorems, we consider something like the substitution instances of $\{\mathbf{F_1}, \ldots, \mathbf{F_n}, \sim\mathbf{B}\}$, i.e., formulas obtained from these by substituting terms for free variables. Thus, the essential idea is to change the framework of the search from formulas in a first-order language to formulas which can be treated as belonging to a propositional language.

We know that:

(1) $\vdash_{\{\mathbf{F_1},\ldots,\mathbf{F_n}\}}\mathbf{B}$ if and only if **B** is true in all models of $\{\mathbf{F_1},\ldots,\mathbf{F_n}\}$.

(2) **B** is true in all models of $\{\mathbf{F_1},\ldots,\mathbf{F_n}\}$ if and only if $\{\mathbf{F_1},\ldots,\mathbf{F_n}, \sim\mathbf{B}\}$ has no models, i.e., is inconsistent.

(3) $\{\mathbf{F_1},\ldots,\mathbf{F_n}, \sim\mathbf{B}\}$ has no models if and only if, by previous sections of this chapter, in particular the corollary to the Skolem theorem, $\{\mathbf{F_1}^{\mathbf{S}},\ldots,\mathbf{F_n}^{\mathbf{S}}, (\sim\mathbf{B})^{\mathbf{S}}\}$ has no models, where $\mathbf{F_i}^{\mathbf{S}}$ and $(\sim\mathbf{B})^{\mathbf{S}}$ are the formulas obtained by taking the prenex normal form, then the conjunctive normal form, and finally the Skolem function form starting from $\mathbf{F_i}$ and $\sim\mathbf{B}$, $1 \le i \le n$.

We also know that it is a fairly simple (at least in principle) mechanical procedure to compute the truth table of a formula and thus ascertain whether or not it is a contradiction. The Herbrand theorem will show us that $\{\mathbf{F_1}^{\mathbf{S}},\ldots,\mathbf{F_n}^{\mathbf{S}}, (\sim\mathbf{B})^{\mathbf{S}}\}$ is consistent if and only if in searching through a certain infinite recursively enumerable set of formulas we can find one formula which is a contradiction. Before we can state the theorem properly and prove it, we must define the "certain recursively enumerable set of formulas." We now turn our attention to that task.

For the purpose of this section, we define the *rank of a term* **t** (write $\rho(\mathbf{t})$) as follows:

1. $\rho(\mathbf{x_i}) = 0$ for all individual variables $\mathbf{x_i}$, and $\rho(\mathbf{a_i}) = 0$ for all names of constants $\mathbf{a_i}$.
2. If **f** is a k-ary function symbol and $\mathbf{t_1}, \ldots, \mathbf{t_k}$ are terms, then

$$\rho(\mathbf{f}(\mathbf{t_1}, \ldots, \mathbf{t_k})) = 1 + \max_{1 \le i \le k} (\rho(\mathbf{t_i})).$$

We will call formulas of the form $(\mathbf{t} = \mathbf{t}')$ *equations* and define the rank of an equation to be $\max(\rho(\mathbf{t}), \rho(\mathbf{t}'))$.

Let Γ be any set of formulas. We say a symbol occurs in Γ if it occurs in some formula belonging to Γ. A term occurs in Γ if it occurs in some formula belonging to Γ. A formula occurs in Γ if either it belongs to Γ or occurs as a subformula of a formula belonging to Γ.

Let Γ be any finite set of quantifier-free formulas. Let $\mathbf{x_1}, \ldots, \mathbf{x_n}$ be a list of all the individual variables which occur in Γ, and let $\mathbf{f_1}, \ldots, \mathbf{f_r}$ be a list of all the function symbols occurring in Γ. We define, inductively, the *Herbrand universe for* Γ, $H(\Gamma)$, as follows:

1. $H_0(\Gamma)$ is the set of rank 0 terms that can be formed from the set of listed symbols; i.e.,

$$H_0(\Gamma) = \{\mathbf{x_1}, \ldots, \mathbf{x_n}, \mathbf{a_1}, \ldots, \mathbf{a_m}\}.$$

2. $H_n(\Gamma)$ is the set of rank n terms that can be formed from the set of listed symbols. For example, if $\mathbf{t_1}, \ldots, \mathbf{t_k} \in H_{n-1}(\Gamma)$ and $\mathbf{f}$ is a k-ary function symbol in the list, then $\mathbf{f(t_1, \ldots, t_k)} \in H_n(\Gamma)$.

Let $H(\Gamma) = \bigcup_{n \in N} H_n(\Gamma)$.

Exercise 13.14 Let Γ be

$$\{\mathbf{P(f(x, a), g(g(b))) \vee \sim Q(f(a, a))}, \quad \mathbf{f(a, g(y)) = f(b, y)}\}$$

Write out $H_0(\Gamma)$, $H_1(\Gamma)$, and $H_2(\Gamma)$. For what n does

$$\mathbf{f(g(g(f(x, a))), f(b, f(g(y), g(a))))}$$

belong to $H_n(\Gamma)$?

A set of formulas Γ is defined to be *equationally closed* if it satisfies the following conditions:

1. If $\mathbf{(t = t')}$ occurs in Γ, then $\mathbf{((t = t') \supset (t' = t))} \in \Gamma$.
2. If $\mathbf{(t = t')}$ and $\mathbf{(t' = t'')}$ occur in Γ, then

$$\mathbf{(t = t') \supset ((t' = t'') \supset (t = t''))} \in \Gamma$$

3. If $\mathbf{(t_1 = t_1')}, \ldots, \mathbf{(t_k = t_k')}$ occur in Γ and $\mathbf{f}$ is a k-ary function symbol occurring in Γ, then

$$\mathbf{((t_1 = t_1') \wedge \cdots \wedge (t_k = t_k')) \supset (f(t_1, \ldots, t_k) = f(t_1', \ldots, t_k'))} \in \Gamma$$

Also, if $\mathbf{P}$ is a k-ary predicate symbol occurring in Γ, then

$$\mathbf{((t_1 = t_1') \wedge \cdots \wedge (t_k = t_k')) \supset (P(t_1, \ldots, t_k) \equiv P(t_1', \ldots, t_k'))} \in \Gamma$$

Referring once more to the finitely axiomatizable theory with nonlogical axioms $\mathbf{F_1}, \ldots, \mathbf{F_n}$, recall that $\{\mathbf{F_1}^{\mathbf{S}}, \ldots, \mathbf{F_n}^{\mathbf{S}}, (\sim\mathbf{B})^{\mathbf{S}}\}$ are all in Skolem function form and conjunctive normal form. Let $\mathbf{C_1}, \ldots, \mathbf{C_p}$ be a list of all the conjuncts that occur there. Obviously, $\{\mathbf{F_1}^{\mathbf{S}}, \ldots, \mathbf{F_n}^{\mathbf{S}}, (\sim\mathbf{B})^{\mathbf{S}}\}$ has a model if and only if $\{\mathbf{C_1}, \ldots, \mathbf{C_p}\}$ does. Henceforth, let χ denote $\{\mathbf{C_1}, \ldots, \mathbf{C_p}\}$. $H(\chi)$ is the Herbrand universe for χ.

We now define, inductively, two sets of formulas, $\Lambda(\chi)$ and $\Gamma(\chi)$, in such a way that $\Lambda(\chi) \subset \Gamma(\chi)$ and $\Gamma(\chi)$ is equationally closed. For each $n \geq 0$, for each $\mathbf{C} \in \chi$, if $\mathbf{C}$ has individual variables $\mathbf{x_1}, \ldots, \mathbf{x_k}$, then for each k-tuple of elements in $H_n(\chi)$, $(\mathbf{h_1}, \ldots, \mathbf{h_k})$, $\Lambda_n(\chi)$ is to include $\mathbf{C}_{(\mathbf{x_1}, \ldots, \mathbf{x_k})}[\mathbf{h_1}, \ldots, \mathbf{h_k}]$. [Notice that $\Lambda_n(\chi) \supset \Lambda_{n-1}(\chi)$.] Let $\Gamma_0(X) = \Lambda_0(\chi)$, and for each $n > 0$, $\Gamma_n(\chi)$ is to include $\Lambda_n(\chi)$ and, in addition:

1. If $(\mathbf{t} = \mathbf{t}')$ is an equation of rank $\leq n$ occurring in $\Gamma_{n-1}(\chi)$, then $\Gamma_n(\chi)$ is to include the formula $((\mathbf{t} = \mathbf{t}') \supset (\mathbf{t}' = \mathbf{t}))$.

2. If $(\mathbf{t} = \mathbf{t}')$ and $(\mathbf{t}' = \mathbf{t}'')$ are equations occurring in $\Gamma_{n-1}(\chi)$ such that $\max(\rho(\mathbf{t}), \rho(\mathbf{t}'), \rho(\mathbf{t}'')) \leq n$, then the formula $((\mathbf{t} = \mathbf{t}') \supset ((\mathbf{t}' = \mathbf{t}'') \supset (\mathbf{t} = \mathbf{t}''))$ is to be in $\Gamma_n(\chi)$.

3. For each k-ary function symbol $\mathbf{f}$ and each k-ary predicate symbol $\mathbf{P}$ occurring in χ, if $(\mathbf{t_1} = \mathbf{t_1}'), \ldots, (\mathbf{t_k} = \mathbf{t_k}')$ are equations occurring in $\Gamma_{n-1}(\chi)$ of rank $\leq n$, then both of the following formulas are members of $\Gamma_n(\chi)$:

$$((\mathbf{t_1} = \mathbf{t_1}') \wedge \cdots \wedge (\mathbf{t_k} = \mathbf{t_k}')) \supset (\mathbf{f}(\mathbf{t_1}, \ldots, \mathbf{t_k}) = \mathbf{f}(\mathbf{t_1}', \ldots, \mathbf{t_k}'))$$
$$((\mathbf{t_1} = \mathbf{t_1}') \wedge \cdots \wedge (\mathbf{t_k} = \mathbf{t_k}')) \supset (\mathbf{P}(\mathbf{t_1}, \ldots, \mathbf{t_k}) \equiv \mathbf{P}(\mathbf{t_1}', \ldots, \mathbf{t_k}'))$$

[By definition, $\Gamma_{n-1}(\chi) \subset \Gamma_n(\chi)$.] Define $\Lambda(\chi)$ to be $\bigcup_{n \in N} \Lambda_n(\chi)$ and $\Gamma(\chi)$ to be $\bigcup_{n \in N} \Gamma_n(\chi)$. Since, for each n, $\Lambda_n(\chi)$ and $\Gamma_n(\chi)$ are finite sets, $\Lambda(\chi)$ and $\Gamma(\chi)$ are recursively enumerable sets. Let $\mathbf{D_1}, \mathbf{D_2}, \ldots$ be a recursive enumeration of the formulas of $\Gamma(\chi)$.

Lemma 13.2 $\Gamma(\chi)$ is equationally closed.

Exercise 13.15 Prove the lemma.

$\Gamma(\chi)$ is the "certain infinite recursively enumerable set of formulas" mentioned earlier, and so we are in a position to state the theorem. Let $\Gamma(\chi)$ be called "the Herbrand open formulas for $\{\mathbf{F_1}, \ldots, \mathbf{F_n}, \sim\mathbf{B}\}$."

Theorem 13.5† (Herbrand) Let $\{\mathbf{F_1}, \ldots, \mathbf{F_n}\}$ be the nonlogical axioms of a (finitely axiomatizable) first-order theory with equality. $\vdash_{\{\mathbf{F_1}, \ldots, \mathbf{F_n}\}} \mathbf{B}$ if and only if some finite conjunction of Herbrand open formulas for

$$\{\mathbf{F_1}, \ldots, \mathbf{F_n}, \sim\mathbf{B}\}$$

is a contradiction.

† This theorem was first proved in Herbrand (1930).

PROOF As mentioned earlier, $\vdash_{\{\mathbf{F_1}, \ldots, \mathbf{F_n}\}} \mathbf{B}$ if and only if $\{\mathbf{C_1}, \ldots, \mathbf{C_p}\}$ is not consistent. Thus, what must be shown is that $\{\mathbf{C_1}, \ldots, \mathbf{C_p}\}$ has no models if and only if there is a finite subset $\{\mathbf{D_{i_1}}, \ldots, \mathbf{D_{i_k}}\}$, of elements of $\Gamma(\chi)$ such that $(\mathbf{D_{i_1}} \wedge \cdots \wedge \mathbf{D_{i_k}})$ is a contradiction. [Abbreviate $(\mathbf{D_{i_1}} \wedge \cdots \wedge \mathbf{D_{i_k}})$ by $\bigwedge_{1 \le j \le k} \mathbf{D_{i_1}}$.] The proof of the theorem continues until the end of this section.

We show first that if there is a set $\{\mathbf{D_{i_1}}, \ldots, \mathbf{D_{i_k}}\}$ such that $\bigwedge_{1 \le j \le k} \mathbf{D_{i_j}}$ is a contradiction, then χ cannot have a model. Suppose χ does have a model, $\mathfrak{A}$. Let φ_0 be any arbitrary, but fixed, assignment of the individual variables of the language to the elements of the universe of $\mathfrak{A}$. Let $\mathbf{A_1}, \ldots, \mathbf{A_m}$ be the atomic formulas occurring in $\bigwedge_{1 \le j \le k} \mathbf{D_{i_j}}$ and consider $V_{\varphi_0}(\mathbf{A_1}), \ldots, V_{\varphi_0}(\mathbf{A_m})$. Since $\bigwedge_{1 \le j \le k} \mathbf{D_{i_j}}$ is a contradiction, for each possible assignment of truth values to $\mathbf{A_1}, \ldots, \mathbf{A_m}$ (i.e., each row of the truth table for $\bigwedge_{1 \le j \le k} \mathbf{D_{i_j}}$), at least one $\mathbf{D_{i_j}}$ has the truth value $\mathfrak{f}$. Further, one of these truth value assignments is the same as $V_{\varphi_0}(\mathbf{A_1}), \ldots, V_{\varphi_0}(\mathbf{A_m})$ and so for the corresponding $\mathbf{D_{i_j}}$, $V_{\varphi_0}(\mathbf{D_{i_j}})$ is $\mathfrak{f}$. Consider that $\mathbf{D_{i_j}}$. By construction, it is one of the following forms:

1. $((\mathbf{t} = \mathbf{t}') \supset (\mathbf{t}' = \mathbf{t}))$.
2. $((\mathbf{t} = \mathbf{t}') \supset ((\mathbf{t}' = \mathbf{t}'') \supset (\mathbf{t} = \mathbf{t}'')))$.
3. $((\mathbf{t_1} = \mathbf{t_1}') \wedge \cdots \wedge (\mathbf{t_k} = \mathbf{t_k}')) \supset (\mathbf{f}(\mathbf{t_1}, \ldots, \mathbf{t_k}) = \mathbf{f}(\mathbf{t_1}', \ldots, \mathbf{t_k}'))$.
4. $((\mathbf{t_1} = \mathbf{t_1}') \wedge \cdots \wedge (\mathbf{t_k} = \mathbf{t_k}')) \supset (\mathbf{P}(\mathbf{t_1}, \ldots, \mathbf{t_k}) \equiv \mathbf{P}(\mathbf{t_1}', \ldots, \mathbf{t_k}'))$.
5. $\mathbf{C_{i(x_1 \ldots, x_k)}}[\mathbf{h_1}, \ldots, \mathbf{h_k}]$ for some $\mathbf{C_i} \in \chi$ and k-tuple of elements in $H(\chi)$.

By the definition of valuation, if $\mathbf{D_{i_j}}$ is any one of the first four forms, $V_{\varphi_0}(\mathbf{D_{i_j}})$ must be $\mathfrak{t}$ in $\mathfrak{A}$, so we consider the fifth possibility. That is,

$$V_{\varphi_0}(\mathbf{C_{i(x_1, \ldots, x_k)}}[\mathbf{h_1}, \ldots, \mathbf{h_k}])$$

is $\mathfrak{f}$. Since $\mathfrak{A}$ is supposed to be a model of χ, it must be the case that for all assignments φ and all $\mathbf{C_j} \in \chi$, $V_\varphi(\mathbf{C_j}) = \mathfrak{t}$ in $\mathfrak{A}$. However, let ψ be an assignment for which $\psi(\mathbf{x_j}) = V_{\varphi_0}(\mathbf{h_j})$, $1 \le j \le k$. By exercise 10.21, $V_\psi(\mathbf{C_i}) = V_{\varphi_0}(\mathbf{D_{i_j}}) = \mathfrak{f}$, which contradicts the assumption that $\mathfrak{A}$ is a model of χ.

To prove the rest of the theorem, we must show that if χ has no model, then there is a finite subset of $\Gamma(\chi)$, $\{\mathbf{D_{i_1}}, \ldots, \mathbf{D_{i_k}}\}$, such that $\bigwedge_{1 \le j \le k} \mathbf{D_{i_j}}$ is a contradiction. Define the following set of conjunctions: $\mathbf{S_1}$ is to be $\mathbf{D_1}$ and $\mathbf{S_n}$ is to be $\mathbf{S_{n-1}} \wedge \mathbf{D_n}$. Certainly, if there is a set of $\mathbf{D_{i_j}}$ such that $\bigwedge_{1 \le j \le k} \mathbf{D_{i_j}}$ is a contradiction, and if $i_1 < i_2 \cdots < i_k$, then $\mathbf{S_{i_k}}$ is also a contradiction. We can complete the proof by showing that there is an n such that $\mathbf{S_n}$ is a contradiction. Suppose there is no such n; then for each n, there is at least one row of the truth table for $\mathbf{S_n}$ for which $\mathbf{S_n}$ has the truth value t. Let

$$\mathbf{A_1}, \ldots, \mathbf{A_{k_{n-1}}}, \ldots, \mathbf{A_{k_n}}$$

be the list of the all the atomic formulas occurring in $\mathbf{S_n}$ (in order of their first occurrence) such that $\mathbf{A_1}, \ldots, \mathbf{A_{k_{n-1}}}$ is the list of the atomic fomulas occurring in $\mathbf{S_{n-1}}$. Notice that if there is an assignment of truth values to $\mathbf{A_1}, \ldots, \mathbf{A_{k_{n-1}}}, \ldots, \mathbf{A_{k_n}}$ for which $\mathbf{S_n}$ has the truth value t, then, for that same assignment restricted to $\mathbf{A_1}, \ldots, \mathbf{A_{k_{n-1}}}$, the truth value for $\mathbf{S_{n-1}}$ is t. With this in mind, we construct a tree $\mathscr{T}$ as follows: (i) $\mathscr{T}$ has a level 0 vertex. (ii) For $n > 0$, each level n vertex is to correspond to an assignment of truth values to the atomic formulas of $\mathbf{S_n}$ in such a way that (a) the corresponding truth value of $\mathbf{S_n}$ is t and (b) the level $n - 1$ vertex connected to that level n vertex corresponds to the same assignment restricted to the atomic formulas occurring in $\mathbf{S_{n-1}}$. (Each level 1 vertex is connected to the level 0 vertex.) Since, for any n, the number of different truth value assignments to the atomic formulas of $\mathbf{S_n}$ is finite, $\mathscr{T}$ is finitely branched. Further, by assumption, for any n, $\mathbf{S_n}$ has at least one assignment of truth values to its atomic formulas for which the corresponding truth value of $\mathbf{S_n}$ is t, so there is no bound on the lengths of the branches of $\mathscr{T}$. Hence, by König's lemma, $\mathscr{T}$ has at least one infinite branch.

Consider henceforth one particular infinite branch of $\mathscr{T}$ and keep in mind that it corresponds to an assignment of truth values to all the atomic formulas occurring in $\Gamma(\chi)$ and that for each $\mathbf{D_i} \in \Gamma(\chi)$, the corresponding truth value is t. We shall refer to this as "the special assignment" and use it in constructing a model of χ. The reader should check that the following defines an equivalence relation on the set $H(\chi)$: If $\mathbf{t}$ and $\mathbf{t'}$ are terms which occur in $\Gamma(\chi)$ and if $(\mathbf{t} = \mathbf{t'})$ is an atomic formula in $\Gamma(\chi)$, then it is assigned a truth value by the special assignment. If it is assigned t, then $\mathbf{t}$ and $\mathbf{t'}$ are to belong to the same equivalence class and we write $[\mathbf{t}] = [\mathbf{t'}]$. The universe of the relational system which we hope will be a model of χ is to be the set of equivalence classes over $H(\chi)$ as just defined. The functions and relations of the relational system are defined just as in the model constructed in the proof of the completeness theorem for first-order theories with equality. That is, for each function symbol $\mathbf{f^k}$, we define the function $\Phi\mathbf{f^k}$ by defining $\Phi\mathbf{f^k}([\mathbf{t_1}], \ldots, [\mathbf{t_k}])$ to have the value $[\mathbf{f^k}(\mathbf{t_1}, \ldots, \mathbf{t_k})]$. We see that this is well defined because if $[\mathbf{t_j}] = [\mathbf{t_j'}]$, $1 \leq j \leq k$, it is because $(\mathbf{t_j} = \mathbf{t_j'})$ is assigned t by the special assignment $(1 \leq j \leq k)$. Further, since $\Gamma(\chi)$ is equationally closed, some $\mathbf{D_j}$ is

$$((\mathbf{t_1} = \mathbf{t_1'}) \wedge \cdots \wedge (\mathbf{t_k} = \mathbf{t_k'})) \supset (\mathbf{f^k}(\mathbf{t_1}, \ldots, \mathbf{t_k}) = \mathbf{f^k}(\mathbf{t_1'}, \ldots, \mathbf{t_k'}))$$

and, as each $\mathbf{D_j}$ must have the truth value t for the special assignment, the formula $(\mathbf{f^k}(\mathbf{t_1}, \ldots, \mathbf{t_k}) = \mathbf{f^k}(\mathbf{t_1'}, \ldots, \mathbf{t_k'}))$ must also be assigned t. For each predicate symbol $\mathbf{P^k}$, we define the relation $\Phi\mathbf{P^k}$ to contain the k-tuple $([\mathbf{t_1}], \ldots, [\mathbf{t_k}])$ if and only if $\mathbf{P^k}(\mathbf{t_1}, \ldots, \mathbf{t_k})$ is assigned t by the special assignment. That $\Phi\mathbf{P^k}$ is well defined follows as in the case for functions. Since each $\mathbf{C} \in \chi$ is $\mathbf{D_j} \in \Gamma(\chi)$ for some j, and since the truth value of every $\mathbf{D_j} \in \Gamma(\chi)$ is t in the special assignment, the relational system just constructed is a model of χ. But then the assumption that χ has no models is contradicted, so the assumption that there is no n such that $\mathbf{S_n}$ is a contradiction must be false. ∎

Exercise 13.16 Prove Herbrand's theorem for finitely axiomatizable first-order theories without equality.

13.6 MECHANICAL THEOREM-PROVING†

The general subject of mechanical theorem-proving includes any way a computer can be used to answer "yes" to the question: "Is this formula provable from these axioms?" The machine might actually try to produce a proof or it might try other ways. One of these other ways is to use techniques developed within the framework of the Herbrand theorem. Given axioms $\{\mathbf{F_1}, \ldots, \mathbf{F_n}\}$ and a formula $\mathbf{B}$, rather than trying to find a proof of $\mathbf{B}$ from $\{\mathbf{F_1}, \ldots, \mathbf{F_n}\}$, we try to show that there can be no model of $\{\mathbf{F_1}, \ldots, \mathbf{F_n}, \sim\mathbf{B}\}$. If we were to use the Herbrand theorem as stated and proved in the last section, we would face two immediate problems. The first is to choose a recursive enumeration of $\Gamma(\chi)$, and to try to choose it so as to lead as quickly as possible to a contradiction, if there is one. The second problem is the size of the truth tables that must be computed—these are soon astronomical. The main investigations in mechanical theorem-proving from the Herbrand point of view have been attempts to overcome these two problems. As we shall see, great improvements have been made, but at the cost of introducing other problems. We will assume throughout the section the same notation as in the last, unless otherwise specified. In particular, χ and the sets associated with it are the same as before. Further, all formulas are assumed to be quantifier-free.

We recall that in the proof of Herbrand's theorem in the last section, successive conjunctions are formed on the basis of *some* recursive enumeration of $\Gamma(\chi)$: $\mathbf{D_1}, \mathbf{D_2}, \ldots$. In the following we will take a slightly different point of view. Let $\{\mathbf{x_1}, \ldots, \mathbf{x_t}\}$ be the set of distinct individual variables occuring in the formulas $\mathbf{F_1}, \ldots, \mathbf{F_n}, \sim\mathbf{B}$. We augment the set of formulas $\{\mathbf{F_1}, \ldots, \mathbf{F_n}, \sim\mathbf{B}\}$ as follows:

† The material in this section was developed from the following papers: Davis (1963), J. A. Robinson (1965a, b; 1967).

1. For each $\mathbf{x_i}$, $\mathbf{x_j}$ and $\mathbf{x_k}$ in $\{\mathbf{x_1}, \ldots, \mathbf{x_t}\}$ add the formulas

$$(\mathbf{x_i} = \mathbf{x_j}) \supset (\mathbf{x_j} = \mathbf{x_i})$$
$$(\mathbf{x_i} = \mathbf{x_j}) \supset ((\mathbf{x_j} = \mathbf{x_k}) \supset (\mathbf{x_i} = \mathbf{x_k}))$$

2. For each k-ary function symbol $\mathbf{f}$ and predicate symbol $\mathbf{P}$ and every $2k$-tuple of variables, $\mathbf{x_{i_1}}, \mathbf{x'_{i_1}}, \ldots, \mathbf{x_{i_k}}, \mathbf{x'_{i_k}}$, in $\{\mathbf{x_1}, \ldots, \mathbf{x_t}\}$ add the formulas

$$((\mathbf{x_{i_1}} = \mathbf{x'_{i_1}}) \wedge \cdots \wedge (\mathbf{x_{i_k}} = \mathbf{x'_{i_k}})) \supset (\mathbf{f}(\mathbf{x_{i_1}}, \ldots, \mathbf{x_{i_k}}) = \mathbf{f}(\mathbf{x'_{i_1}}, \ldots, \mathbf{x'_{i_k}}))$$

and

$$((\mathbf{x_{i_1}} = \mathbf{x'_{i_1}}) \wedge \ldots \wedge (\mathbf{x_{i_k}} = \mathbf{x'_{i_k}})) \supset (\mathbf{P}(\mathbf{x_{i_1}}, \ldots, \mathbf{x_{i_k}}) \equiv \mathbf{P}(\mathbf{x'_{i_1}}, \ldots, \mathbf{x'_{i_k}}))$$

The augmented set is $\{\mathbf{F_1}, \ldots, \mathbf{F_n}, \ldots, \mathbf{F_{n'}}, \sim\mathbf{B}\}$. Let χ^+ be the set of conjuncts obtained from the formulas of the augmented set by taking the prenex normal form, then the conjunctive normal form and finally the Skolem function form of each. Obviously, $\chi \subset \chi^+$. Let the elements of χ^+ be denoted by $\mathbf{C_1}', \ldots, \mathbf{C'_{k'}}$. The individual variables occuring χ^+ are the same as those in χ. It is easy to see that $H(\chi^+) = H(\chi)$. Let $\mathbf{z_1}, \mathbf{z_2}, \ldots$ be an infinite set of individual variables disjoint from those in χ^+ [and hence from those in $H(\chi)$]. Consider an infinite set of conjunctions obtained from $\mathbf{C_1}' \wedge \cdots \wedge \mathbf{C'_{k'}}$ as follows. Let $\mathbf{V_1}$ be the result of substituting $\mathbf{z_1}, \ldots, \mathbf{z_t}$ for the t free variables in $\mathbf{C_1}' \wedge \cdots \wedge \mathbf{C'_{k'}}$; inductively, $\mathbf{V_{n+1}}$ is to be the result of substituting $\mathbf{z_{nt+1}}, \ldots, \mathbf{z_{(n+1)t}}$ for the t free variables in $\mathbf{C_1}' \wedge \cdots \wedge \mathbf{C'_{k'}}$. Define $\mathbf{S_1}' = \mathbf{V_1}'$ and $\mathbf{S'_{n+1}} = \mathbf{S_n}' \wedge \mathbf{V_{n+1}}$.

Lemma 13.3 $\vdash_{\{\mathbf{F_1}, \ldots, \mathbf{F_n}\}} \mathbf{B}$ if and only if there exists an n and an nt-tuple of elements of $H(\chi)$ such that the formula obtained by replacing the individual variables in $\mathbf{S'_n}$ by the elements of that tuple is a contradiction.

Exercise 13.17 Prove the lemma.

We will consider substitutions of terms for individual variables in formulas as mappings from formulas to formulas. These will be denoted by $\theta, \sigma, \lambda, \ldots$. In order to indicate exactly which term is to be substituted for which individual variable, we sometimes write $\theta = (\mathbf{x_1}/\mathbf{t_1}, \ldots, \mathbf{x_k}/\mathbf{t_k})$. For each i, $1 \le i \le k$, $\mathbf{x_i}/\mathbf{t_i}$ is the (x_i)th *component of* θ. Thus, for any formula $\mathbf{A}$, $\theta\mathbf{A}$ is defined to be the formula obtained from $\mathbf{A}$ by simultaneously replacing all (if any) occurrences of $\mathbf{x_i}$ in $\mathbf{A}$ by $\mathbf{t_i}$, $1 \le i \le k$, i.e., $\theta\mathbf{A}$ is $\mathbf{A}_{(\mathbf{x_1}, \ldots, \mathbf{x_k})}[\mathbf{t_1}, \ldots, \mathbf{t_k}]$. If Γ is a set of formulas $\{\mathbf{A_1}, \mathbf{A_2}, \ldots\}$, then $\theta\Gamma$ is $\{\theta\mathbf{A_1}, \theta\mathbf{A_2}, \ldots\}$. Similarly for a set of terms. Substitution maps can be composed in the obvious way, that is, if λ is a substitution map and θ is as above, then $\lambda\theta$ is $(\mathbf{x_1}/\lambda\mathbf{t_1}, \ldots, \mathbf{x_k}/\lambda\mathbf{t_k})$ and so $(\lambda\theta)\mathbf{A}$ is $\lambda(\theta\mathbf{A})$. The substitution map like θ but without the (x_i)th component will be written $\theta - (\mathbf{x_i}/\mathbf{t_i})$. If θ has no $\mathbf{y}$ component, the substitution map obtained from θ by adding the component $(\mathbf{y}/\mathbf{t}')$ will be written $\theta + (\mathbf{y}/\mathbf{t}')$. The identity substitution is $(\mathbf{x_1}/\mathbf{x_1}, \ldots, \mathbf{x_k}/\mathbf{x_k})$ for any k individual variables.

Lemma 13.4 (a) If σ has no $\mathbf{y}$ component, then $\sigma + (\mathbf{y}/\sigma\mathbf{t}) = \sigma(\mathbf{y}/\mathbf{t})$. (b) If $\theta = \sigma + (\mathbf{y}/\theta\mathbf{t})$, where there is no $\mathbf{y}$ component in σ and no occurrence of $\mathbf{y}$ in t, then $\theta = \sigma + (\mathbf{y}/\sigma\mathbf{t})$. (c) For substitution maps λ, σ, θ, $\lambda(\sigma\theta) = (\lambda\sigma)\theta$.

Exercise 13.18 Prove the lemma.

Exercise 13.19 Let $\chi = \{\mathbf{C_1}, \mathbf{C_2}\}$, where $\mathbf{C_1}$ is $\mathbf{P(x, f(u, a))}$ and $\mathbf{C_2}$ is $\sim \mathbf{P(g(y), f(g(z), a))}$. Let $\sigma_1 = (\mathbf{x/g(y)})$, $\sigma_2 = (\mathbf{x/g(y), u/g(z)})$, $\theta = (\mathbf{x/g(y), z/f(a), u/g(f(a))})$, $\lambda = (\mathbf{z/f(a)})$. Show (a) $\mathbf{C_1} \wedge \mathbf{C_2}$ is not a contradiction. (b) $\sigma_1\mathbf{C_1} \wedge \sigma_1\mathbf{C_2}$ is not a contradiction. (c) $\sigma_2\mathbf{C_1} \wedge \sigma_2\mathbf{C_2}$ is a contradiction. (d) $\theta\mathbf{C_1} \wedge \theta\mathbf{C_2}$ is a contradiction. (e) $\sigma_2\mathbf{C_i} \neq \theta\mathbf{C_i}$, $i = 1, 2$. (f) $\theta\mathbf{C_i} = \lambda\sigma_2\mathbf{C_i}$, $i = 1, 2$.

If a set of formulas $\alpha = \{\mathbf{A_1}, \ldots, \mathbf{A_t}\}$ is divided up into subsets, for whatever reason, so that $\alpha = \alpha_1 \cup \cdots \cup \alpha_r$ and $\alpha_i \cap \alpha_j = \varnothing$, for all $i \neq j$, we say we have *a partition P^α of* α and each subset α_i, $1 \leq i \leq r$, is an *element of the partition.* (The superscript will sometimes be dropped.) Notice that a partition determines an equivalence relation by defining $\mathbf{A_i}$ to be equivalent to $\mathbf{A_j}$ if and only if $\mathbf{A_i}$ and $\mathbf{A_j}$ belong to the same subset α_k. (The reader must check to see that it really is an equivalence relation.) Denote the equivalence classes by $[\mathbf{A_i}]_{P^\alpha}$; that is, $\mathbf{A_j} \in [\mathbf{A_i}]_{P^\alpha}$ if and only if $\mathbf{A_j}$ is equivalent to $\mathbf{A_i}$. *Two partitions of* α, P^α and P'^α, are defined to be *equal* if for all i, j, $\mathbf{A_j} \in [\mathbf{A_i}]_{P^\alpha}$ if and only if $\mathbf{A_j} \in [\mathbf{A_i}]_{P'^\alpha}$, i.e., $[\mathbf{A_i}]_{P^\alpha} = [\mathbf{A_i}]_{P'^\alpha}$. For partitions of α, P_1^α and P_2^α, if for all i, j, $1 \leq i, j \leq k$, $[\mathbf{A_i}]_{P_1^\alpha} = [\mathbf{A_j}]_{P_1^\alpha}$ implies that $[\mathbf{A_i}]_{P_2^\alpha} = [\mathbf{A_j}]_{P_2^\alpha}$, then we say that P_1^α *is less than* P_2^α and write $P_1^\alpha \leq P_2^\alpha$.

Given a substitution map θ and any set of atomic formulas $\alpha = \{\mathbf{A_1}, \ldots, \mathbf{A_t}\}$, there is a partition P_θ^α of α defined by: $\mathbf{A_i}$ and $\mathbf{A_j}$ belong to the same subset α_k if and only if $\theta\mathbf{A_i}$ is the same formula as $\theta\mathbf{A_j}$; i.e., $\mathbf{A_j} \in [\mathbf{A_i}]_{P_\theta^\alpha}$ if and only if $\theta\mathbf{A_i} \equiv \theta\mathbf{A_j}$. For example, in exercise 13.19, if $\mathbf{A_1}$ is $\mathbf{C_1}$ and $\mathbf{A_2}$ is $\mathbf{C_2}$ without the negation symbol, $[\mathbf{A_1}]_{P_\theta^\alpha} = [\mathbf{A_2}]_{P_\theta^\alpha}$. Notice also in the exercise that $P_{\sigma_2}^\alpha = P_\theta^\alpha$ and there is a λ such that $\theta = \lambda\sigma_2$. Having seen that every substitution map determines a partition of a set α of atomic formulas, we might ask if, given a partition P^α, there is a θ such that $P_\theta^\alpha = P^\alpha$. When there is such a θ, we say that θ *unifies* P^α.

Let $\mathbf{F}$ be a formula, $\alpha = \{\mathbf{A_1}, \ldots, \mathbf{A_k}\}$ the set of all atomic formulas occurring in $\mathbf{F}$, and P^α some partition of α. Let q be the number of equivalence classes of α determined by P^α and denote them by $[\mathbf{A_{i_1}}]_{P^\alpha}, \ldots, [\mathbf{A_{iq}}]_{P^\alpha}$. Let $\mathbf{p_1}, \ldots, \mathbf{p_q}$ be q distinct propositional variables. Define the formula $\mathbf{F}$ mod P^α to be the formula obtained from $\mathbf{F}$ by substituting (into $\mathbf{F}$) $\mathbf{p_j}$ for all occurrences of all the atomic formulas in the equivalence class $[\mathbf{A}_{\mathbf{i_j}}]_{P^\alpha}$, for all j, $1 \leq j \leq q$. The following lemma is immediate from the definition.

Lemma 13.5 Let α be a set of atomic formulas. Let θ and σ be substitution maps such that $P_\theta^\alpha = P_\sigma^\alpha$. Then, (1) **F** mod P_θ^α is the same formula as **F** mod P_σ^α and (2) θ**F** is a contradiction if and only if **F** mod P_θ^α a contradiction.

Lemma 13.6 Let **F** be a formula. Let P_1 and P_2 be two partitions of the atomic formulas occuring in **F** and assume that $P_1 \leq P_2$. If **F** mod P_1 is a contradiction, then **F** mod P_2 is a contradiction,

Exercise 13.20 Prove the lemma.

Now we can easily appreciate the question at the end of the previous paragraph. Consider the conjunction $\mathbf{S_n}'$ mentioned earlier in the section. For each partition P of the atomic formulas occurring in $\mathbf{S_n}'$, we can compute the truth table (or otherwise check) to see if $\mathbf{S_n}'$ mod P is a contradiction. *If* it is and *if* there is a θ such that $P_\theta = P$, then $\theta\mathbf{S_n}'$ is a contradiction.

In fact, there is an algorithm (due to Prawitz †) which, when applied to any partition of a finite set of atomic formulas, will either give a substitution map which unifies the partition or will indicate that no unifying map exists. If α is a finite, nonempty set of atomic formulas, *the disagreement set of* α is obtained by looking at each atomic formula, symbol by symbol, from left to right. At the first position in which they do not all have the same symbol, take from each formula the term (or whole formula) occurring in it which begins at that position. It is this set of terms and formulas which is the disagreement set of α. (The reader must reassure himself, inductively, that this makes sense in terms of the formation of atomic formulas). Thus, unless the atomic formulas disagree in the leftmost symbol, the predicate symbol, the disagreement set is a set of terms. Take a fixed ordering of the terms occurring in α with all the individual variables coming before any other terms. We will always assume that the disagreement set of α or of any subset of α is ordered according to the fixed ordering.

The Unification Algorithm

Let α be a finite set of atomic formulas and P^α a partition of α into $\alpha_1 \cup \cdots \cup \alpha_t$. We apply a series of substitution maps σ_i, $i = 0, 1, \ldots,$ which will be defined by the algorithm.

† The presentation here follows the version given by J. A. Robinson (1965a).

Step 0: Set $i = 0$ and let σ_0 be the identity substitution map.

Step 1: If $\sigma_i \alpha_j$, for all j, $1 \leq j \leq t$, are all singleton sets, stop, and let σ^* be σ_i. If not, let k be the smallest j such that $\sigma_i \alpha_j$ is not a singleton and look at the disagreement set of $\sigma_i \alpha_j$ as ordered by the fixed ordering. Let $\mathbf{t_1}$ and $\mathbf{t_2}$ be the first two terms of the ordered disagreement set. Go to step 2.

Step 2: If $\mathbf{t_1}$ is an individual variable and it does not occur in $\mathbf{t_2}$, set $\sigma_{i+1} = (\mathbf{t_1}/\mathbf{t_2})\sigma_i$ and go back to step 1. Otherwise, stop.

To be sure that the unification algorithm is an algorithm, we must show that it stops for any set α.

Lemma 13.7 If α is a set of atomic formulas, the unification algorithm, applied to α, terminates.

PROOF Suppose m distinct individual variables occur in α. For any cycle k, the unification algorithm stops in step 2 unless the first two members of the disagreement set, $\mathbf{t_1}$ and $\mathbf{t_2}$, satisfy the conditions that $\mathbf{t_1}$ is an individual variable, say $\mathbf{y}$, and that it does not occur in $\mathbf{t_2}$. In that case, $\sigma_{k+1} = (\mathbf{y}/\mathbf{t_2})\sigma_k$ and the unification algorithm goes back to step 1 for the $k+1$ cycle. Notice that $\mathbf{y}$ as well as the individual variables occurring in $\mathbf{t_2}$ are among the m individual variables of α. If $\sigma_k = (\mathbf{x_1}/\mathbf{t_1}', \ldots, \mathbf{x_k}/\mathbf{t_k}')$, then $\sigma_{k+1} = (\mathbf{x_1}/(\mathbf{y}/\mathbf{t_2})\mathbf{t_1}', \ldots, \mathbf{x_k}/(\mathbf{y}/\mathbf{t_2})\mathbf{t_k}')$. Thus, $\sigma_{k+1}\alpha$ contains no occurrence of $\mathbf{y}$ and no individual variables not occurring in $\sigma_k \alpha$. Since there are only m distinct individual variables to start with and since the number of individual variables occurring in $\sigma_{k+1}\alpha$ is less than the number in $\sigma_k \alpha$, the algorithm must terminate. ∎

It is clear from the statement of the algorithm that if it stops in step 1, σ^* unifies the partition.

Theorem 13.6 Let $\alpha = \alpha_1 \cup \cdots \cup \alpha_t$ be a finite, nonempty set of atomic formulas. If there is a substitution map θ such that $\theta\alpha_1, \ldots, \theta\alpha_t$ are all singletons, then the unification algorithm stops in step 1 with a σ^* for which there exists a λ such that $\theta = \lambda\sigma^*$.

PROOF We will show, by induction on i, that (1) the unification algorithm cannot stop in step 2 and therefore by the lemma it must stop in step 1 with $\sigma^* = \sigma_k$ for some k, and (2) for all i, $0 \leq i \leq k$, there exists a λ_i such that $\theta = \lambda_i \sigma_i$. If $i = 0$, then σ_0 is the identity substitution, so $\lambda_0 = \theta$. Inductively, assume that θ unifies α and that we are considering σ_i; so, by the induction hypothesis, for each $j \leq i$, there exists a λ_j such that $\theta = \lambda_j \sigma_j$.

Case 1: For all m, $1 \leq m \leq t$, $\sigma_k \alpha_m$ is a singleton. Then the algorithm stops in step 1 and $\sigma^* = \sigma_i$ (i.e., $k = i$), so, by the induction hypothesis, $\lambda_i \sigma^* = \theta$.

Case 2: There is an m, $1 \leq m \leq t$, such that $\sigma_i \alpha_m$ is not a singleton. The algorithm goes into step 2. As we are given that θ does unify α, we have that $\theta\alpha_m$ is a singleton and, by the induction hypothesis, that $\lambda_i \sigma_i \alpha_m = \theta\alpha_m$, so λ_i unifies the disagreement set of $\sigma_i \alpha_m$; call the disagreement set δ. Since δ is unifiable, some of its terms must be individual variables and so come first in the ordering of δ determined by the fixed ordering. That is, $\mathbf{t_1}$ is an individual variable, say $\mathbf{y}$. From the facts that $\lambda_i \delta$ is a singleton, $\mathbf{y} \neq \mathbf{t_2}$, and $\lambda_i \mathbf{y} = \lambda_i \mathbf{t_2}$, we know that $\mathbf{y}$ does not occur in $\mathbf{t_2}$; also, that $(\mathbf{y}/\lambda_i \mathbf{y}) = (\mathbf{y}/\lambda_i \mathbf{t_2})$. Therefore, the unification algorithm must return to step 1 with $\sigma_{i+1} = (\mathbf{y}/\mathbf{t_2})\sigma_i$. So it remains to be shown that there is a λ_{i+1} such that $\theta = \lambda_{i+1}\sigma_{i+1}$. Let $\lambda_{i+1} = \lambda_i - (\mathbf{y}/\lambda_i \mathbf{y})$. Then,

$$\begin{aligned}
\lambda_i &= \lambda_{i+1} + (\mathbf{y}/\lambda_i \mathbf{y}) \\
&= \lambda_{i+1} + (\mathbf{y}/\lambda_i \mathbf{t_2}) \\
&= \lambda_{i+1} + (\mathbf{y}/\lambda_{i+1} \mathbf{t_2}) && \text{since } \mathbf{y} \text{ is not in } \mathbf{t_2} \\
&= (\lambda_{i+1}(\mathbf{y}/\mathbf{t_2})) && \text{(refer to Lemma 13.4)}
\end{aligned}$$

$$\begin{aligned}
\theta = \lambda_i \sigma_i &= (\lambda_{i+1}(\mathbf{y}/\mathbf{t_2}))\sigma_i \\
&= \lambda_{i+1}((\mathbf{y}/\mathbf{t_2})\sigma_i) && \text{by Lemma 13.4} \\
&= \lambda_{i+1}\sigma_{i+1} && \text{by the definition of } \sigma_{i+1} \text{ above } \blacksquare
\end{aligned}$$

Given α and θ as in the hypothesis of the theorem, the substitution map σ^* obtained by the unification algorithm is called a *most general unifier of* θ. If α is a set of atomic formulas, $\mathbf{y_1}, \ldots, \mathbf{y_m}$ are the individual variables occurring in α, and σ is a substitution map obtained by applying the unification algorithm to a partition of α, then σ is of the form $(\mathbf{y_1}/\mathbf{t_1}, \ldots, \mathbf{y_m}/\mathbf{t_m})$ where the variables occurring in the $\mathbf{t_i}$ belong to $\{\mathbf{y_1}, \ldots, \mathbf{y_m}\}$. Suppose that by means of the unification algorithm we find a σ such that $\sigma\mathbf{S_n}'$ is a contradiction. What does this have to do with there being an nt-tuple of elements of $H(\chi)$ such that the formula obtained by substituting that tuple into $\mathbf{S_n}'$ is a contradiction? For each $n \geq 1$ let $H(\mathbf{S_n}')$ be the Herbrand universe of $\mathbf{S_n}'$.

Lemma 13.8 There is an nt-tuple of elements of $H(\chi)$ such that the formula obtained by replacing the individual variables of $\mathbf{S_n}'$ by those elements is a contradiction if and only if there is an nt-tuple of elements of $H(\mathbf{S_n}')$ such that the formula obtained by replacing the individual variables of $\mathbf{S_n}'$ by those elements is a contradiction.

PROOF Let θ_χ denote a substitution map which substitutes elements of $H(\chi)$ for individual variables in $\{\mathbf{z_1}, \mathbf{z_2}, \ldots\}$; and let θ_n denote a substitution map which substitutes elements of $H(\mathbf{S_n}')$ for individual variables in $\{\mathbf{z_1}, \mathbf{z_2}, \ldots\}$. Restating the lemma in these terms we have: There exists a θ_χ such that $\theta_\chi\mathbf{S_n}'$ is a contradiction if and only if there is a θ_n such that $\theta_n\mathbf{S_n}'$ is a contradiction. Define the map μ: $\{\mathbf{x_1}, \ldots, \mathbf{x_t}\} \to \{\mathbf{z_1}, \ldots, \mathbf{z_t}\}$ by $\mu(\mathbf{x_i}) = \mathbf{z_i}$; and define the map ν: $\{\mathbf{z_1}, \mathbf{z_2}, \ldots\} \to \{\mathbf{z_1}, \ldots, \mathbf{z_t}\}$ by: for all $k \geq 0$ and all j, $1 \leq j \leq t$, $\nu(\mathbf{z_{kt+j}}) = \mathbf{z_j}$. If there is a θ_χ such that $\theta_\chi\mathbf{S_n}'$ is a contradiction then, obviously, $\mu\theta_\chi\mathbf{S_n}'$ is a contradiction and $\mu\theta_\chi$ is the necessary θ_n. On the other hand, suppose there is a θ_n such that $\theta_n\mathbf{S_n}'$ is a contradiction. Of course, θ_n defines a partition of the atomic formulas of $\mathbf{S_n}'$ so that $\mathbf{A_i}$ and $\mathbf{A_j}$ belong to the same equivalence class if and only if $\theta_n\mathbf{A_i} \equiv \theta_n\mathbf{A_j}$. But then $\nu\theta_n\mathbf{A_i} \equiv \nu\theta_n\mathbf{A_j}$ and so $P_{\theta_n} \leq P_{\nu\theta_n}$. By Lemma 13.6, if $\theta_n\mathbf{S_n}'$ is a contradiction then $\nu\theta_n\mathbf{S_n}'$ is a contradiction. It follows that $\mu^{-1}\nu\vartheta_n$ is the θ_χ required by the lemma. ▌

Exercise 13.21 Outline a search procedure based on Lemmas 13.3 and 13.8 and the unification algorithm.

Such procedures invite improvements in the areas of hunting for partitions and computing truth tables. The truth table computation can be improved along lines suggested by Davis (1963). Define a *literal* to be an atomic formula or an atomic formula preceded by a negation symbol. Then a *clause* is defined to be a disjunction of a set of literals. Define the conjunction of a set of clauses to be a *linked conjunction* if it is the case that whenever an atomic formula **A** (or $\sim$**A**) occurs in some clause, then $\sim$**A** (or **A**) occurs in some other clause. ("Other" because if **A** and $\sim$**A** occur in the same clause, we can simplify by removing that clause altogether since it is true under all truth value assignments.)

Theorem 13.7 Let $\mathbf{E_1}, \ldots, \mathbf{E_t}$ be a set of clauses. $\bigwedge_{1 \leq i \leq t} \mathbf{E_i}$ is a contradiction if and only if there is a subset $\mathbf{E}_{i_1}, \ldots, \mathbf{E_{i_m}}$ such that $\bigwedge_{1 \leq j \leq m} \mathbf{E_{i_j}}$ is a linked conjunction and a contradiction.

Exercise 13.22 Prove the theorem. (Hint: Show that if $\mathbf{E_j}$ is a clause in which **A** (or $\sim$**A**) occurs and no clause contains $\sim$**A** (or **A**), then $\bigwedge_{1 \leq i \leq t} \mathbf{E_i}$ is a contradiction if and only if $\bigwedge_{1 \leq i \leq j-1} \mathbf{E_i} \wedge \bigwedge_{j+1 \leq i \leq t} \mathbf{E_i}$ is a contradiction.)

Using the discussion so far in this section, we can state another version of the Herbrand theorem:

Theorem 13.5D $\vdash_{\{\mathbf{F_1}, \ldots, \mathbf{F_n}\}} \mathbf{B}$ if and only if there is an n and a θ such that the conjunction of some subset of the clauses of $\theta\mathbf{S}_n'$ is a contradiction and a linked conjunction.

Lemma 13.9 Let θ be a substitution map and $\mathbf{S}$ a conjunction of clauses. $\theta\mathbf{S}$ is a linked conjunction if and only if $\mathbf{S}$ mod P_θ is.

Combining these observations with the unification algorithm, we have the following search procedure:

Step 0: Set $n = 0$.

Step 1: Take a partition P of the atomic formulas of $\mathbf{S}_n'$, form $\mathbf{S}_n'$ mod P, and go to step 2.

Step 2: Check through the clauses of $\mathbf{S}_n'$ mod P to see if a linked conjunction can be formed. If not, go to step 5; if so, go to step 3.

Step 3: Is the linked conjunction a contradiction? If so, go to step 4. If not, go to step 5.

Step 4: Does P have a unifying substitution map? If so, done. If not, go to step 5.

Step 5: Have all the partitions of $\mathbf{S}_n'$ been tried? If not, go back to step 1 to try a new one. If so, increase n by 1 and start the procedure for this new value of n back at step 1.

There is no improvement over the Prawitz-type procedures as far as choosing partitions is concerned, but some improvement with regard to computing truth tables. Other improvements, especially with regard to computing truth tables, are also discussed by Davis (1963).

The last improvement that we will consider is along lines developed by J. A. Robinson. He calls his general methods *resolutions procedures*; we will consider just one particular kind. In general, resolution procedures substitute symbol checking in place of computing truth tables, but still leave open the question of how "best" to hunt for useful partitions. If Σ is a set of clauses, we will say "Σ is a contradiction" as an abbreviation for "the conjunction of the clauses in Σ is a contradiction." We assume that a given clause contains at most one occurrence of a given atomic formula. A *positive clause* is a clause in which no negation symbol occurs; i.e., a disjunction of atomic formulas. A *negative clause* is a clause in which at least one negation symbol occurs. A positive clause $\mathbf{P}$ and a negative clause $\mathbf{N}$ *clash* if at least one of the atomic formulas of $\mathbf{P}$ occurs in $\mathbf{N}$ preceded by a negation symbol. Suppose $\mathbf{A}_1, \ldots, \mathbf{A}_r$, $r \geq 1$, are atomic formulas. A positive clause $\mathbf{P}$ and a negative clause $\mathbf{N}$ *clash over* $\mathbf{A}_1, \ldots, \mathbf{A}_r$ if for each i, $1 \leq i \leq r$, $\mathbf{A}_i$ occurs in $\mathbf{P}$ and $\sim \mathbf{A}_i$ occurs in $\mathbf{N}$. For $\mathbf{P}$ and $\mathbf{N}$ which clash over $\mathbf{A}_1, \ldots, \mathbf{A}_r$, let $\mathbf{P}^0$ be the clause obtained from $\mathbf{P}$ by deleting $\mathbf{A}_1, \ldots, \mathbf{A}_r$ and let $\mathbf{N}^0$ be the clause obtained from $\mathbf{N}$ by deleting $\sim \mathbf{A}_1, \ldots, \sim \mathbf{A}_r$ as well as any literal that occurs in both

P and **N**. Then the clause ($\mathbf{P^0} \vee \mathbf{N^0}$) is defined to be the P_1*-resolvent* of **P** and **N**; denote it by [**P**, **N**]. Of course, if **P** is just **A** and **N** is just $\sim$**A**, then [**P**, **N**] is empty; the empty clause is denoted by $\square$. A P_1-resolvent of a set Σ of clauses is a P_1-resolvent of a pair, positive and negative, of clauses in Σ. A P_1*-deduction* from a set Σ of clauses is a sequence of sets of clauses such that

$$\begin{aligned}
\Sigma_0 &= \Sigma \\
\Sigma_1 &= \Sigma_0 \cup \{\mathbf{R_0}\} \\
&\vdots \\
\Sigma_n &= \Sigma_{n-1} \cup \{\mathbf{R_{n-1}}\}
\end{aligned}$$

where $\mathbf{R_i}$ is a P_1-resolvent of Σ_i and $\mathbf{R_i} \notin \Sigma_i$. A clause **C** is P_1*-deducible* in n steps from a set Σ of clauses if there is a P_1-deduction from Σ for which $\mathbf{R_{n-1}}$ is **C**. Thus, $\square$ is P_1-deducible from a set Σ if there is such a sequence and for some n, $\square$ is a P_1-resolvent of Σ_{n-1}.

Lemma 13.10 Let Σ be a finite set of clauses. If $\Sigma_0, \Sigma_1, \ldots$ is a P_1-deduction from Σ, then there exists a k such that there is no P_1-resolvent of Σ_k which does not belong to Σ_k.

Exercise 13.23 Prove the lemma. (Hint: Think about how many literals a clause in a deduction can contain.)

Lemma 13.11 Let Σ be a finite set of clauses. If Σ is a contradiction and Σ does not contain $\square$, then there is a P_1-resolvent of Σ which is not Σ.

PROOF Let Π be the subset of Σ containing all the positive clauses. Consider all assignments of truth values to the atomic formulas of Σ for which (i) the corresponding truth values of the clauses in Π are $\mathfrak{t}$ and (ii) no other assignment satisfying (i) has fewer $\mathfrak{t}$'s. Call such assignments "minimal truth value assignments." Since Σ is a contradiction, for any minimal truth value assignment there is a clause belonging to $\Sigma - \Pi$ which has the truth value $\mathfrak{f}$ under that assignment. For each minimal truth value assignment, consider all clauses of $\Sigma - \Pi$ that are assigned $\mathfrak{f}$ and pick those in which the fewest negation symbols occur; call these the "minimal negative clauses" corresponding to the assignment. Claim that for all minimal truth value assignments and all corresponding minimal negative clauses **K**, if the atomic formula **A** occurs preceded by a negation symbol in **K**, then there is a clause $\mathbf{J} \in \Pi$ such that **A** occurs in **J** and, by the assignment, $\mathfrak{f}$ is assigned to all other atomic formulas in **J**. First notice that there is a **J** in which **A** occurs, for otherwise we need not assign $\mathfrak{t}$ to **A**, contradicting the minimality of the assignment. Take any one of the minimal truth value assignments and some corresponding minimal negative clause **K** in which $\sim$**A** occurs. Look at all clauses of Π in which **A** occurs. If there is one in which all other atomic formulas are assigned $\mathfrak{f}$ by this assignment, then the claim is proved. So suppose that in all positive clauses in which **A** occurs, at least one other atomic formula is assigned $\mathfrak{t}$. But in that case, **A** need not be assigned $\mathfrak{t}$ to make all those positive clauses have the truth value $\mathfrak{t}$, thus contradicting the minimality of the assignment. For such a pair **J** and **K**, let $\mathbf{R} = [\mathbf{J}, \mathbf{K}]$, the P_1-resolvent of **J** and **K**. Notice that for the fixed minimal assignment on which **J** and **K** depend, **R** is assigned $\mathfrak{f}$. We now show that **R** is not in Σ. If **R** is $\square$, then $\mathbf{R} \notin \Sigma$ by the hypothesis of the lemma. If **R** is positive and $\mathbf{R} \in \Sigma$, then $\mathbf{R} \in \Pi$, but all clauses of Π are assigned $\mathfrak{t}$ by any minimal assignment; contradiction. If **R** is negative, then **R** contains one less negation symbol than any minimal negative clause associated with the fixed minimal assignment and so $\mathbf{R} \notin \Sigma - \Pi$. ▮

Theorem 13.8 Let Σ be a finite set of clauses. The conjunction of the clauses of Σ is a contradiction if and only if there is a P_1-deduction of $\square$ from Σ.

PROOF *Part I.* (a) We show by induction on n that if $\Sigma_0, \ldots, \Sigma_n$ is a P_1-deduction from Σ and Σ_n is a contradiction, then Σ is a contradiction. The case for $n = 0$ is trivial. Suppose $\Sigma_n - \Sigma_{n-1} \cup [\mathbf{N}, \mathbf{P}]$ is a contradiction. If $[\mathbf{N}, \mathbf{P}]$ is assigned $\mathfrak{f}$, then $\mathbf{N} \wedge \mathbf{P}$ is assigned $\mathfrak{f}$, therefore Σ_{n-1} is assigned $\mathfrak{f}$. If $[\mathbf{N}, \mathbf{P}]$ is assigned $\mathfrak{t}$, then since Σ_n is a contradiction, Σ_{n-1} must be assigned $\mathfrak{f}$. (b) By induction on n, we show that if there is a P_1-deduction of $\square$ from Σ, then Σ is a contradiction. If $n = 1$, $\Sigma_1 = \Sigma_0 \cup \square$, where $\square$ is $[\mathbf{A}, \sim\mathbf{A}]$ and the formulas **A** and $\sim$**A** are clauses in Σ_0 (i.e., Σ). So Σ is a contradiction. Similarly, if $\Sigma_n = \Sigma_{n-1} \cup \square$, then Σ_{n-1} is a contradiction, so by part (a), Σ is also a contradiction.

Part II. Consider a P_1-deduction from Σ: Σ_0, $\Sigma_0 \cup \{\mathbf{R_0}\}$, ..., $\Sigma_{n-1} \cup \{\mathbf{R_{n-1}}\}$, ... If Σ_0 is a contradiction, so is every Σ_i. By Lemma 13.10, there is a k such that there is no P_1-resolution of Σ_k not belonging to Σ_k. Further, Σ_k must contain $\square$ because the conjunction of clauses in Σ_k is a contradiction and so by Lemma 13.11, either Σ_k contains $\square$ or there is a P_1-resolvent of Σ_k which does not belong to Σ_k. But the latter is impossible. ▮

Corollary Let θ be a substitution map. For some n, let Σ be the clauses occurring in $\mathbf{S'_n}$. $\theta\mathbf{S'_n}$ is a contradiction if and only if there is a P_1-deduction of $\square$ from $\theta\Sigma$.

It is now easy to verify a third version of the Herbrand theorem, namely:

Theorem 13.5R $\vdash_{\{\mathbf{F_1}, \ldots, \mathbf{F_n}\}} \mathbf{B}$ if and only if there is an n and a partition P

of the atomic formulas in $\mathbf{S}_{\mathbf{n}}'$ such that there is a P_1-deduction of □ from Σ mod P, where Σ is the set of clauses occurring in $\mathbf{S}_{\mathbf{n}}'$, and P is unifiable

Exercise 13.24 Outline a search procedure based on Theorem 13.5R.

Although a search procedure based on Theorem 13.5R completely eliminates computing truth-tables, it does not help us choose an efficient ordering of the partitions. As one might imagine, considerable attention has been given to this problem. Notice also that, for a given Σ, if □ is P_1-deducible from Σ then there is a smallest n such that □ is P_1-deducible from Σ in n steps. In the discussion above no consideration was given to how to find a P_1-deduction of □ with the minimal number of steps. The reader can consult the papers of Robinson mentioned at the beginning of the section for some discussion of these problems.

The number of papers concerned with mechanical theorem-proving is growing very rapidly and no attempt is made in this book to give a complete bibliography. A few papers are listed here for the interested reader; to really know what is going on, he will have to consult the latest issues of the various computing journals. The papers listed at the beginning of this section contain more information than we covered here; in addition to these, the following papers are suggested†: Allen and Luckham (1970), Anderson and Bledsoe (1970), Chang (1970), Friedman (1963), Loveland (1970a, b), Prawitz (1970), Robinson and Wos (1969), and Slagle (1967).

† I am very grateful to Professors Donald Loveland and Martin Davis for helping me make this selection.

Chapter XIV

DECIDABLE AND UNDECIDABLE THEORIES

14.1 NUMBER THEORY AND PEANO'S AXIOMS

Let $\mathfrak{N} = \langle N; ', +, \cdot; =\rangle$. The subject usually called number theory is the study of the relational system $\mathfrak{N}$. Let $\mathsf{L}(N)$ be a first-order language with equality with one unary function symbol **S**, two binary function symbols **+** and **·**, and one 0-ary function symbol (i.e., constant) **0**. Then, $\mathfrak{N}$ is a relational symbol for $\mathsf{L}(N)$ in the obvious way. Notice that for every $n \in N$, there is a corresponding closed term $\underbrace{\mathbf{SS} \cdots \mathbf{S}}_{n}(\mathbf{0})$ in $\mathsf{L}(N)$. We abbreviate that term by $\underline{n}$ and call all such terms *numerals*. Let $\mathcal{N}$ denote the set of sentences of $\mathsf{F}(N)$ which are true in $\mathfrak{N}$. We are interested in finding a recursively axiomatizable theory which is equal to $\mathcal{N}$, that is, which is complete and thus also decidable. How to choose a set of nonlogical axioms which might serve our purpose is by no means obvious. One cannot settle on a suitable set of axioms until one has settled the more fundamental problem of what a number really is. In the late 19th and early 20th centuries, there were very serious studies of this basic question.† In particular, Frege, Dedekind, and Peano directed their efforts toward defining "number." As a consequence of these efforts, certain sets of axioms were proposed as simultaneously defining the arithmetic objects, namely numbers, and as being true in arithmetic. Subsequently, when logicians attempted to formalize arithmetic, one of those sets of axioms, that formulated by Peano, was adapted for that purpose. Let P be the first-order theory with the following nonlogical axioms:

*P*1. $(\mathbf{x} = \mathbf{y}) \supset (\mathbf{S}(\mathbf{x}) = \mathbf{S}(\mathbf{y}))$

*P*2. $(\mathbf{S}(\mathbf{x}) = \mathbf{S}(\mathbf{y})) \supset (\mathbf{x} = \mathbf{y})$

*P*3. $\mathbf{S}(\mathbf{x}) \neq \mathbf{0}$

† For a brief discussion of these developments, see Kneale and Kneale (1962, chapter VII, in particular, section 5).

*P*4. $\mathbf{x} + \mathbf{0} = \mathbf{x}$
*P*5. $\mathbf{x} + \mathbf{S}(\mathbf{x}) = \mathbf{S}(\mathbf{x} + \mathbf{y})$
*P*6. $\mathbf{x} \cdot \mathbf{0} = \mathbf{0}$
*P*7. $\mathbf{x} \cdot \mathbf{S}(\mathbf{y}) = (\mathbf{x} \cdot \mathbf{y}) + \mathbf{x}$
*P*8. For any formula **A** with free variable **x** in F(*N*),

$$\mathbf{A}_{(\mathbf{x})}[\mathbf{0}] \supset ((\forall \mathbf{x})(\mathbf{A} \supset \mathbf{A}_{(\mathbf{x})}[\mathbf{S}(\mathbf{x})]) \supset (\forall \mathbf{x})\mathbf{A})$$

Notice that *P*1 is just an instance of the equality axiom and is given only for emphasis. *P*8 is known as the "first-order induction axiom." Notice that since *P*8 is an axiom scheme and there are countably infinitely many formulas **A**, the set of axioms for P is infinite but recursive. $\mathfrak{N}$ is called the *standard model* for P; of course, P has many other models.

The theory P was established in an effort to formalize $\mathfrak{N}$. Naturally, one purpose in formalizing $\mathfrak{N}$ is to establish a uniform standard of rigor for number theory. But there are at least two further goals of considerable interest. (1) It was proposed by Hilbert that mathematics (number theory, analysis, etc.) be formalized with the idea that the consistency of mathematics could be proved in a rigorous manner. (2) If the formal theory were complete and decidable, then verifying the truth of a mathematical statement would become a routine or mechanical computation. That the latter goal turned out to be unachievable is the meaning of the Gödel incompleteness theorem that we will be discussing. That the former goal is also unachievable in a naive way is the outcome of another theorem of Gödel's (known as "Gödel's second theorem"), which we will not discuss.†

Exercise 14.1 Can P have any finite models?

Exercise 14.2 What is the cardinality of the set of all number-theoretic relations? What is the cardinality of the set of all formulas of L(*N*)? Is axiom *P*8 as powerful as the usual, informal, induction axiom?

It is already clear, from the observations about cardinalities in exercise 14.2, that some aspects of $\mathfrak{N}$ are not included when looking at $\mathfrak{N}$ from the point of view of what is provable in P. Along this line, one should wonder what the relations and functions (not named by $\mathfrak{N}$) on the natural numbers have to do with what is provable in P. This requires a reasonable and formal definition of "having to do with." If T is a theory on a language which includes $=$, S, and **0**, and in which *P*1, *P*2, and *P*3 are theorems, then we say that *N is expressible in* T. Let T be a theory in which *N* is expressible. A *number-theoretic relation* R^k is defined to be *numeralwise-representable* (write

† For a discussion of this result, see Mendelson (1964, pp.148–149).

n-representable) *in* T if and only if there is a formula **R** with k free variables such that

$$\text{if} \quad (n_1, \ldots, n_k) \in R^k, \quad \text{then} \quad \vdash_\mathsf{T} \mathbf{R}(\underline{n}_1, \ldots, \underline{n}_k)$$

and

$$\text{if} \quad (n_1, \ldots, n_k) \notin R^k, \quad \text{then} \quad \vdash_\mathsf{T} \sim \mathbf{R}(\underline{n}_1, \ldots, \underline{n}_k)$$

A *number-theoretic function* f^k is defined to be *numeralwise-representable in* T if and only if there is a formula **F** with $k + 1$ free variables such that

$$\text{if} \quad f^k(n_1, \ldots, n_k) = n_{k+1}, \quad \text{then} \quad \vdash_\mathsf{T} \mathbf{F}(\underline{n}_1, \ldots, \underline{n}_k, \underline{n}_{k+1}) \qquad (*)$$

$$\text{if} \quad f^k(n_1, \ldots, n_k) \neq n_{k+1}, \quad \text{then} \quad \vdash_\mathsf{T} \sim \mathbf{F}(\underline{n}_1, \ldots, \underline{n}_k, \underline{n}_{k+1}) \qquad (**)$$

$$\vdash_\mathsf{T} (\forall \mathbf{x})(\forall \mathbf{y})(\mathbf{F}(\underline{n}_1, \ldots, \underline{n}_k, \mathbf{x}) = \mathbf{F}(\underline{n}_1, \ldots, \underline{n}_k, \mathbf{y}) \supset \mathbf{x} = y) \qquad (***)$$

We say "**R** represents R in T" and "**F** represents f in T."

Exercise 14.3 Show that in the definition of *n*-representable for functions, the second condition, (**), follows from the other two and therefore need not be explicitly stated in the definition.

Exercise 14.4 Consider the following alternative definition of *n*-representable function f^k: the formula **F** with $k + 1$ free variables represents the function f^k in T if

$$\vdash_\mathsf{T} (\mathbf{F}(\underline{n}_1, \ldots, \underline{n}_k, \mathbf{x}) \equiv (\mathbf{x} = \underline{f^k(n_1, \ldots, n_k)}))$$

Prove that this definition is equivalent to the one just above.

Exercise 14.5 Let $R^k \subset N^k$ and $\bar{R}^k = N^k - R^k$. Prove that for a theory T, R^k is *n*-representable in T if and only if $\bar{R}^k$ is *n*-representable in T.

Exercise 14.6 Let **R** be a formula with k free variables. Let R be the k-ary number-theoretic relation defined by

$$(n_1, \ldots, n_k) \in R \quad \text{if and only if} \quad V(\mathbf{R}(\underline{n}_1, \ldots, \underline{n}_k)) = \mathrm{t} \qquad \text{in } \mathfrak{N}$$

Show that if R is not *n*-representable in P, then there exists a k-tuple $(n_1, \ldots, n_k)$ such that neither $\vdash_\mathsf{P} \mathbf{R}(\underline{n}_1, \ldots, \underline{n}_k)$ nor $\vdash_\mathsf{P} \sim \mathbf{R}(\underline{n}_1, \ldots, \underline{n}_k)$.

It is natural to compare the classes of functions and relations studied in Part I of this book to the functions and relations which are *n*-representable in P. We state for use now the following theorem; it will be proved in section 14.4.

Theorem 14.1 If f is a recursive function, it is *n*-representable in P.

Exercise 14.7 Show that a number-theoretic relation is n-representable in P if and only if its characteristic function is n-representable in P.

Exercise 14.8 Use Church's thesis to show that if T is a theory with a recursively enumerable set of theorems and R is n-representable in T, then R is a recursive relation.

Exercise 14.9 Prove that if the set S (i.e., unary relation) and the function f are n-representable in theory T, then so is the set $\{x \mid f(x) \in S\}$.

Corollary to Theorem 14.1 A number-theoretic relation is n-representable in P if and only if it is recursive.

Since the objects in $\mathfrak{N}$ are just exactly the natural numbers, if there is some statement which is true about each natural number, then, obviously, we cannot truthfully say there is some object in $\mathfrak{N}$ for which the negation of the statement is true. However, this would not be the case for the models of P of cardinality greater than $\aleph_0$. A theory T in which the natural numbers are expressible is *ω-consistent* if the following is the case: for each formula **A** with free variable **x**, if for every $n \in N$ $\vdash_{\mathsf{T}} \mathbf{A}_{(\mathbf{x})}[\underline{n}]$, then *not* $\vdash_{\mathsf{T}} (\exists \mathbf{x}) \sim \mathbf{A}$. Thus, if P is ω-consistent, then what its theorems tell us about its models is in some sense restricted to the natural number part of those models.†

Exercise 14.10 Show that if a theory is ω-consistent, then it is consistent.

Recall that the formulas of L(N) can be Gödel numbered, and so by thinking of the Gödel number of a formula as its name, we have in $\mathfrak{N}$ the names, so to speak, of all the formulas of L(N). From a similar point of view, we have the names of all the proofs. This view raises the possibility of proving in P a formula of the form $\mathbf{A}(\underline{n})$ whose truth in $\mathfrak{N}$ says "something interesting" about the formula with Gödel number n. One of the "interesting things" about a formula **A** with Gödel number n is whether or not $\vdash_{\mathsf{P}} \mathbf{A}$. To be able to do this would require a relation linking the Gödel numbers of formulas to the Gödel numbers of proofs.

Let "$G\#$" be an abbreviation for "Gödel number." Define the binary relation P_{P} by:

† A related concept is ω-completeness. A theory T with the property that for any formula A with free variable x, if $\vdash_{\mathsf{T}} \mathbf{A}_{(\mathbf{x})}[\underline{n}]$ for all $n \in N$, then also $\vdash_{\mathsf{T}} (\forall \mathbf{x})\mathbf{A}$, is said to be ω-complete. The reader can easily show that if a theory is both consistent and ω-complete, then it is ω-consistent. For a brief discussion, the reader can consult Kneale and Kneale (1962, p. 722); for a lengthy discussion, consult "Some observations on the concepts of ω-consistency and ω-completeness," in Tarski (1956, pp. 279–295).

$(m, n) \in P_{\mathsf{P}}$ if and only if m is the $G\#$ of a formula $\mathbf{A}$ with free variable $\mathbf{x}_1$ and n is the $G\#$ of a proof in P of $\mathbf{A}_{(\mathbf{x}_1)}[\underline{m}]$.

Thus, $\vdash_{\mathsf{P}} \mathbf{A}_{(\mathbf{x}_1)}[\underline{m}]$ if and only if $G\#(\mathbf{A}) = m$ and there is an n such that $(m, n) \in P_{\mathsf{P}}$. Further, should P_{P} be a recursive relation, then by Theorem 14.1, $\vdash_{\mathsf{P}} \mathbf{A}_{(\mathbf{x}_1)}[\underline{m}]$ if and only if $G\#(\mathbf{A}) = m$ and there is an n such that $\vdash_{\mathsf{P}} \mathbf{P}_{\mathsf{P}}(\underline{m}, \underline{n})$, where $\mathbf{P}_{\mathsf{P}}$ represents P_{P}. In this way, there would be theorems in P giving information about theorems in P.

Lemma 14.1 The relation P_{P} is a (primitive) recursive relation.

The formal proof, which consists of a long and fairly complicated computation, will not be given in this book.† However, the reader's understanding of recursive procedures should be such that he can give a proof using Church's thesis.

Exercise 14.11 Think about the similaritics between this relation P_{P} and the Kleene T-predicate.

14.2 THE INCOMPLETENESS THEOREM

If P is consistent, then certainly $\mathfrak{N}$ is one of its models. So if P is not complete, there is a sentence $\mathbf{A}$ such that neither $\vdash_{\mathsf{P}} \mathbf{A}$ nor $\vdash_{\mathsf{P}} \sim \mathbf{A}$; but one of them, $\mathbf{A}$ or $\sim\mathbf{A}$, is true in $\mathfrak{N}$. So for whichever it is, say $\mathbf{A}$, the theory $\mathsf{P} \cup \{\mathbf{A}\}$ also has $\mathfrak{N}$ as a model. One could then ask if $\mathsf{P} \cup \{\mathbf{A}\}$ is complete.

Theorem 14.2 (Gödel's Incompleteness Theorem)‡ There is a sentence $\mathbf{G}$ such that

(a) if P is consistent, then not $\vdash_{\mathsf{P}} \mathbf{G}$, and
(b) if P is ω-consistent, then not $\vdash_{\mathsf{P}} \sim \mathbf{G}$.

PROOF Recall the relation P_{P} of the last section and the formula $\mathbf{P}_{\mathsf{P}}$ representing it. To be specific, let $\mathbf{P}_{\mathsf{P}}$ have free variables $\mathbf{x}_1$ and $\mathbf{x}_2$. For any n, consider the formula

$$(\forall \mathbf{x}_2) \sim \mathbf{P}_{\mathsf{P}(\mathbf{x}_1)}[\underline{n}]$$

† Among the places to find the formal proof are Mendelson (1964, pp. 128–134) and Shoenfield (1967, pp. 115–126).

‡ The thcorem was first proved in Gödel (1931). The theorem with consistent replacing ω-consistent in (b) was subsequently proved by Rosser. For a proof, consult Rosser (1936) or Mendelson (1964, pp. 145–146).

In terms of $\mathfrak{N}$, it means "no $\mathbf{x}_2$ is the $G\#$ of a proof of a formula $\mathbf{A}_{(\mathbf{x}_1)}[\underline{n}]$ where n is the $G\#$ of $\mathbf{A}$ with free variable $\mathbf{x}_1$." The formula $(\forall \mathbf{x}_2) \sim \mathbf{P}_{\mathsf{P}}$ has $\mathbf{x}_1$ as its one free variable and it has a Gödel number, say m. Let $\mathbf{G}$ be the sentence

$$(\forall \mathbf{x}_2) \sim \mathbf{P}_{\mathsf{P}_{(\mathbf{x}_1)}}[\underline{m}]$$

$\mathbf{G}$ means "no $\mathbf{x}_2$ is the Gödel number of a proof of $(\forall \mathbf{x}_2) \sim \mathbf{P}_{\mathsf{P}_{(\mathbf{x}_1)}}[\underline{m}]$," or, roughly, "I am not provable." [We emphasize that this meaning for $\mathbf{G}$ depends on the particular m, i.e., the $G\#$ of $(\forall \mathbf{x}_2) \sim \mathbf{P}_{\mathsf{P}}$.] We will now show that $\mathbf{G}$ has the properties claimed by the theorem.

(a) Assume that P is consistent and that

$$\vdash_{\mathsf{P}} (\forall \mathbf{x}_2) \sim \mathbf{P}_{\mathsf{P}_{(\mathbf{x}_1)}}[\underline{m}] \qquad (*)$$

Then there must be a proof; let k be the $G\#$ of that proof. Thus, $(m, k) \in P_{\mathsf{P}}$. But since P_{P} is n-representable in P,

$$\vdash_{\mathsf{P}} \mathbf{P}_{\mathsf{P}_{(\mathbf{x}_1, \mathbf{x}_2)}}[\underline{m}, \underline{k}] \qquad (**)$$

$$\vdash_{\mathsf{P}} (\forall \mathbf{x}_2) \sim \mathbf{P}_{\mathsf{P}_{(\mathbf{x}_1)}}[\underline{m}] \supset \sim \mathbf{P}_{\mathsf{P}_{(\mathbf{x}_1, \mathbf{x}_2)}}[\underline{m}, \underline{k}]$$

with $(*)$, by mp,

$$\vdash_{\mathsf{P}} \sim \mathbf{P}_{\mathsf{P}_{(\mathbf{x}_1, \mathbf{x}_2)}}[\underline{m}, \underline{k}]$$

But this contradicts $(**)$, thus contradicting the consistency of P.

(b) Assume that P is ω-consistent and

$$\vdash_{\mathsf{P}} \sim (\forall \mathbf{x}_2) \sim \mathbf{P}_{\mathsf{P}_{(\mathbf{x}_1)}}[\underline{m}] \qquad (***)$$

As ω-consistency implies consistency, we know that

$$\text{not} \quad \vdash_{\mathsf{P}} (\forall \mathbf{x}_2) \sim \mathbf{P}_{\mathsf{P}_{(\mathbf{x}_1)}}[\underline{m}]$$

That is, no n is the $G\#$ of a proof of

$$(\forall \mathbf{x}_2) \sim \mathbf{P}_{\mathsf{P}_{(\mathbf{x}_1)}}[\underline{m}]$$

Thus, for all n, $(m, n) \notin P_{\mathsf{P}}$. Then, by the n-representability of P_{P}, we have, for every n,

$$\vdash_{\mathsf{P}} \sim \mathbf{P}_{\mathsf{P}_{(\mathbf{x}_1, \mathbf{x}_2)}}[\underline{m}, \underline{n}]$$

Hence, by the ω-consistency of P,

$$\text{not} \quad \vdash_{\mathsf{P}} (\exists \mathbf{x}_2) \sim (\sim \mathbf{P}_{\mathsf{P}_{(\mathbf{x}_1)}}[\underline{m}]), \qquad \text{i.e.,} \qquad \text{not} \quad \vdash_{\mathsf{P}} (\exists \mathbf{x}_2) \mathbf{P}_{\mathsf{P}_{(\mathbf{x}_1)}}[\underline{m}]$$

But this contradicts the assumption $(***)$. ▮

Corollary **G** is true in $\mathfrak{N}$.

Remark: The phrase T *decides* **A** is sometimes used to mean that for a sentence **A**, either $\vdash_T$ **A** or $\vdash_T \sim$ **A**. Obviously, if T is not complete, there is a sentence which T does not decide and such a sentence is said to be an *undecidable sentence*. Thus, **G** is undecidable with respect to P. We only mention this terminology so that the reader confronted with it elsewhere will not confuse it with the usual meaning for decidable and undecidable.

Exercise 14.12 (a) Using Theorem 14.1 but without reference to Theorem 14.2, show that P is not complete. (Hint: Use the Kleene predicate, the unsolvability of the halting problem, etc.) (b) Show that the set of $G\#$'s of sentences provable in P is a creative set. (Hint: Show that the set of $G\#$'s of formulas not provable in P is a productive set, notice that the set $\{(\exists \mathbf{y})\mathbf{T}(\underline{m}, \underline{n}, \mathbf{y}) \mid \text{all } n, m \in N\}$, where **T** represents the T-predicate, is recursive; use exercise 7.32.)

We know from Theorem 14.2 that $\sim$**G** is not provable in P and hence (on the assumption that P is consistent) the theory P $\cup$ {**G**} is consistent; similarly, P $\cup$ {$\sim$**G**} is also consistent. Further, P $\cup$ {**G**} is recursively axiomatizable. Let $P_{\mathsf{P} \cup \{\mathbf{G}\}}$ be the relation depending on P $\cup$ {**G**} analogous to the relation relation P_{P} which depends on P. $P_{\mathsf{P} \cup \{\mathbf{G}\}}$ is a primitive recursive relation for the same reasons that P_{P} is. Also, all recursive functions and relations are n-representable in P $\cup$ {**G**}, since they are in P. By exercise 14.8, these are the only relations representable in P $\cup$ {**G**}. Thus, Theorem 14.2 can be restated in terms of the theory P $\cup$ {**G**} and a new sentence **G**′ related to P $\cup$ {**G**} in the same way that **G** is related to P. The proof is exactly the same. And so this process can be repeated any number of times. That is, starting with the recursive set of axioms of P, one can keep adding axioms in such a way that each expanded set is recursive; but the result is never a complete theory.

Exercise 14.13 Show that P $\cup$ {$\sim$**G**} is not ω-consistent even though it is consistent if P is consistent.

Having reached this impasse in our attempts to find a recursively axiomatizable complete theory for $\mathfrak{N}$, we might still hope that P or P $\cup$ {**G**, **G**′, ..., $\mathbf{G}^{(n)}$} for some n, is decidable. We might also reexamine the axioms of P in an effort to find some sort of culprit. Of these, probably the induction axiom scheme has the guiltiest air about it. Perhaps if we traded it in for something simpler, the rather negative sorts of results we have just obtained for P could be avoided. However, later, we will investigate two sets of axioms for which $\mathfrak{N}$ is a model, neither containing the induction axiom, and find that we obtain the same kinds of results as for P. A further result that we will be considering is that the induction axiom may not be quite as mischievous as it appears; for Presburger has exhibited a theory without the function symbol $\cdot$ and with all the axioms of P except *P*6 and *P*7 which is decidable and complete.

14.3 THE UNDECIDABILITY OF P AND $\mathcal{N}$

In setting out to show that P is not decidable, we will actually get a stronger result which leads to various significant conclusions.

Lemma 14.2 There is a recursive function d such that if n is the $G\#$ of a formula $\mathbf{A}$ with one free variable $\mathbf{x}_1$, then $d(n)$ is the $G\#$ of $\mathbf{A}_{(\mathbf{x}_1)}[\underline{n}]$.

Again we omit the formal proof and ask the reader to convince himself of the truth of the lemma by means of Church's thesis. Of course, d is a "diagonal function."

For a first-order theory T, define

$$PR_{\mathsf{T}} = \{n \mid n \text{ is the } G\# \text{ of a sentence } \mathbf{A} \text{ and } \vdash_{\mathsf{T}} \mathbf{A}\}$$

Obviously, T is decidable if and only if PR_{T} is a recursive set. For any theory T in which exactly the recursive relations are n-representable, we can investigate the decidability of T by looking into whether or not PR_{T} is n-representable in T. PR_{P} is just the set proved to be creative in exercise 14.12. It follows that PR_{P} is not recursive and hence P is not decidable. However, the proof of the exercise depends on results about Tm's. We wish to carry out the proof of the undecidability of P without reference to the Tm results precisely because we will see that in the apparently different setting of formal logic there are exactly analogous results.

Theorem 14.3 If T is a consistent first-order theory, then it is impossible for both the function d and the relation PR_{T} to be n-representable in T.

PROOF† Assume that both d and PR_{T} are n-representable in T; then by exercises 14.5 and 14.9, the set $\{x \mid d(x) \notin PR_{\mathsf{T}}\}$ is also n-representable. Let $\mathbf{M}$ be the formula with one free variable $\mathbf{x}_1$ which represents that set. Suppose m is the $G\#$ of $\mathbf{M}$. We then have the following:

$\vdash_{\mathsf{T}} \mathbf{M}_{(\mathbf{x}_1)}[\underline{m}]$ if and only if, by the definition of $\mathbf{M}$,
$d(m) \notin PR_{\mathsf{T}}$ if and only if, by the definition of d,
$G\#(\mathbf{M}_{(\mathbf{x}_1)}[\underline{m}]) \notin PR_{\mathsf{T}}$, if and only if, by the definition of PR_{T},
not $\vdash_{\mathsf{T}} \mathbf{M}_{(\mathbf{x}_1)}[\underline{m}]$.

That is, $\vdash_{\mathsf{T}} \mathbf{M}_{(\mathbf{x}_1)}[\underline{m}]$ if and only if not $\vdash_{\mathsf{T}} \mathbf{M}_{(\mathbf{x}_1)}[\underline{m}]$, a clear-cut contradiction. ∎

† This tidy proof has been contributed by Mitsuru Yasuhara.

Corollary 1 Let T be a theory in which all the recursive functions are n-representable. Then T is not decidable.

Corollary 2 P is not decidable.

Corollary 3 $\mathcal{N}$ is not decidable.

A k-ary number-theoretic relation R^k is defined to be an *arithmetic relation* if there is a formula **R** in F(N) with exactly k free variables such that $(n_1, \ldots, n_k) \in R^k$ if and only if $V(\mathbf{R}(\underline{n}_1, \ldots, \underline{n}_k)) = \mathrm{t}$ in $\mathfrak{N}$. Certainly, the recursive and recursively enumerable relations are included in the set of arithmetic relations. [In fact, the set of arithmetic relations is just exactly the set of relations in the Kleene–Mostowski hierarchy; see section 8.3. For a proof, see Davis (1958, chapter 9, section 7) or Hermes (1965, section 29).] Since the axioms of $\mathcal{N}$ are all sentences of F(N) that are true in $\mathfrak{N}$, $n \in PR_{\mathcal{N}}$ if and only if n is the $G\#$ of a sentence true in $\mathfrak{N}$. It is clear from the definitions that all arithmetic relations are n-representable in $\mathcal{N}$.

Corollary 4 to Theorem 14.3 The relation $PR_{\mathcal{N}}$ is not arithmetic.

Thus, whatever we mean by "truth" with regard to first-order sentences *about* arithmetic, we cannot formalize it *in* first-order arithmetic.

Corollary 5 to Theorem 14.3 The set of all sentences in F(N) that are true in $\mathfrak{N}$ is not a recursively enumerable set.

Exercise 14.14 Corollary 5 can be proved without reference to arithmetic relations as follows: (a) Prove that if the set of theorems of a first-order theory is a recursively enumerable set, then the theory is recursively axiomatizable. (Hint: An amusing trick, due to Craig—let $\mathbf{A}_1, \mathbf{A}_2, \ldots$ be a recursive enumeration of the nonlogical axioms and think about $\mathbf{A}_1$, $\mathbf{A}_2 \wedge \mathbf{A}_2$, $\mathbf{A}_3 \wedge \mathbf{A}_3 \wedge \mathbf{A}_3, \ldots$.) (b) Prove that the set of all sentences of F(N) that are true in $\mathfrak{N}$ is not a recursively enumerable set.

14.4 TWO SUBTHEORIES OF $\mathcal{N}$: R AND Q†

L(N) will be the language for this section. Let Q be the theory with nonlogical axioms

$Q1.$ $(\mathbf{S}(\mathbf{x}) = \mathbf{S}(\mathbf{y})) \supset (\mathbf{x} = \mathbf{y})$

$Q2.$ $\mathbf{0} \neq \mathbf{S}(\mathbf{y})$

† This section is based on Tarski *et al.* (1953, Part II, sections 3 and 4).

$Q3.$ $\mathbf{x \neq 0 \supset (\exists y)(x = S(y))}$

$Q4.$ $\mathbf{x + 0 = x}$

$Q5.$ $\mathbf{x + S(y) = S(x + y)}$

$Q6.$ $\mathbf{x \cdot 0 = 0}$

$Q7.$ $\mathbf{x \cdot S(y) = x \cdot y + x}$

Notice, in particular, that Q has only a finite number of axioms.

Let R be the theory with nonlogical axioms

$R1.$ $\underline{m} + \underline{n} = \underline{m+n}$ for all $m, n \in N$.

$R2.$ $\underline{m} \cdot \underline{n} = \underline{mn}$ for all $m, n \in N$.

$R3.$ $\underline{m} \neq \underline{n}$ for all $m, n \in N$ such that $m \neq n$.

$R4.$ $\mathbf{x} \leq \underline{n} \supset (\mathbf{x} = \mathbf{0} \vee \mathbf{x} = \underline{1} \vee \cdots \vee \mathbf{x} = \underline{n})$ for all $n \in N$.

$R5.$ $\mathbf{x} \leq \underline{n} \vee \underline{n} \leq \mathbf{x}$ for all $n \in N$.

As before, $\underline{m} + \underline{n}$ denotes the term $(\underbrace{\mathbf{SS \cdots S}}_{m}\mathbf{(0)} + \underbrace{\mathbf{SS \cdots S}}_{n}\mathbf{(0)})$, and $\underline{m} \cdot \underline{n}$ denotes the term $(\underbrace{\mathbf{SS \cdots S}}_{m}\mathbf{(0)} \cdot \underbrace{\mathbf{SS \cdots S}}_{n}\mathbf{(0)})$. Further, $\underline{m+n}$ denotes the term $\underbrace{\mathbf{SS \cdots S}}_{m+n}\mathbf{(0)}$ and $\underline{mn}$ denotes $\underbrace{\mathbf{SS \cdots S}}_{n}\mathbf{(0)}$. Thus, in any model of R, the object corresponding to $\underline{m} + \underline{n}$ is also the object corresponding to $\underline{m+n}$; similarly $\underline{m} \cdot \underline{n}$ and $\underline{mn}$ correspond to the same object. Each axiom of R is an axiom scheme and so R has an infinite number of axioms, but is still recursively axiomatizable. Notice that the induction axiom of P is not present in either set of axioms. This section will be devoted to showing that R is a subtheory of Q, that Q is a subtheory of P, and that all recursive functions are n-representable in R. It then follows that the recursive functions are n-representable in P; the proof for Theorem 14.1 stated in section 14.1 is thus supplied. It will then also follow from Theorem 14.3 that R and Q and any extensions of these are undecidable.

The language L(N) does not contain the symbol $\leq$, but we will use $\underline{m} \leq \underline{n}$ as an abbreviation for $(\exists \mathbf{x})(\mathbf{x} + \underline{m} = \underline{n})$.

Lemma 14.3 If $m \leq n$, then $\vdash_{\mathsf{R}} \underline{m} \leq \underline{n}$.

PROOF $m \leq n$ implies there is a $k \in N$ such that $k + m = n$. So by $R1$, $\vdash_{\mathsf{R}} \underline{k} + \underline{m} = \underline{k+m}$, which is the same as $\vdash_{\mathsf{R}} \underline{k} + \underline{m} = \underline{n}$ and therefore $\vdash_{\mathsf{R}} (\exists \mathbf{x})(\mathbf{x} + \underline{m} = \underline{n})$. ∎

Theorem 14.4 R is a subtheory of Q.

PROOF We must show that each of the axioms of R is a theorem of Q. *R*2 and *R*3 are left to the reader; the main steps showing that *R*1, *R*4, and *R*5 are provable will be given. In each case, it must be shown that for each numeral, $\underline{n}$, the axiom with that numeral $\underline{n}$ is provable in Q. Sometimes we will do an induction on $n \in N$ to get the result for the corresponding numeral $\underline{n}$. Notice that this kind of induction is just the usual, informal induction we have used many times. It has nothing to do with *P*8, the induction axiom in P. In the proof of Theorem 14.5, *P*8 will be used and then the difference between usual, informal induction and formal, *P*8, induction should be clear.

*R*1: Show that for each $m, n \in N$, $\vdash_{\mathrm{Q}} \underline{m} + \underline{n} = \underline{m+n}$. By induction on n with m arbitrary,

$n = 0$:	$\vdash \mathbf{x} + \mathbf{0} = \mathbf{x}$	*Q*4
	$\vdash (\forall \mathbf{x})(\mathbf{x} + \mathbf{0} = \mathbf{x}) \supset (\underline{m} + \mathbf{0} = \underline{m})$	
	$\vdash \underline{m} + \mathbf{0} = \underline{m}$,	i.e., $\underline{m}$ is $\underline{m+0}$
$n = k + 1$:	$\vdash \underline{m} + \underline{k} = (\underline{m+k})$	induction hypothesis
	$\vdash \mathrm{S}(\underline{m} + \underline{k}) = \mathrm{S}(\underline{m+k})$	
	$\vdash \underline{m} + \mathrm{S}(\underline{k}) = \mathrm{S}(\underline{m} + \underline{k})$	*Q*5
	$\vdash \underline{m} + \mathrm{S}(\underline{k}) = \mathrm{S}(\underline{m+k})$	

*R*4: Show that for each $n \in N$, $\vdash_{\mathrm{Q}} \mathbf{x} \leq \underline{n} \supset (\mathbf{x} = \mathbf{0} \vee \mathbf{x} = \underline{1} \vee \cdots \vee \mathbf{x} = \underline{n})$. (The justifications for the less obvious steps are given at the right; in some cases, there will just be a number which refers to a list at the end of the proof. DT indicates deduction theorem. Parentheses will be omitted by writing $\mathbf{A} \vee \mathbf{B} \supset \mathbf{C}$ for $((\mathbf{A} \vee \mathbf{B}) \supset \mathbf{C})$ and $\mathbf{A} \supset \mathbf{B} \vee \mathbf{C}$ for $(\mathbf{A} \supset (\mathbf{B} \vee \mathbf{C}))$; similarly with $\wedge$ in place of $\vee$.)

For $n = 0$:

$\mathbf{z} + \mathbf{x} = \mathbf{0}, \mathbf{x} = \mathrm{S}(\mathbf{y}) \vdash \mathbf{x} = \mathrm{S}(\mathbf{y}) \wedge \mathbf{z} + \mathrm{S}(\mathbf{y}) = \mathbf{0}$	by 1
$\vdash \mathbf{z} + \mathrm{S}(\mathbf{y}) = \mathrm{S}(\mathbf{x} + \mathbf{y})$	*Q*5
$\mathbf{z} + \mathbf{x} = \mathbf{0}, \mathbf{x} = \mathrm{S}(\mathbf{y}) \vdash \mathbf{x} = \mathrm{S}(\mathbf{y}) \wedge \mathrm{S}(\mathbf{z} + \mathbf{y}) = \mathbf{0}$	by 1
$\mathbf{z} + \mathbf{x} = \mathbf{0}, \mathbf{x} = \mathrm{S}(\mathbf{y}) \vdash (\exists \mathbf{y})(\mathbf{x} = \mathrm{S}(\mathbf{y}) \wedge \mathrm{S}(\mathbf{z} + \mathbf{y}) = \mathbf{0})$	by 2
$\mathbf{z} + \mathbf{x} = \mathbf{0} \vdash (\exists \mathbf{y})(\mathbf{x} = \mathrm{S}(\mathbf{y})) \supset (\exists \mathbf{y})(\mathbf{x} = \mathrm{S}(\mathbf{y}) \wedge \mathrm{S}(\mathbf{z} + \mathbf{y}) = \mathbf{0})$	DT and 7
$\mathbf{x} \neq \mathbf{0} \vdash (\exists \mathbf{y})(\mathbf{x} = \mathrm{S}(\mathbf{y}))$	*Q*3
$\mathbf{z} + \mathbf{x} = \mathbf{0}, \mathbf{x} \neq \mathbf{0} \vdash (\exists \mathbf{y})(\mathbf{x} = \mathrm{S}(\mathbf{y}) \wedge \mathrm{S}(\mathbf{z} + \mathbf{y}) = \mathbf{0})$	mp
$\vdash \mathbf{z} + \mathbf{x} = \mathbf{0} \wedge \mathbf{x} \neq \mathbf{0} \supset (\exists \mathbf{y})(\mathbf{x} = \mathrm{S}(\mathbf{y}) \wedge \mathrm{S}(\mathbf{z} + \mathbf{y}) = \mathbf{0})$	DT
$\vdash (\forall \mathbf{v})((\mathbf{v} \neq \mathbf{0} \wedge \mathbf{x} + \mathbf{v} = \mathbf{0}) \supset (\exists \mathbf{y})(\mathbf{v} = \mathrm{S}(\mathbf{y}) \wedge \mathrm{S}(\mathbf{x} + \mathbf{y}) = \mathbf{0}))$	ug
$\vdash (\exists \mathbf{v})(\mathbf{v} \neq \mathbf{0} \wedge \mathbf{x} + \mathbf{v} = \mathbf{0}) \supset (\exists \mathbf{v})(\exists \mathbf{y})(\mathbf{v} = \mathrm{S}(\mathbf{y}) \wedge \mathrm{S}(\mathbf{x} + \mathbf{y}) = \mathbf{0})$	by 3
$\mathbf{x} \leq \mathbf{0} \vdash \mathbf{x} = \mathbf{0} \vee \mathbf{x} < \mathbf{0}$	

which is

$\mathbf{x \le 0 \vdash x = 0 \lor (\exists v)(v \ne 0 \land x + v = 0)}$
$\mathbf{x \le 0 \vdash x = 0 \lor (\exists v)(\exists y)(v = S(y) \land S(x + y) = 0)}$

or equivalently

$\mathbf{x \le 0 \vdash x = 0 \lor \sim(\forall v)(\forall y)(v \ne S(y) \lor S(x + y) \ne 0)}$
$\mathbf{\vdash S(x + y) \ne 0 \lor v \ne S(y)}$ — $Q2$
$\mathbf{\vdash (\forall v)(\forall y)(S(x + y) \ne 0 \lor v \ne S(y))}$ — ug
$\mathbf{x \le 0 \vdash x = 0}$ — by 8
$\mathbf{\vdash x \le 0 \supset x = 0}$ — DT

Suppose $R4$ is provable in Q for $n = k$ and show that it is for $n = k + 1$: By methods similar to those just used, we have

$\mathbf{\vdash z + x = S(\underline{k}) \land x \ne 0 \supset (\exists y)(x = S(y) \land S(z + y) = S(\underline{k}))}$ (*)
$\mathbf{\vdash S(z + y) = S(\underline{k}) \supset z + y = \underline{k}}$ — $Q1$
$\mathbf{\vdash (\forall z)(x = S(y) \land S(z + y) = S(\underline{k}) \supset x = S(y) \land z + y = \underline{k})}$ — If $\vdash \mathbf{A \supset B}$, then $\vdash \mathbf{C \land A \supset C \land B}$
$\mathbf{\vdash x = S(y) \land (\exists z)(S(z + y) = S(\underline{k})) \supset x = S(y) \land (\exists z)(z + y = \underline{k})}$ — by 3 and 4
$\mathbf{\vdash (\exists y)(x = S(y) \land (\exists z)(S(z + y) = S(\underline{k}))) \supset (\exists y)(x = S(y) \land y \le \underline{k})}$ — by ug and 3
$\mathbf{\vdash (\forall z)(*)}$ — ug on (*)
$\mathbf{\vdash (\exists z)(z + x = S(\underline{k})) \land x \ne 0 \supset (\exists y)(x = S(y) \land (\exists z)(S(z + y) = S(\underline{k})))}$ — by 3 and 5
$\mathbf{\vdash x \le S(\underline{k}) \supset x = 0 \lor (\exists y)(x = S(y) \land y \le \underline{k})}$ (**) — by 6 and def. of $\le$
$\mathbf{\vdash x = S(y) \land y \le \underline{k} \supset x = S(y) \land (y = 0 \lor \cdots \lor y = \underline{k})}$ — by the induction hypothesis, etc.
$\mathbf{\vdash x = S(y) \land y \le \underline{n} \supset (x = S(y) \land y = 0) \lor \cdots \lor (x = S(y) \land y = \underline{k})}$
For each i, $0 \le i \le n$,
$\mathbf{\vdash y = \underline{i} \supset (x = S(y) \equiv x = \underline{i+1})}$ — equality axiom and def. $\underline{i+1}$
$\mathbf{\vdash (x = S(y) \land y = \underline{i}) \supset (x = \underline{i+1})}$
$\mathbf{\vdash (x = S(y) \land y \le \underline{k}) \supset (x = \underline{1}) \lor \cdots \lor (x = \underline{k+1})}$
$\mathbf{\vdash (\exists y)(x = S(y) \lor y \le \underline{k}) \supset ((x = \underline{1}) \lor \cdots \lor (x = \underline{k+1}))}$ — by 7
$\mathbf{\vdash x \le S(\underline{k}) \supset (x = 0 \lor x = \underline{1} \lor \cdots \lor x = \underline{k+1})}$ — transitivity with (**)

List of justifications:

1. Equality axioms and propositional calculus.
2. $\mathbf{A(y) \supset (\exists y)A(y)}$.
3. $\mathbf{(\forall x)(A \supset B) \supset ((\exists x)A \supset (\exists x)B)}$.
4. $\mathbf{(\exists x)(A \wedge B) \supset (A \wedge (\exists x)B)}$, where $\mathbf{x}$ is not free in $\mathbf{A}$.
5. $\mathbf{(\exists x)(\exists y)A \equiv (\exists y)(\exists x)A}$.
6. Transitivity and if $\vdash \mathbf{A} \wedge \sim\mathbf{C} \supset \mathbf{B}$, then $\vdash \mathbf{A} \supset \mathbf{C} \vee \mathbf{B}$.
7. If $\vdash \mathbf{A} \supset \mathbf{B}$, then $\vdash (\exists \mathbf{y})\mathbf{A} \supset \mathbf{B}$, where $\mathbf{y}$ is not free in $\mathbf{B}$.
8. If $\mathbf{A} \vdash \mathbf{B} \vee \sim\mathbf{C}$ and $\vdash\mathbf{C}$, then $\mathbf{A} \vdash \mathbf{B}$.

*R*5: Show that $\vdash_{Q} \underline{n} \leq \mathbf{x} \vee \mathbf{x} \leq \underline{n}$, by induction on n. For $n = 0$:

$$\vdash \mathbf{x + 0 = x}$$
$$\vdash (\exists \mathbf{z})(\mathbf{z + 0 = x}),$$

which is

$$\vdash \mathbf{0 \leq x} \qquad \text{by 9 below}$$
$$\vdash \mathbf{0 \leq x \vee x \leq 0}$$

The following five little results are needed to complete the induction step.

(i) By induction on n, show that for all $n \in N$, $\vdash_{Q} \mathbf{S(x)} + \underline{n} = \mathbf{x} + \underline{n+1}$.

$n = 0$:
$$\vdash \mathbf{x + S(0) = S(x) + 0}$$
$$\vdash \mathbf{S(x) + 0 = S(x)}$$
$$\vdash \mathbf{x + S(0) = S(x)}$$
$$\vdash \mathbf{x + 0 = x}$$
$$\vdash \mathbf{x + S(0) = S(x + 0)}$$
$$\vdash \mathbf{S(x) + 0 = S(x + 0)}$$

$n = k + 1$:
$$\vdash \mathbf{S(x)} + \mathbf{S}(\underline{k}) = \mathbf{S}(\mathbf{S(x)} + \underline{k})$$
$$\vdash \mathbf{S(x)} + \underline{k} = \mathbf{x} + (\underline{k+1}) \qquad \text{induction hypothesis}$$
$$\vdash \mathbf{S(x)} + \mathbf{S}(\underline{k}) = \mathbf{S}(\mathbf{x} + (\underline{k+1}))$$
$$\vdash \mathbf{x} + \mathbf{S}(\underline{k+1}) = \mathbf{S}(\mathbf{x} + (\underline{k+1}))$$
$$\vdash \mathbf{S(x)} + \mathbf{S}(\underline{k}) = \mathbf{x} + \mathbf{S}(\underline{k+1})$$
$$\vdash \mathbf{S(x)} + (\underline{k+1}) = \mathbf{x} + \mathbf{S}(\underline{k+1})$$

(ii) Show that for all $n \in N$, $\vdash_{Q} \underline{n} \leq (\underline{n+1})$:

$$\vdash \underline{n} + \underline{1} = (\underline{n+1}) \qquad \text{already proved}$$
$$\vdash (\exists \mathbf{z})(\underline{n} + \mathbf{z} = \underline{n+1}),$$

that is

$$\vdash \underline{n} \leq (\underline{n+1}) \qquad \text{by 9 below}$$

(iii) Show that for all $n \in N$, $\vdash_Q (\mathbf{x} \leq \underline{n}) \supset (\mathbf{x} \leq \underline{(n+1)})$. For all $i \leq n+1$, there is a k_i such that $k_i + i = (n+1)$. So, for each i,

$$\vdash \underline{k}_i + \underline{i} = \underline{(n+1)}$$
$$\vdash (\exists \mathbf{y})(\mathbf{y} + \underline{i} = \underline{(n+1)}),$$

that is (*)

$$\vdash \underline{i} \leq \underline{(n+1)}$$
$$\vdash \mathbf{x} \leq \underline{n} \supset (\mathbf{x} = 0 \vee \cdots \vee \mathbf{x} = \underline{i} \vee \cdots \vee \mathbf{x} = \underline{n}) \qquad \text{already proved}$$

Then, for each $i \leq (n+1)$,

$$\vdash (\mathbf{x} = \underline{i}) \supset (\mathbf{x} \leq \underline{(n+1)} \equiv \underline{i} \leq \underline{(n+1)})$$
$$\vdash (\mathbf{x} \leq \underline{n}) \supset \bigvee_{0 \leq i \leq n} (\mathbf{x} \leq \underline{(n+1)} \equiv \underline{i} \leq \underline{(n+1)}) \qquad (**)$$
$$\vdash \mathbf{x} \leq \underline{n} \supset \mathbf{x} \leq \underline{(n+1)} \qquad \text{from } (*) \text{ and } (**)$$

(iv) Show that for all $n \in N$, $\vdash_Q \mathbf{z} + \underline{n} = \mathbf{x} \wedge \mathbf{z} \neq 0 \supset (\underline{(n+1)} \leq \mathbf{x})$:

$$\vdash S(\mathbf{y}) = \mathbf{z} \supset (\mathbf{z} + \underline{n} = \mathbf{x} \equiv S(\mathbf{y}) + \underline{n} = \mathbf{x})$$
$$\vdash S(\mathbf{y}) + \underline{n} = \mathbf{y} + \underline{(n+1)} \qquad \text{already proved}$$
$$\vdash (S(\mathbf{y}) + \underline{n} = \mathbf{y} + \underline{(n+1)}) \supset (S(\mathbf{y}) + \underline{n} = \mathbf{x} \equiv \mathbf{y} + \underline{(n+1)} = \mathbf{x})$$
$$\vdash (\forall \mathbf{y})(S(\mathbf{y}) = \mathbf{z} \supset (\mathbf{z} + \underline{n} = \mathbf{x} \equiv \mathbf{y} + \underline{(n+1)} = \mathbf{x}))$$
$$\vdash (\exists \mathbf{y})(S(\mathbf{y}) = \mathbf{z} \supset (\mathbf{z} + \underline{n} = \mathbf{x} \equiv (\exists \mathbf{y})(\mathbf{y} + \underline{(n+1)} = \mathbf{x})))$$
$$\vdash \mathbf{z} + \underline{n} = \mathbf{x} \wedge \mathbf{z} \neq 0 \supset (\exists \mathbf{y})(S(\mathbf{y}) = \mathbf{z})$$
$$\vdash \mathbf{z} + \underline{n} = \mathbf{x} \wedge \mathbf{z} \neq 0 \supset (\mathbf{z} + \underline{n} = \mathbf{x} \equiv (\exists \mathbf{y})(\mathbf{y} + \underline{(n+1)} = \mathbf{x}))$$
$$\vdash \mathbf{z} + \underline{n} = \mathbf{x} \wedge \mathbf{z} \neq 0 \supset (\exists \mathbf{y})(\mathbf{y} + \underline{(n+1)} = \mathbf{x}),$$

that is

$$\vdash \mathbf{z} + \underline{n} = \mathbf{x} \wedge \mathbf{z} \neq 0 \supset (\underline{(n+1)} \leq \mathbf{x}) \qquad (*)$$

(v) Show that for all $n \in N$, $\vdash_Q \underline{n} \leq \mathbf{x} \supset \mathbf{x} = \underline{n} \vee \underline{(n+1)} \leq \mathbf{x}$:

$$\underline{n} \leq \mathbf{x} \vdash \underline{n} = \mathbf{x} \vee \underline{n} < \mathbf{x},$$

that is

$$\underline{n} \leq \mathbf{x} \vdash \underline{n} = \mathbf{x} \vee (\exists \mathbf{z})(\mathbf{z} + \underline{n} = \mathbf{x} \wedge \mathbf{z} \neq 0)$$
$$\vdash \underline{n} \leq \mathbf{x} \supset (\underline{n} = \mathbf{x} \vee \underline{(n+1)} \leq \mathbf{x}) \qquad \text{by } (*) \text{ in (iv), 7, and DT}$$

Finally, for the induction step proving that $R5$ is a theorem in **Q**, we assume it true for $n = k$ and show for $n = k + 1$.

$$\vdash \mathbf{x} \leq \underline{k} \vee \underline{k} \leq \mathbf{x} \qquad \text{by the induction hypothesis}$$
$$\vdash (\mathbf{x} \leq \underline{k}) \vee (\mathbf{x} = \underline{k} \vee \underline{(k+1)} \leq \mathbf{x}) \qquad \text{using (v)}$$
$$\vdash \mathbf{x} \leq \underline{(k+1)} \vee (\mathbf{x} = \underline{k} \vee \underline{(k+1)} \leq \mathbf{x}) \qquad \text{using (iii)}$$

$\vdash \mathbf{x} = \underline{k} \supset (\mathbf{x} \leq \underline{(k+1)} \equiv \underline{k} \leq \underline{(k+1)})$
$\vdash \underline{k} \leq \underline{(k+1)}$
$\vdash \mathbf{x} = \underline{k} \supset \mathbf{x} \leq \underline{(k+1)}$
$\vdash \mathbf{x} \leq \underline{(k+1)} \vee (\mathbf{x} \leq \underline{(k+1)} \vee \underline{(k+1)} \leq \mathbf{x})$
$\vdash \mathbf{x} \leq \underline{(k+1)} \vee \underline{(k+1)} \leq \mathbf{x}.$

(Justification 9 is $\vdash \mathbf{A}_{(\mathbf{x})}[\mathbf{t}] \supset (\exists \mathbf{x})\mathbf{A}$, $\mathbf{t}$ free for $\mathbf{x}$ in $\mathbf{A}$.) ▮

Theorem 14.5 Q is a subtheory of P.

PROOF Since all the axioms of Q except *Q*3 are axioms of P, we need only show that *Q*3 is provable in P. We will use *P*8 with $\mathbf{A}$ the formula $(\mathbf{x} = \mathbf{0} \vee (\exists \mathbf{y})(\mathbf{S}(\mathbf{y}) = \mathbf{x}))$. That is, we have, by *P*8,

$$\vdash (\mathbf{0} = \mathbf{0} \vee (\exists \mathbf{y})(\mathbf{S}(\mathbf{y}) = \mathbf{0}))\supset$$
$$((\forall \mathbf{x})((\mathbf{x} = \mathbf{0} \vee (\exists \mathbf{y})(\mathbf{S}(\mathbf{y}) = \mathbf{x})) \supset (\mathbf{S}(\mathbf{x}) = \mathbf{0} \vee (\exists \mathbf{y})(\mathbf{S}(\mathbf{y}) = \mathbf{S}(\mathbf{x}))))$$
$$\supset (\forall \mathbf{x})(\mathbf{x} = \mathbf{0} \vee (\exists \mathbf{y})(\mathbf{S}(\mathbf{y}) = \mathbf{x})))$$

Thus, we must show the two following:

$$\vdash \mathbf{0} = \mathbf{0} \vee (\exists \mathbf{y})(\mathbf{S}(\mathbf{y}) = \mathbf{0}) \qquad (*)$$
$$\vdash (\forall \mathbf{x})((\mathbf{x} = \mathbf{0} \vee (\exists \mathbf{y})(\mathbf{S}(\mathbf{y}) = \mathbf{x}))\supset (\mathbf{S}(\mathbf{x}) = \mathbf{0} \vee (\exists \mathbf{y})(\mathbf{S}(\mathbf{y}) = \mathbf{S}(\mathbf{x})))) \qquad (**)$$

Since $\vdash \mathbf{0} = \mathbf{0}$, $(*)$ follows immediately. From $\vdash \mathbf{S}(\mathbf{x}) = \mathbf{S}(\mathbf{y})$, we have $\vdash (\exists \mathbf{y})(\mathbf{S}(\mathbf{y}) = \mathbf{S}(\mathbf{x}))$, and then $\vdash \mathbf{S}(\mathbf{x}) = \mathbf{0} \vee (\exists \mathbf{y})(\mathbf{S}(\mathbf{y}) = \mathbf{S}(\mathbf{x}))$. And hence we have $(**)$.

Using $(*)$ and $(**)$ in *P*8 and mp twice, we have $\vdash \mathbf{x} = \mathbf{0} \vee (\exists \mathbf{y})(\mathbf{S}(\mathbf{y}) = \mathbf{x})$, which is equivalent to $\vdash \mathbf{x} \neq \mathbf{0} \supset (\exists \mathbf{y})(\mathbf{S}(\mathbf{y}) = \mathbf{x})$. In this case, the proof is by formal induction in the theory P. That is, by using *P*8, we never need to refer to *N* "outside" of P. The transfer of the formula from $\mathbf{x}$ to $\mathbf{S}(\mathbf{x})$ is entirely inside the formal theory P. ▮

The alternative formulation of recursive functions given in section 6.3 will be used in showing that the recursive functions are *n*-representable in R. We will be using both the original and the alternative definitions of *n*-representable that were given in section 14.1.

Theorem 14.6 If $f \in \mathscr{R}'$, then f is *n*-representable in R.

PROOF The proof is by induction on the formation of f. The reader can show for himself that the successor function S is *n*-representable. Notice that if g is a recursive function then $\underline{g(n)}$ denotes $\underbrace{\mathbf{SS}\cdots\mathbf{S}}_{g(n)}(\mathbf{0})$.

1. Show that E, the excess over a square function, is n-representable. We will show that the formula

$$(\exists \mathbf{x})(\mathbf{x} \leq \mathbf{u} \wedge \mathbf{v} = \mathbf{x} + \mathbf{x} \wedge \mathbf{u} = \mathbf{x} \cdot \mathbf{x} + \mathbf{v})$$

denoted by $\mathbf{E}(\mathbf{u}, \mathbf{v})$, represents E in R.

(a) Show that $\vdash \mathbf{v} = \underline{E(n)} \supset \mathbf{E}(\underline{n}, \mathbf{v})$. By the definition of E, for each n, there must be an m such that $m^2 + E(n) = n$ and $n < (m+1)^2$. Thus, $E(n) \leq 2m$ and $m^2 \leq n$, so $m \leq n$. Therefore, by Lemma 14.3 and $R1$, $\vdash \underline{m} \leq \underline{n} \wedge \underline{E(n)} \leq \underline{m} + \underline{m} \wedge \underline{n} = \underline{m} \cdot \underline{m} + \underline{E(n)}$. Combining this with an instance of the equality axiom and the propositional calculus, we have

$$\vdash \mathbf{v} = \underline{E(n)} \supset (\underline{m} \leq \underline{n} \wedge \mathbf{v} \leq \underline{m} + \underline{m} \wedge \underline{n} = \underline{m} \cdot \underline{m} + \mathbf{v})$$

therefore

$$\vdash \mathbf{v} = \underline{E(n)} \supset (\exists \mathbf{x})(\mathbf{x} \leq \underline{n} \wedge \mathbf{v} \leq \mathbf{x} + \mathbf{x} \wedge \underline{n} = \mathbf{x} \cdot \mathbf{x} + \mathbf{v})$$

i.e.,

$$\vdash \mathbf{v} = \underline{E(n)} \supset \mathbf{E}(\underline{n}, \mathbf{v})$$

(b) It takes a little longer to show $\vdash \mathbf{E}(\underline{n}, \mathbf{v}) \supset \mathbf{v} = \underline{E(n)}$:

$$\vdash \big((\mathbf{x} \leq \underline{n}) \wedge (\mathbf{v} \leq \mathbf{x} + \mathbf{x}) \wedge (\underline{n} = \mathbf{x} \cdot \mathbf{x} + \mathbf{v})\big) \supset (\mathbf{x} = \mathbf{0} \vee \mathbf{x} = \underline{1} \vee \cdots \vee \mathbf{x} = \underline{n})$$

from $R4$

For all j, $0 \leq j \leq n$,

$$\vdash (\mathbf{x} = \underline{j}) \supset \big((\mathbf{v} \leq \underline{j} + \underline{j} \wedge \underline{n} = \underline{j} \cdot \underline{j} + \mathbf{v}) \equiv (\mathbf{v} \leq \mathbf{x} + \mathbf{x} \wedge \underline{n} = \mathbf{x} \cdot \mathbf{x} + \mathbf{v})\big)$$

$$\vdash \big((\mathbf{x} \leq \underline{n}) \wedge (\mathbf{v} \leq \mathbf{x} + \mathbf{x}) \wedge (\underline{n} = \mathbf{x} \cdot \mathbf{x} + \mathbf{v})\big)$$
$$\supset \big((\mathbf{v} \leq \mathbf{0} + \mathbf{0} \wedge \underline{n} = \mathbf{0} \cdot \mathbf{0} + \mathbf{v}) \vee \cdots \vee (\mathbf{v} \leq \underline{j} + \underline{j} \wedge \underline{n} = \underline{j} \cdot \underline{j} + \mathbf{v})$$
$$\cdots \vee (\mathbf{v} \leq \underline{n} + \underline{n} \wedge \underline{n} = \underline{n} \cdot \underline{n} + \mathbf{v})\big)$$

For all j, $0 \leq j \leq n$,

$$\vdash \big((\mathbf{v} \leq \underline{j} + \underline{j}) \wedge (\underline{n} = \underline{j} \cdot \underline{j} + \mathbf{v})\big)$$
$$\supset (\mathbf{v} = \mathbf{0} \vee \mathbf{v} = \underline{1} \vee \cdots \vee \mathbf{v} = \underline{p} \vee \cdots \vee \mathbf{v} = \underline{j} + \underline{j}) \qquad \text{from } R4$$

For all p, $0 \leq p \leq j + j$,

$$\vdash \mathbf{v} = \underline{p} \supset \big((\underline{n} = \underline{j} \cdot \underline{j} + \mathbf{v}) \equiv (\underline{n} = \underline{j} \cdot \underline{j} + \underline{p})\big)$$

For all j, $0 \leq j \leq n$,

$$\vdash (\mathbf{v} \leq \underline{j} + \underline{j} \wedge \underline{n} = \underline{j} \cdot \underline{j} + \mathbf{v}) \supset \bigvee_{0 \leq p \leq j + j} (\mathbf{v} = \underline{p} \wedge \underline{n} = \underline{j} \cdot \underline{j} + \underline{p})$$

$$\vdash \big((\mathbf{x} \leq \underline{n}) \wedge (\mathbf{v} \leq \mathbf{x} + \mathbf{x}) \wedge (\underline{n} = \mathbf{x} \cdot \mathbf{x} + \mathbf{v})\big) \qquad \text{combine steps 3 and 7}$$
$$\supset \bigvee_{0 \leq j \leq n} \left(\bigvee_{0 \leq p \leq j + j} (\mathbf{v} = \underline{p} \wedge \underline{n} = \underline{j} \cdot \underline{j} + \underline{p}) \right) \qquad \text{with transitivity}$$

By the nature of the function $E(n)$, for a given n, there is a unique p such that $n = j^2 + p$ and $p \leq 2j$ for some $j \leq n$. Therefore, *with one exception*, namely $p = E(n)$, it is the case that $n \neq j^2 + p$ for $0 \leq p \leq j + j$ and $0 \leq j \leq n$. Hence, by $R3$,

$$\vdash \underline{n} \neq \underline{j} \cdot \underline{j} + \underline{p}$$

for all j, p, $0 \leq p \leq 2j$ and $0 \leq j \leq n$ *except* $p = E(n)$;

$$\vdash ((\mathbf{x} \leq \underline{n}) \wedge (\mathbf{v} \leq \mathbf{x} + \mathbf{x}) \wedge (\underline{n} = \mathbf{x} \cdot \mathbf{x} + \mathbf{v})) \supset (\mathbf{v} = \underline{E(n)} \wedge \underline{n} = \underline{j} \cdot \underline{j} + \underline{E(n)})$$

$$\vdash (\exists \mathbf{x})((\mathbf{x} \leq \underline{n}) \wedge (\mathbf{v} \leq \mathbf{x} + \mathbf{x}) \wedge (\underline{n} = \mathbf{x} \cdot \mathbf{x} + \mathbf{v})) \supset (\mathbf{v} = \underline{E(n)})$$

2. Show that if the function $g \in \mathscr{R}'$ and g is onto N and the formula $\mathbf{G}$ represents g, then the formula $\mathbf{H}(\mathbf{u}, \mathbf{v})$ given by

$$\mathbf{G}(\mathbf{v}, \mathbf{u}) \wedge (\forall \mathbf{y})(\mathbf{G}(\mathbf{y}, \mathbf{u}) \supset \mathbf{v} \leq \mathbf{y})$$

represents g^{-1}.

(a) Show $\vdash \mathbf{H}(\underline{n}, \mathbf{v}) \supset \mathbf{v} = \underline{g^{-1}(n)}$.

$$\vdash \mathbf{G}(\underline{m}, \mathbf{v}) \equiv (\mathbf{v} = \underline{g(m)}) \qquad \text{induction hypothesis} \qquad (*)$$

$$\vdash (\mathbf{G}(\mathbf{v}, \underline{n}) \wedge \mathbf{v} \leq \underline{g^{-1}(n)}) \supset (\mathbf{v} = \mathbf{0} \vee \cdots \vee \mathbf{v} = \underline{i} \vee \cdots \vee \mathbf{v} = \underline{g^{-1}(n)}) \qquad R4$$

For all m, if $m < g^{-1}(n)$, then $g(m) \neq n$, so for all $m < g^{-1}(n)$, $\vdash \underline{g(m)} \neq \underline{n}$. Then, by the contrapositive of $(*)$ and mp, for all $m < g^{-1}(n)$, $\vdash \sim\mathbf{G}(\underline{m}, \underline{n})$. For all i, $0 \leq i \leq g^{-1}(n)$,

$$\vdash (\mathbf{v} = \underline{i}) \supset (\mathbf{G}(\mathbf{v}, \underline{n}) \equiv \mathbf{G}(\underline{i}, \underline{n}))$$

So,

$$\begin{aligned}&\vdash (\mathbf{G}(\mathbf{v}, \underline{n}) \wedge \mathbf{v} \leq \underline{g^{-1}(n)})\\ &\quad \supset ((\mathbf{v} = \mathbf{0} \wedge \mathbf{G}(\mathbf{0}, \underline{n})) \vee \cdots \vee (\mathbf{v} = \underline{i} \wedge \mathbf{G}(\underline{i}, \underline{n}))\\ &\qquad\qquad \vee \cdots \vee (\mathbf{v} = \underline{g^{-1}(n)} \wedge \mathbf{G}(\underline{g^{-1}(n)}, \underline{n})))\end{aligned}$$

$$\vdash (\mathbf{G}(\mathbf{v}, \underline{n}) \wedge (\mathbf{v} \leq \underline{g^{-1}(n)})) \supset (\mathbf{v} = \underline{g^{-1}(n)} \wedge \mathbf{G}(\underline{g^{-1}(n)}, \underline{n}))$$

$$\vdash (\mathbf{G}(\mathbf{v}, \underline{n}) \wedge \mathbf{v} \leq \underline{g^{-1}(n)}) \supset (\mathbf{v} = \underline{g^{-1}(n)})$$

$$\vdash \mathbf{G}(\underline{n}, \underline{g(n)}) \qquad \text{induction hypothesis}$$

$$\begin{aligned}&\vdash (\mathbf{G}(\mathbf{v}, \underline{n}) \wedge (\forall \mathbf{y})(\mathbf{G}(\mathbf{y}, \underline{n}) \supset (\mathbf{v} \leq \mathbf{y})))\\ &\quad \supset (\mathbf{G}(\mathbf{v}, \underline{n}) \wedge (\mathbf{G}(\underline{g^{-1}(n)}, \underline{n})) \supset \mathbf{v} \leq \underline{g^{-1}(n)})\end{aligned}$$

$$\vdash (\mathbf{G}(\mathbf{v}, \underline{n}) \wedge (\forall \mathbf{y})(\mathbf{G}(\mathbf{y}, \underline{n}) \supset \mathbf{v} \leq \mathbf{y})) \supset (\mathbf{v} = \underline{g^{-1}(n)}),$$

i.e.,

$$\vdash \mathbf{H}(\underline{n}, \mathbf{v}) \supset (\mathbf{v} = \underline{g^{-1}(n)})$$

(b) To show $\vdash \mathbf{v} = \underline{g^{-1}(n)} \supset \mathbf{H}(\underline{n}, \mathbf{v})$:

$$\mathbf{G}(\mathbf{y}, \underline{n}), \mathbf{y} \leq \underline{g^{-1}(n)} \vdash \mathbf{y} = \underline{g^{-1}(n)}$$

from (∗) of part (a), therefore

$$\mathbf{G}(\mathbf{y}, \underline{n}), \mathbf{y} \leq \underline{g^{-1}(n)} \vdash \mathbf{y} = \underline{g^{-1}(n)} \vee \underline{g^{-1}(n)} < \mathbf{y}$$

i.e.,

$$\mathbf{G}(\mathbf{y}, \underline{n}), \mathbf{y} \leq \underline{g^{-1}(n)} \vdash \underline{g^{-1}(n)} \leq \mathbf{y}$$

Also,

$\mathbf{G}(\mathbf{y}, \underline{n}), \mathbf{y} \leq \underline{g^{-1}(n)} \vdash \mathbf{y} \leq \underline{g^{-1}(n)}$

$\vdash \mathbf{y} \leq \underline{g^{-1}(n)} \vee \underline{g^{-1}(n)} \leq \mathbf{y}$ by *R*5

$\vdash \mathbf{G}(\mathbf{y}, \underline{n}) \supset \underline{g^{-1}(n)} \leq \mathbf{y}$

$\vdash \mathbf{G}(\underline{g^{-1}(n)}, \underline{n}) \wedge (\forall \mathbf{y})(\mathbf{G}(\mathbf{y}, \underline{n}) \supset \underline{g^{-1}(n)} \leq \mathbf{y})$ induction hypothesis and ug on the previous step

i.e.,

$\vdash \mathbf{H}(\underline{n}, \underline{g^{-1}(n)})$

$\vdash \mathbf{v} = \underline{g^{-1}(n)} \supset (\mathbf{H}(\underline{n}, \mathbf{v}) \equiv \mathbf{H}(\underline{n}, \underline{g^{-1}(n)}))$

$\vdash \mathbf{v} = \underline{g^{-1}(n)} \supset \mathbf{H}(\underline{n}, \mathbf{v})$. ∎

Exercise 14.15 To finish the proof of the theorem, show the following: Let $\mathbf{G}(\mathbf{u}, \mathbf{x})$ represent g and $\mathbf{H}(\mathbf{u}, \mathbf{y})$ represent h; (a) if f is defined by $f(x) = g(x) + h(x)$, then the formula $(\exists \mathbf{x})(\exists \mathbf{y})(\mathbf{x} + \mathbf{y} = \mathbf{v} \wedge \mathbf{G}(\mathbf{u}, \mathbf{x}) \wedge \mathbf{H}(\mathbf{u}, \mathbf{y}))$ represents f, and (b) if f is defined by $f(x) = g(h(x))$, then the formula

$$(\exists \mathbf{z})(\mathbf{H}(\mathbf{u}, \mathbf{z}) \wedge \mathbf{G}(\mathbf{z}, \mathbf{v}))$$

represents f.

Theorem 14.7 If f is a recursive function, then f is n-representable in R.

Exercise 14.16 Prove the theorem. (Hint: Recall Theorem 6.4. J depends on functions in $\mathscr{R}'$, $u + v$, and composition, and the recursive functions depend on $\mathscr{R}'$, J, and composition. Hence, we need only show that $u + v$ is n-representable and that if $g_1, \ldots, g_k$, h are n-representable, then f defined by composition on the g_i and h is also n-representable.)

Thus, three attempts (P, Q, and R) to formalize $\mathfrak{N}$ by means of a recursively axiomatizable theory which is complete and/or decidable have failed. We have learned further that such attempts are doomed. The three attempts share at least two properties: (i) every consistent extension is undecidable and (ii) all the recursive functions are *n*-representable.

If T is an undecidable theory such that every consistent extension of T is also undecidable, then T is said to be *essentially undecidable*. The following then obviously holds.

Theorem 14.8 R, Q, and P are essentially undecidable.

Theorem 14.9 The set of sentences in the language L(N) that are universally valid [that is, $\mathscr{V}(N)$] is not a recursive set.

Theorem 14.10 If T is a theory in which all recursive functions are *n*-representable, then T is essentially undecidable.

Exercise 14.17 Prove the theorems. (Hint for 14.9: Recall exercise 11.9b and use Q.)

Exercise 14.18 Prove that if T_1 and T_2 are two theories such that T_1 is a consistent extension of T_2 and if T_2 is essentially undecidable, then T_1 is essentially undecidable.

Section 1.7 included a short discussion of first-order arithmetic. Let $\mathfrak{Z}$ be the relational system $\langle Z; +, \cdot; = \rangle$. $\mathfrak{Z}$ is a relational system for the language L(A) given there. To say that "the truth problem for first-order arithmetic is unsolvable" is a rough way of saying that the set of sentences of L(A) that are true in $\mathfrak{Z}$ is not a recursive set. The reader is now in a position to follow a proof of that fact. There is a proof in Tarski *et al.* (1953), section II.6). Similar methods ("interpretability") are used to prove that the pure predicate calculus is undecidable. A proof of that can be found in Mendelson (1964, p. 55).

14.5 PRESBURGER'S THEORY OF ADDITION OF NATURAL NUMBERS; ELIMINATION OF QUANTIFIERS†

Let L(n) be the same as L(N) but without the binary function symbol •. Let Ps be the first-order theory on L(n) with all the axioms of P except $P6$ and $P7$. The numerals are expressible in Ps and are to be denoted by $\underline{1}, \underline{2}, \ldots, \underline{n}, \ldots$, as usual. $\mathfrak{N}$ is a relational system for L(n) in the obvious way. Notice that for

† The theorems of this section were first proved by Presburger (1929). The presentation here is based on that paper and notes made by Mitsuru Yasuhara from the presentation in Hilbert-Bernays (1934, pp. 359–368 or 1968, pp. 368–377).

any assignment φ and numeral $\underline{n}$, $V_\varphi(\underline{n}) = n$. We will use **r**, **s**, **t** with or without subscripts or primes as variables for terms and $\underline{k}$, $\underline{m}$, $\underline{n}$ with or without subscripts or primes as variables for numerals. Some examples of terms and formulas in F(n) are:

1. $\underbrace{\mathbf{x} + \cdots + \mathbf{x}}_{n}$
2. $\underbrace{\underline{k} + \cdots + \underline{k}}_{n}$
3. $(\exists \mathbf{x})(\mathbf{s} + \mathbf{x} = \mathbf{t})$
4. $(\exists \mathbf{x})((\mathbf{r} = \mathbf{s} + \underbrace{\mathbf{x} + \cdots + \mathbf{x}}_{n}) \vee (\mathbf{s} = \mathbf{r} + \underbrace{\mathbf{x} + \cdots + \mathbf{x}}_{n}))$

We introduce some symbols, not in L(n), to abbreviate these as follows:

1′. Abbreviate 1 by $\underline{n}\mathbf{x}$.
2′. Abbreviate 2 by $\underline{n}\underline{k}$.
3′. Abbreviate 3 by $\mathbf{s} < \mathbf{t}$.
4′. Abbreviate 4 by $\mathbf{r} \equiv \mathbf{s}(\mathbf{mod}\ \underline{n})$.

For any terms **s** and **t** and numeral $\underline{n}$, define $(\mathbf{s} < \mathbf{t})$ and $(\mathbf{s} \equiv \mathbf{t}(\mathbf{mod}\ \underline{n}))$ to be *elementary atomic formulas*. Then, any elementary atomic formula is defined to be an *elementary formula* and if **A** and **B** are elementary formulas, then so are $(\mathbf{A} \wedge \mathbf{B})$ and $(\mathbf{A} \vee \mathbf{B})$. Let $\mathscr{C}$ be the set of closed elementary formulas of F(n).

Theorem 14.11 The set of closed elementary formulas of F(n) that are true in $\mathfrak{N}$ is decidable.

Exercise 14.19 Prove the theorem.

The principal goal of this section is to show that the set of sentences of F(n) that are true in $\mathfrak{N}$ is a recursive set, and secondarily that it is equal to the set of sentences provable in Ps. Thus, Ps is a complete and decidable theory. As a means of reaching this goal, a general method known as "elimination of quantifiers" will be introduced. Once the goals have been reached, we will look into the elimination-of-quantifiers method a little further. Let $\mathscr{S}$ be the set of all closed formulas in F(n). We will define an effective mapping $*: \mathscr{S} \to \mathscr{C}$ with the property that for $\mathbf{A} \in \mathscr{S}$, **A** is true in $\mathfrak{N}$ if and only if $\mathbf{A}^* \in \mathscr{C}$ is true in $\mathfrak{N}$. A fair amount of space will now be required to define the mapping. It is the mapping to be described that comes under the general heading of "elimination of quantifiers." This procedure is distinctly different from the method of eliminating quantifiers by introducing Skolem functions that was used in chapter XIII. In that case, there was no reference to a relational system and the language was enlarged to include the new functions; in this case, the language stays the same and the quantifiers are eliminated by referring to what we know about the relational system $\mathfrak{N}$.

Given a formula **(∃x)B**, where **B** is quantifier-free, the effective procedure will describe how to obtain a formula **C** in which **x** does not occur such that **((∃x)B ≡ C)** is true in $\mathfrak{N}$. Once this is established, the procedure can be used roughly as follows on any sentence **A**. There is a sentence **A′** in prenex normal form which is logically equivalent to **A**; there is a sentence **A″** logically equivalent to **A′** in which no universal quantifiers appear [i.e., **A″** looks like **A′** except all **(∀x$_i$)** are replaced by **∼(∃x$_i$)∼**]; there is a sentence **A‴** logically equivalent to **A″** with the same prefex of quantifiers such that the scope of the rightmost (existential) quantifier is in disjunctive normal form.† For a sentence **A**, let **A‴** as just described be called the "existential normal form of **A**". Thus, **A‴** is of the form **(···(∃x)B)** and so the process can be applied to **(∃x)B** with the result **C**, in which **x** does not occur and **((∃x)B ≡ C)** is true in $\mathfrak{N}$. Hence, **(···(∃x)B ≡ (···C))** is true in $\mathfrak{N}$ and **(···C)** contains one less quantifier than **(···(∃x)B)**, namely **(∃x)**. Given the existential normal form of **(···C)**, say **(···(∃x′)B′)**, the procedure can be applied to **((∃x′)B′)**. Then the process is repeated as many times as there are quantifiers in **A** so that the final result is a quantifier-free closed formula; that formula is **A***.

For the purposes of this section, define two formulas **A** and **B** to be $\mathfrak{N}$-*equivalent* if **(A ≡ B)** is true in $\mathfrak{N}$. Recall that a formula is defined to be true in a relational system if its universal closure is true there. Notice also that **A** and **B** are $\mathfrak{N}$-equivalent if and only if for all assignments φ, $V_\varphi(\mathbf{A}) = V_\varphi(\mathbf{B})$ in $\mathfrak{N}$; this fact will be useful hereafter.

Lemma 14.4 If **A** and **B** are $\mathfrak{N}$-equivalent, then **(Q$_1$x$_1$) ··· (Q$_n$x$_n$)A** is true in $\mathfrak{N}$ if and only if **(Q$_1$x$_1$) ··· (Q$_n$x$_n$)B** is true in $\mathfrak{N}$, where the **Q$_i$** are either existential or universal quantifiers.

Exercise 14.20 Prove the lemma.

Lemma 14.5 If (i) **A** and **B** are $\mathfrak{N}$-equivalent formulas, (ii) **B** is a subformula of formula **C**, and (iii) **D** is a formula obtained from **C** by replacing an occurrence of **A** by **B**, then **C** and **D** are $\mathfrak{N}$-equivalent.

Exercise 14.21 Prove the lemma.

† In this section disjunctive normal form is to be taken with elementary atomic and atomic formulas in place of just atomic formulas.

Exercise 14.22 (Do-it-yourself number theory) In order to define the mapping $*$ and prove that it has the required properties, a little knowledge about congruences is required. By doing all the parts of this exercise, the reader will probably be adequately prepared. The *least common multiple* (l.c.m.) of two numbers m and n is defined to be the smallest number k such that $m|k$ and $n|k$. Recall that the greatest common divisor of m and n, written (m, n), is the largest number k such that $k|m$ and $k|n$. It is a theorem of number theory that there exist *integers* a and b such that $am + bn = (m, n)$.†

(a) Show that if $k|mn$ and k and m are mutually prime, then $k|n$.

(b) Show that the l.c.m. of m and n is $mn/(m, n)$.

(c) Show that $h \in N$ is a solution of the congruence $kx + m \equiv m'(\text{mod } n)$ if and only if it is a solution of the congruence $kx \equiv (m' + (n-1)m)(\text{mod } n)$.

(d) Show that $h \in N$ is a solution of the pair of congruences $k_1 x \equiv m_1 (\text{mod } n_1)$ and $k_2 x \equiv m_2(\text{mod } n_2)$ if and only if it is a solution of the pair $n_2' k_1 x \equiv n_2' m_1(\text{mod } n)$ and $n_1' k_2 x \equiv n_1' m_2$ $(\text{mod } n)$, where n is the l.c.m. of n_1 and n_2 and $n_i = n_i'(n_1, n_2)$, $i = 1, 2$.

(e) Show that $h \in N$ is a solution of the pair of congruences $k_1 x + m_1 \equiv m_1'(\text{mod } n)$ and $k_2 x + m_2 \equiv m_2'(\text{mod } n)$ if and only if it is a solution of the pair $k_1 + m_1 \equiv m_1'(\text{mod } n)$ and $(k_2 - k_1)x + m_2 + m_1' \equiv (m_2' + m_1)(\text{mod } n)$, where $k_2 > k_1$.

(*f*) Show that for any pair of congruences $k_i x \equiv m_i(\text{mod } n)$, $i = 1, 2$, there is a pair $kx + m_i' \equiv m_i''(\text{mod } n)$, $i = 1, 2$, such that $h \in N$ is a solution of the first pair if and only if it is a solution of the second pair.

(g) Show that $h \in N$ is a solution of the pair of congruences $kx + m_i' \equiv m_i''(\text{mod } n)$, $i = 1, 2$, if and only if it is a solution of the congruence $kx + m_1' \equiv m_1''(\text{mod } n)$ and the congruence $(m_1' + m_2'') \equiv (m_1'' + m_2')(\text{mod } n)$ holds.

(h) Show that the equation $kx + ny = m$ has a solution if and only if $(k, n)|m$. [Hint: Remember that there are integers a and b such that $ak + bn = (k, n)$.]

(i) Show that the congruence $kx = m(\text{mod } n)$ has a solution if and only if $(k, n)|m$.

(j) Give an algorithm which operates on a set of congruences

$$C = \{k_i x \equiv m_i(\text{mod } n_i) \mid 1 \leq i \leq q\}$$

so as to end in one of two ways: (1) if C has a solution, the algorithm produces another set Q of q congruences, one of them in the form $kx \equiv m(\text{mod } n)$ and the $(q-1)$ others in the form $k_i' \equiv m_i'(\text{mod } n_i')$ so that $h \in N$ is a solution of the set C if and only if it is a solution of $kx \equiv m(\text{mod } n)$ and $k_i' \equiv m_i'$

† If the reader is not able to prove this, he can consult Herstein (1964, p. 18) or any elementary number theory or modern algebra text.

(mod n_i') is true, for $i = 1, \ldots, q - 1$; or (2) if C has no solution, the algorithm says so.

Given $\mathbf{A} \in \mathscr{S}$, to obtain $\mathbf{A^*}$, find the existential normal form of $\mathbf{A}$; it is in the form $(\cdots(\exists \mathbf{x})\mathbf{B})$. Carry out the following sequence of steps as many times as there are quantifiers in $\mathbf{A}$.

Step 1. (a) In $\mathbf{B}$, make the following replacements: replace $(\mathbf{r} = \mathbf{s})$ by $(\mathbf{r} < \mathbf{S(s)} \wedge \mathbf{s} < \mathbf{S(r)})$; replace $\sim(\mathbf{r} = \mathbf{s})$ by $(\mathbf{r} < \mathbf{s} \vee \mathbf{s} < \mathbf{r})$; replace $\sim(\mathbf{r} < \mathbf{s})$ by $(\mathbf{s} < \mathbf{S(r)})$; replace $\sim(\mathbf{r} \equiv \mathbf{s}(\mathbf{mod}\ \underline{k}))$ by

$$(\mathbf{r} \equiv (\mathbf{s} + \mathbf{S(0)}(\mathbf{mod}\ \underline{k})) \vee \cdots \vee \mathbf{r} \equiv (\mathbf{s} + \mathbf{S^{k+1}(0)})(\mathbf{mod}\ \underline{k}))$$

(b) Let the disjunctive normal form of the formula obtained from $\mathbf{B}$ by making the replacements of part (a) be denoted by $\mathbf{B}'$.

Exercise 14.23 Show that no negation symbols occur in $\mathbf{B}'$ and that $\mathbf{B}'$ is a disjunction of conjunctions of elementary atomic formulas.

Exercise 14.24 Show that $\mathbf{B}$ and $\mathbf{B}'$ are $\mathfrak{N}$-equivalent.

Step 2. (a) If $\mathbf{t}$ is a term in $\mathbf{B}'$ in which $\mathbf{x}$ occurs, then replace $\mathbf{t}$ by a term of the form $(\underline{k}\mathbf{x} + \mathbf{r})$ or $(\underline{k}\mathbf{x})$ so that for all assignments φ,

$$V_\varphi(\mathbf{t}) = \begin{cases} V_\varphi(\underline{k}\mathbf{x} + \mathbf{r}) & \text{or} \\ V_\varphi(\underline{k}\mathbf{x}) \end{cases}$$

Exercise 14.25 Show that this can be done, i.e., that there are such $\underline{k}$ and $\mathbf{r}$.

(b) Replace formulas of the form $(\underline{k}\mathbf{x} + \mathbf{r} < \underline{k}\mathbf{x} + \mathbf{s})$ by $(\mathbf{r} < \mathbf{s})$. Replace formulas of the form $(\underline{m}\mathbf{s} + \mathbf{r} < \underline{n}\mathbf{x} + \mathbf{s})$ by $\underline{(m - n)}\mathbf{x} + \mathbf{r} < \mathbf{s}$ if $m > n$ or by $\mathbf{r} < \underline{(n - m)}\mathbf{x}$ if $n \leq m$.

Exercise 14.26 Show that in Step 2(b), the replaced formula and the replacing formula are $\mathfrak{N}$-equivalent.

(c) Replace formulas of the form $(\underline{k}\mathbf{x} + \mathbf{r} \equiv \mathbf{s}(\mathbf{mod}\ \underline{n}))$ by $(\underline{k}\mathbf{x} \equiv \mathbf{t}(\mathbf{mod}\ \underline{n}))$, where $\mathbf{t}$ is $(\mathbf{s} + \underline{(n - 1)}\mathbf{r})$. Replace formulas of the form $(\underline{k}\mathbf{x} \equiv \mathbf{t}(\mathbf{mod}\ \underline{n}))$, where $\mathbf{t}$ is not a numeral, by

$$(\mathbf{t} \equiv \mathbf{0}(\mathbf{mod}\ \underline{n}) \wedge \underline{k}\mathbf{x} \equiv \mathbf{0}(\mathbf{mod}\ \underline{n})) \vee \cdots \vee (\mathbf{t} \equiv \underline{n - 1}(\mathbf{mod}\ \underline{n}) \wedge \underline{k}\mathbf{x} \equiv \underline{n - 1}(\mathbf{mod}\ \underline{n}))$$

Exercise 14.27 Show that $(\underline{k}\mathbf{x} + \mathbf{r} \equiv \mathbf{s}(\mathbf{mod}\ \underline{n}))$ and $(\underline{k}\mathbf{x} \equiv (\mathbf{s} + \underline{(n-1)}\mathbf{r})(\mathbf{mod}\ \underline{n}))$ are $\mathfrak{N}$-equivalent, and that, for $\mathbf{t}$ not a numeral, $(\underline{k}\mathbf{x} \equiv \mathbf{t}(\mathbf{mod}\ \underline{n}))$ and

$$(\mathbf{t} \equiv \mathbf{0}(\mathbf{mod}\ \underline{n}) \wedge \underline{k}\mathbf{x} \equiv \mathbf{0}(\mathbf{mod}\ \underline{n})) \vee \cdots$$
$$\vee (\mathbf{t} \equiv \underline{(n-1)}(\mathbf{mod}\ \underline{n}) \wedge \underline{k}\mathbf{x} \equiv \underline{(n-1)}(\mathbf{mod}\ \underline{n})$$

are $\mathfrak{N}$-equivalent.

(d) Let $\mathbf{B}''$ be the disjunctive normal form of the formula obtained from $\mathbf{B}'$ by carrying out steps (a), (b), and (c).

Exercise 14.28 Prove that any conjunct of $\mathbf{B}''$ in which $\mathbf{x}$ occurs is in one of the three following forms: (i) $\underline{k}\mathbf{x} + \mathbf{r} < \mathbf{s}$, (ii) $\mathbf{s} < \underline{k}\mathbf{x} + \mathbf{r}$, or (iii) $\underline{k}\mathbf{x} \equiv \underline{m}(\mathbf{mod}\ \underline{n})$, where in (i) or (ii), $\mathbf{r}$ may not appear.

Exercise 14.29 Show that $\mathbf{B}'$ and $\mathbf{B}''$ are $\mathfrak{N}$-equivalent.

Step 3. $\mathbf{B}''$ is a disjunction of conjunctions of elementary atomic formulas of one of the three forms listed in exercise 14.28.

(a) For each conjunction of congruences $\bigwedge_{1 \le i \le q} (\underline{k}_i\mathbf{x} \equiv \underline{m}_i(\mathbf{mod}\ \underline{n}_i))$ apply the algorithm of exercise 14.22 to the corresponding set $C = \{k_i x \equiv m_i(\text{mod}\ n_i) \mid 1 \le i \le q\}$. If the result of the algorithm is that C does not have a solution, replace the conjunction of congruences by $\mathbf{0} < \mathbf{0}$. If the result of the algorithm is a set of congruences Q, replace the conjunction of congruences $\bigwedge_{1 \le i \le q} (\underline{k}_i\mathbf{x} \equiv \underline{m}_i(\mathbf{mod}\ \underline{n}_i))$ by the single congruence $(\underline{k}\mathbf{x} \equiv \underline{m}(\mathbf{mod}\ \underline{n}))$, where $(kx \equiv m(\text{mod}\ n)) \in Q$ with the property that $h \in N$ is a solution if and only if h is a solution of the set C.

Exercise 14.30 (i) Show that the formula

$$(\underline{k}_1\mathbf{x} + \mathbf{r_1} < \mathbf{s_1} \wedge \underline{k}_2\mathbf{x} + \mathbf{r_2} < \mathbf{s_2})$$

is $\mathfrak{N}$-equivalent to the formula

$$(\underline{k}_1\underline{k}_2\mathbf{x} + (\mathbf{r_1}\underline{k}_2 + \mathbf{r_2}\underline{k}_1) < (\mathbf{s_1}\underline{k}_2 + \mathbf{r_2}\underline{k}_1)$$
$$\wedge\ \underline{k}_1\underline{k}_2\mathbf{x} + (\mathbf{r_1}\underline{k}_2 + \mathbf{r_2}\underline{k}_1) < (\mathbf{s_2}\underline{k}_1 + \mathbf{r_1}\underline{k}_2))$$

(ii) Generalize part (i) to conjunctions with more than two conjuncts.

(b) Replace each conjunction (occurring in $\mathbf{B}''$) of inequalities $\bigwedge_{1 \le i \le q} (\underline{k}_i\mathbf{x} + \mathbf{r_i} < \mathbf{s_i})$ by the conjunction $\bigwedge_{1 \le i \le q} (\underline{k}\mathbf{x} + \mathbf{r} < \mathbf{t_i})$ where $\underline{k}$, $\mathbf{r}$, and $\mathbf{t_i}$ are the terms which correspond to $\{k_i, r_i, s_i \mid 1 \le i \le q\}$ according to part (ii) of exercise 14.30. Conjunctions of inequalities $\bigwedge_{1 \leq i \leq q} (\mathbf{s_i} < \underline{k}_i \mathbf{x} + \mathbf{r_i})$ are replaced by a conjunction $\bigwedge_{1 \leq i \leq q} (\mathbf{t_i} < \underline{k}\mathbf{x} + \mathbf{r})$ which corresponds in an analogous way.

(c) Let the formula resulting from **B″** according to steps (a) and (b) be denoted by **B‴**.

Exercise 14.31 Prove that

$$\left(\bigwedge_{1 \leq i \leq q} (\underline{k}\mathbf{x} + \mathbf{r} < \mathbf{t_i})\right) \qquad \text{and} \qquad \left(\bigvee_{1 \leq j \leq q} \left(\bigwedge_{1 \leq i \leq q} (\mathbf{t_j} < \mathbf{S}(\mathbf{t_i})) \wedge \underline{k}\mathbf{x} + \mathbf{r} < \mathbf{t_j}\right)\right)$$

are $\mathfrak{N}$-equivalent.

Step 4. In **B‴**, replace any conjunction

$$\bigwedge_{1 \leq i \leq q} (\underline{k}\mathbf{x} + \mathbf{r} < \mathbf{t_i}) \qquad \text{by} \qquad \bigvee_{1 \leq j \leq q} \left(\bigwedge_{1 \leq i \leq q} (\mathbf{t_j} < \mathbf{S}(\mathbf{t_i})) \wedge \underline{k}\mathbf{x} + \mathbf{r} < \mathbf{t_j}\right)$$

similarly, replace any

$$\bigwedge_{1 \leq i \leq q} (\mathbf{t_i} < \underline{k}\mathbf{x} + \mathbf{r}) \qquad \text{by} \qquad \bigvee_{1 \leq i \leq q} \left(\bigwedge_{1 \leq j \leq q} (\mathbf{t_j} < \mathbf{S}(\mathbf{t_i})) \wedge \mathbf{t_j} < \underline{k}\mathbf{x} + \mathbf{r}\right)$$

Denote by $\mathbf{B}^{(4)}$ the disjunctive normal form of the formula obtained from **B‴** by making these replacements.

Exercise 14.32 Show that **B″** and $\mathbf{B}^{(4)}$ are $\mathfrak{N}$-equivalent.

Step 5. From $(\exists\mathbf{x})\mathbf{B}$, we now have $(\exists\mathbf{x})\mathbf{B}^{(4)}$, where $\mathbf{B}^{(4)}$ is in the form $\mathbf{D_1} \vee \cdots \vee \mathbf{D_p}$. So $(\exists\mathbf{x})\mathbf{B}^{(4)}$ is logically equivalent to

$$(\exists\mathbf{x})\mathbf{D_1} \vee \cdots \vee (\exists\mathbf{x})\mathbf{D_p} \qquad (*)$$

Further, each $\mathbf{D_i}$ is a conjunction of elementary atomic formulas in some of which **x** occurs. So (*) is logically equivalent to

$$(\exists\mathbf{x})\mathbf{D_1}' \vee \cdots \vee (\exists\mathbf{x})\mathbf{D_r}' \vee \mathbf{D}'_{\mathbf{r+1}} \qquad (**)$$

where **x** occurs in each conjunct of each $\mathbf{D_i}'$, $1 \leq i \leq r$, and $\mathbf{D}'_{\mathbf{r+1}}$ is a disjunction of conjunctions of elementary atomic formulas in none of which does **x** occur. Replace $(\exists\mathbf{x})\mathbf{B}^{(4)}$ by (**).

Exercise 14.33 Show that for each i, $1 \leq i \leq r$, $\mathbf{D_i}'$ is a conjunction of the form

$$(\underline{k}\mathbf{x} \equiv \underline{m}(\mathbf{mod}\ \underline{n}) \wedge \mathbf{r} < \underline{k}'\mathbf{x} + \mathbf{s} \wedge \underline{k}'\mathbf{x} + \mathbf{s} < \mathbf{t}) \qquad (1)$$

or the same form with one or two conjuncts missing.

Step 6. (a) For each $\mathbf{D_i}'$ in which two conjuncts are missing, i.e., $\mathbf{D_i}'$ is simply $\underline{k}\mathbf{x} \equiv \underline{m}(\mathbf{mod}\ \underline{n})$, or $\mathbf{r} < \underline{k}'\mathbf{x} + \mathbf{s}$, or $\underline{k}'\mathbf{x} + \mathbf{s} < \mathbf{t}$, do the following: (i) if $\mathbf{D_i}'$ is $\underline{k}\mathbf{x} \equiv \underline{m}(\mathbf{mod}\ \underline{n})$, replace $(\exists \mathbf{x})\mathbf{D_i}'$ by $(\mathbf{0} < \mathbf{S(0)})$ if $(k, n)|m$ and by $(\mathbf{0} < \mathbf{0})$ if not; (ii) if $\mathbf{D_i}'$ is $\mathbf{r} < \underline{k}'\mathbf{x} + \mathbf{s}$, replace $(\exists \mathbf{x})\mathbf{D_i}'$ by $(\mathbf{0} < \mathbf{S(0)})$; (iii) if $\mathbf{D_i}'$ is $\underline{k}'\mathbf{x} + \mathbf{s} < \mathbf{t}$, replace $(\exists \mathbf{x})\mathbf{D_i}'$ by $(\mathbf{s} < \mathbf{t})$.

(b) For each $\mathbf{D_i}'$ in which only one conjunct is missing, do the following: (i) If $\mathbf{r} < \underline{k}'\mathbf{x} + \mathbf{s}$ is lacking, replace $\underline{k}'\mathbf{x} + \mathbf{s} < \mathbf{t}$ by $(\mathbf{0} < \underline{k}\mathbf{x} + \mathbf{S(s)} \wedge \underline{k}'\mathbf{x} + \mathbf{S(s)} < \mathbf{S(t)})$; (ii) if $\underline{k}'\mathbf{x} + \mathbf{s} < \mathbf{t}$ is missing, replace $(\exists \mathbf{x})\mathbf{D_i}'$ by $(\mathbf{0} < \mathbf{S(0)})$ if $(k, n)|m$ and by $(\mathbf{0} < \mathbf{0})$ if not; (iii) if it is the congruence which is missing in $\mathbf{D_i}'$, we distinguish two cases: if $\underline{k}'$ is $\mathbf{S(0)}$, replace $(\exists \mathbf{x})\mathbf{D_i}'$ by $(\mathbf{S(r)} < \mathbf{t} \wedge \mathbf{s} < \mathbf{t})$, and if $\underline{k}'$ is $\mathbf{S^j(0)}$ for $j > 1$, replace $\mathbf{D_i}'$ by

$$\big(\mathbf{x} \equiv \mathbf{0}(\mathbf{mod}\ \underline{k}') \wedge \mathbf{r} < \mathbf{x} + \mathbf{s} \wedge \mathbf{x} + \mathbf{s} < \mathbf{t}\big)$$

In each of the three cases the resulting formula is in form (1) above with all three conjuncts present. Notice in some cases $\mathbf{D_i}'$ was replaced and in some others $(\exists \mathbf{x})\mathbf{D_i}'$ was replaced. So the resulting formula is of the form

$$(\exists \mathbf{x})\mathbf{D}_1'' \vee \cdots \vee (\exists \mathbf{x})\mathbf{D}_{s-1}'' \vee \mathbf{D}_s'' \qquad (***)$$

where $\mathbf{D}_s''$ includes all of $\mathbf{D}_{r+1}'$ of $(**)$ and any formulas resulting by this step from the replacement of $(\exists \mathbf{x})\mathbf{D_i}'$, and each $\mathbf{D}_i''$ results from $\mathbf{D_j}'$, for some j by the replacements made in this step. Further, each $\mathbf{D}_i''$ is of the form (1) above with all three conjuncts present.

Exercise 14.34 Show that $(**)$ and $(***)$ are $\mathfrak{N}$-equivalent. (Hint: Use exercise 14.22.)

Step 7. Each $\mathbf{D}_i''$ is of form (1). Notice that the formulas $(\exists \mathbf{x})(1)$ and

$$(\exists \mathbf{x})\big(\underline{k}'\underline{k}\mathbf{x} \equiv \underline{k}'\underline{m}(\mathbf{mod}\ \underline{k}'\underline{n}) \wedge \underline{k}\mathbf{r} < \underline{k}\underline{k}'\mathbf{x} + \underline{k}\mathbf{s} \wedge \underline{k}\underline{k}'\mathbf{x} + \underline{k}\mathbf{s} < \underline{k}\mathbf{t}\big)$$

are $\mathfrak{N}$-equivalent. So replace each $\mathbf{D}_i''$ by the appropriate conjunction of the *form*

$$\big(\mathbf{x} \equiv \underline{m}(\mathbf{mod}\ \underline{n}) \wedge \mathbf{r} < \mathbf{x} + \mathbf{s} \wedge \mathbf{x} + \mathbf{s} < \mathbf{t}\big) \qquad (2)$$

Step 8. Finally we are ready to eliminate the existential quantifier $(\exists \mathbf{x})$. By the steps so far, we need only consider formulas of the form $(\exists \mathbf{x})(2)$. Replace each such formula by

$$(\mathbf{r} < \mathbf{s} \wedge \mathbf{s} + \underline{m} < \mathbf{t}) \vee \bigvee_{1 \le j \le n} \big(\mathbf{s} < \mathbf{S(r)} \wedge \mathbf{s} + \underline{j} < \mathbf{t} \wedge \mathbf{r} + \underline{j} \equiv \mathbf{s} + \underline{m}(\mathbf{mod}\ \underline{n})\big) \qquad (3)$$

We must show that formulas $(\exists \mathbf{x})(2)$ and (3) are $\mathfrak{N}$-equivalent.

Case 1: Suppose $V_\varphi(\mathbf{r}) < V_\varphi(\mathbf{s})$. Now, $V_\varphi(\mathbf{r}) < V_\varphi(\mathbf{s})$ implies that for any

$x \in N$, $V_\varphi(\mathbf{r}) < x + V_\varphi(\mathbf{r})$, so if $x = m + pn$ (for some p) and $V_\varphi(\underline{m} + \overline{pn} + \mathbf{s}) < V_\varphi(\mathbf{t})$, then $V_\varphi(\mathbf{s} + \underline{m}) < V_\varphi(\mathbf{t})$; so the first disjunct of (3) holds. On the other hand, if $V_\varphi(\mathbf{r}) < V_\varphi(\mathbf{s})$, then only the first disjunct of (3) can be true. That is, for the case $V_\varphi(\mathbf{r}) < V_\varphi(\mathbf{s})$, $V_\varphi((3)) = \mathrm{t}$ implies that $V_\varphi(\mathbf{r} < \mathbf{s} \wedge \mathbf{s} + \underline{m} < \mathbf{t}) = \mathrm{t}$ and so $V_\varphi(\mathbf{s} + \underline{m} < \mathbf{t}) = \mathrm{t}$. Now, $V_\varphi((\exists \mathbf{x})(2)) = \mathrm{t}$ if and only if there is a φ' differing from φ at most at $\mathbf{x}$ such that $V_{\varphi'}((2)) = \mathrm{t}$. If $\varphi'(\mathbf{x}) = m = \varphi(\underline{m})$ and $\varphi' = \varphi$ elsewhere, then φ' satisfies that requirement.

Case 2: $V_\varphi(\mathbf{r}) \geq V_\varphi(\mathbf{s})$. Then $V_\varphi(\mathbf{s}) < (V_\varphi(\mathbf{r}) + 1)$. To show that $V_\varphi((\exists \mathbf{x})(2)) = \mathrm{t}$ implies $V_\varphi((3)) = \mathrm{t}$. Since $V_\varphi((\exists \mathbf{x})(2)) = \mathrm{t}$, there is an assignment φ' differing from φ at most at $\mathbf{x}$ such that for some $p \in N$, $\varphi'(\mathbf{x}) = (m + pn)$. Of course, $V_{\varphi'}(\underline{m} + \overline{pn})$ also is $(m + pn)$. So $V_{\varphi'}(\mathbf{r}) < V_{\varphi'}(\underline{m} + \overline{pn} + \mathbf{s})$ and $V_{\varphi'}(\underline{m} + \overline{pn} + \mathbf{s}) < V_{\varphi'}(\mathbf{t})$. By the first of these, there is a $j \in N$ such that $V_{\varphi'}(\mathbf{r}) + j = V_{\varphi'}(\underline{m} + \overline{pn} + \mathbf{s})$, and therefore, by the second, $V_{\varphi'}(\mathbf{r}) + j < V_{\varphi'}(\mathbf{t})$ and $V_{\varphi'}(\mathbf{r}) + j \equiv (m + V_{\varphi'}(\mathbf{s}))(\mathrm{mod}\, n)$. If there is such a j where $1 \leq j \leq n$, then we will have shown that $V_\varphi((3)) = \mathrm{t}$. Since there is some j as just described, if it is greater than n, then $j = j' + hn$ for some h and $1 \leq j' \leq n$, i.e., $j \equiv j'(\mathrm{mod}\ n)$ and $j' < j$. Since $V_{\varphi'}(\mathbf{s}) + j < V_{\varphi'}(\mathbf{t})$ and $V_\varphi'(\mathbf{s}) + j \equiv V_{\varphi'}(\mathbf{t}) + m(\mathrm{mod}\, n)$, we have that $V_{\varphi'}(\mathbf{s}) + j' < V_{\varphi'}(\mathbf{t})$ and $V_{\varphi'}(\mathbf{s}) + j' < V_{\varphi'}(\mathbf{t}) + m(\mathrm{mod}\, n)$. To show that $V_\varphi((3)) = \mathrm{t}$ implies $V_\varphi((\exists \mathbf{x})(2)) = \mathrm{t}$. $V_\varphi((3)) = \mathrm{t}$ implies there is a $j \in N$, $1 \leq j \leq n$, and an $h \in N$ such that $V_\varphi(\mathbf{r}) + j = V_\varphi(\mathbf{s}) + m + hn$ and $V_\varphi(\mathbf{s}) + j < V_\varphi(\mathbf{t})$. Consider the assignment φ' such that $\varphi' = \varphi$ except at $\mathbf{x}$ and $\varphi'(\mathbf{x}) = m + hn$. Then $\varphi'(\mathbf{x}) \equiv m\ (\mathrm{mod}\ n)$, so $V_{\varphi'}(\mathbf{x} \equiv \underline{m}\ (\mathbf{mod}\ \underline{n})) = \mathrm{t}$. Further, $V_{\varphi'}(\mathbf{r}) + j = V_{\varphi'}(\mathbf{x}) + V_{\varphi'}(\mathbf{s})$ and $V_{\varphi'}(\mathbf{r}) + j < V_{\varphi'}(\mathbf{t})$ and so $V_{\varphi'}(\mathbf{r} < \mathbf{x} + \mathbf{s}) = \mathrm{t}$ and $V_{\varphi'}(\mathbf{x} + \mathbf{s} < \mathbf{t}) = \mathrm{t}$. Hence, $V_\varphi((\exists \mathbf{x})(2)) = \mathrm{t}$.

Step 9. Find the disjunctive normal form of the formula obtained by the replacements of Step 8. Call the resulting formula **C**. It follows from Lemma 14.5 that by replacing **(∃x)B** in **A‴**, the existential normal form of **A**, the resulting formula is 𝔑-equivalent to **A‴**, and in fact, to **A**.

Theorem 14.12 The set of sentences of F(n) that are true in 𝔑 is a recursive set.

PROOF Follows from all the preceding.

Theorem 14.13 The theory Ps is complete.

Exercise 14.35 Prove the theorem. [Hints: (i) Show that a closed elementary formula is true in 𝔑 if and only if it is provable in Ps. (ii) Show that in each place above where two formulas **C** and **D** are shown to be 𝔑-equivalent, it is also true that $\vdash_{\mathrm{Ps}}(\mathbf{C} \equiv \mathbf{D})$. (iii) Save this exercise for a very rainy week or two.]

Corollary Ps is a decidable theory.

Usually, we choose a language with a particular relational system in mind —e.g., $\mathfrak{N}$, a group, a field, etc. Given a relational system and a language for it, two of the obvious investigations are (1) to find a recursively axiomatizable theory which has the given relational system as a model, if possible so that the set of theorems is decidable and/or complete, and (2) to find out if the set of sentences of the language that are true in the relational system is a recursive (or at least r.e.) set. The method of eliminating quantifiers has turned out to be extremely useful in both kinds of endeavors. In the Presburger system, we did the second problem and from there went to the first. The first kind of problem can be approached more directly by using the fact that for a theory T and a sentence **A**, $\vdash_T \mathbf{A}$ if and only if **A** is true in all models of T. Thus, in trying to prove T decidable, it may be possible to decide whether or not $\vdash_T \mathbf{A}$ by deciding whether or not **A** is true in all models of T and, in turn, it may be possible to do that by means of some quantifier-elimination method. An example of this approach is given in exercise 14.36 just below. Although "quantifier-elimination method" cannot be defined explicitly, the general idea is that there is an effective process which when applied to any sentence (of the language in question) produces a quantifier-free formula (in the same language) belonging to a particular set of formulas in the language. Then, depending on the situation, the intersection of that particular set of formulas with (a) the set of provable sentences or (b) the set of sentences true in a particular model, is recursive.

Exercise 14.36 Let L(O) be a first-order language with equality and the binary predicate symbol $<$. (L(O) had no constant or function symbols.) Let DL be the first-order theory with axioms those of $\mathsf{V}({}_{O}{}^{=})$ plus

$$(\forall x)(\forall y)(x < y \vee x = y \vee y < x)$$

$$(\forall x)(\forall y)(\forall z)(x < y \wedge y < z \supset x < z)$$

$$(\forall x) \sim (x < x)$$

$$(\forall x)(\forall y)(\exists z)\big(x < y \supset (x < z \wedge z < y)\big)$$

Any relational system in which the first three axioms are true is called a *linearly ordered system* and if the fourth axiom is also true in that system then it is called a *dense, linearly ordered system.* Familiar examples of linearly ordered systems are the integers, the rationals, the reals, or any interval of these. The rationals and the reals or any interval of these are also examples of dense, linearly ordered systems. Notice that the axioms do not specify anything about the existence of end points. Thus, the closed interval [0, 1], the open interval (0, 1), or the half-open intervals (0, 1] and [0, 1) over the rationals or reals are all examples of dense, linearly ordered systems. Let DLO be the first-order theory with the axioms of DL plus:

$$(\forall x)(\exists y)(\exists z)(y < x \wedge x < z)$$

Models of DLO are *dense, linearly ordered systems without end points*. The reals, the rationals, and any open interval of these are examples. As we know, to decide of an arbitrary sentence **A** whether or not $\vdash_{DLO}$ **A** is equivalent to deciding whether or not **A** is true in all models of DLO. We will approach the problem of showing that DLO is a complete and decidable theory by means of this equivalence. A quantifier-elimination scheme can be used to do this. The three main differences between what is to be done here and what is done in the Presburger case are: (i) there are closed terms in L(n), there are no closed terms in L(O); and (ii) in order to eliminate quantifiers in the Presburger method, closed terms relating to the specific relational system $\mathfrak{N}$ replace bound variables so that the result of eliminating quantifiers is a closed quantifier-free formula; in this case, there is no reference to a specific model of DLO and the result of eliminating quantifiers is a quantifier-free formula in which no constants appear; and (iii) in the Presburger case, a formula and the formula obtained from it by eliminating quantifiers are $\mathfrak{N}$-equivalent for the specific model for Ps, $\mathfrak{N}$; in this case, a formula and the formula obtained from it by eliminating quantifiers are $\mathfrak{D}$-equivalent for *all* models $\mathfrak{D}$ of DLO. Let the elementary atomic formulas be $(x = x)$ and $(x < x)$, for any variable **x**, and define the elementary formulas to be the elementary atomic formulas and $(A \wedge B)$ and $(A \vee B)$ for any elementary formulas **A** and **B**. Let $\mathscr{E}$ be the set of elementary formulas. Let $\mathscr{S}$ be the set of sentences of F(O).

(a) Show that for any elementary formula **A**, if **A** is true in one model of DLO, then it is true in all models of DLO.

(b) Show that the set of elementary formulas that are true in all models of DLO is a recursive set.

(c) Define an effective mapping $*: \mathscr{S} \to \mathscr{E}$ such that $(A \equiv A^*)$ is true in all models of DLO and $((\sim A)^* \equiv \sim(A^*))$ is universally valid. [Hints: Set up a quantifier-elimination scheme using existential normal form. Notice that the following pairs of formulas are $\mathfrak{D}$-equivalent for all models $\mathfrak{D}$ of DLO: (i) $\sim(x = y)$ and $(x < y \vee y < x)$, (ii) $\sim(x < y)$ and $(x = y \vee y < x)$, (iii) $(\exists x)(y < x \wedge x < z)$ and $(y < z)$, (iv) $(\exists x)(x < y \wedge y < x)$ and $(x < x)$, (v) $(\exists x)(x < y \wedge x < z)$ and $(x = x)$.]

(d) Show that for any sentence **A**, if **A** is true in one model of DLO, then it is true in all models of DLO.

(e) Show that DLO is a decidable theory.

(f) Show that DLO is a complete theory.

14.6 FINITE AUTOMATA AND THE WEAK MONADIC SECOND-ORDER THEORY OF SUCCESSOR†

Finally, we turn our attention to second-order languages; that is, we can quantify over predicate variables as well as individual variables. For example, if **P** is a unary predicate variable and = a predicate constant, then

$$(\mathbf{x} = \mathbf{y}) \equiv (\forall \mathbf{P})(\mathbf{P}(\mathbf{x}) \equiv \mathbf{P}(\mathbf{y}))$$

is a second order formula. Obviously this formula is universally valid and therefore we do not even need the predicate constant =.‡ This is in sharp contrast to the situation for first-order languages. Recall that the differences between first-order languages without = and first-order languages with = are so great that the theories about the two languages were developed separately. One particular difference is that if a first-order theory (on a language) without = has a (finite) model of cardinality n, it has a model of cardinality $n + 1$, whereas this is not the case for first-order theories with =. Nor, obviously, is it the case for second-order theories. But the second-order languages differ strongly even from first-order languages with = since, as one should suspect, the Löwenheim–Skolem and Tarski theorems (see chapter XII) are false for second-order theories.

By means of a first-order language, we are able to "talk about" the elements of a universe but not about the subsets of a universe. So, for example, if one is interested in the real numbers, one can express that they are densely ordered by the first-order formula

$$(\forall \mathbf{x})(\forall \mathbf{y})(\exists \mathbf{z})(\mathbf{x} \neq \mathbf{y} \supset ((\mathbf{x} < \mathbf{y} \wedge \mathbf{x} < \mathbf{z} \wedge \mathbf{z} < \mathbf{y}) \vee (\mathbf{y} < \mathbf{x} \wedge \mathbf{y} < \mathbf{z} \wedge \mathbf{z} < \mathbf{x})))$$

However, there does not seem to be (and it can be proved that there is not) a first-order formula that expresses the fact that every set of real numbers that has an upper bound has a least upper bound. But it is easy to write a second-order formula that does. To say that y is an upper bound of the set S we have the (first-order) formula

$$(\forall \mathbf{x})(\mathbf{S}(\mathbf{x}) \supset \mathbf{y} > \mathbf{x})$$

which we denote by **Bd(y, S)**. Then, for the least upper bound statement, we have

$$(\forall \mathbf{S})(\exists \mathbf{x})(\mathbf{Bd}(\mathbf{x}, \mathbf{S}) \supset (\forall \mathbf{y})(\mathbf{Bd}(\mathbf{y}, \mathbf{S}) \supset \mathbf{y} \geq \mathbf{x}))$$

† The material in this section is based on Elgot's paper (1961).

‡ In section 10.5, there was a brief discussion of what it means to say that a relation of a relational system is definable in a first-order language. What we are saying here is that the equality relation for any relational system is definable in any second-order language which has unary predicate variables.

For the last subject in this book, we consider which sets of second-order sentences true about the natural numbers are undecidable and which are decidable. Of course, if the second-order language includes symbols for successor, addition, and multiplication, then the set of all sentences true about the natural numbers is undecidable. Even without these specific symbols, if the second-order language has binary predicate variables and a symbol for successor, the set of all sentences true about N is still undecidable, as shown in R. M. Robinson (1958). (The theorem will be proved in the next section.) So, if we are to find a decidable set of sentences, we must restrict the language to one containing only unary (monadic) predicate variables and a symbol for successor. And, indeed, the set of sentences of this language that are true about N has been shown to be decidable by Büchi (1962). The method of proof used in that paper is to correlate the sentences of the language and finite automata.

In this section, we present the decidability of the "weak monadic second-order theory of successor." That is, the language is exactly the same as that used for the Büchi result—there are monadic predicate variables and a symbol for successor—but we think of the predicate variables as ranging only over the *finite* subsets of N. This simpler result is obtained by methods similar to, but less complicated than those used by Büchi. References to some related results will be mentioned at the end of the section.

The rest of the section is divided into three subsections. The first is to get a little more concrete understanding of monadic second-order languages. In the second, finite automata are defined carefully and the basic theorems about them that will be used later are presented. In the third subsection, the machinery necessary to prove the theorem is developed and the theorem is proved. None of these subsections present thorough treatments of the theories involved; however the setting has been made as general as possible to make it easier for the reader to go on to read some of the allied papers.

The Monadic Second-Order Language with Successor

We are interested in the relational system $\mathfrak{S}$ made up of the set of natural numbers and the successor function. Let $\mathsf{L}(S)$ be the monadic second-order language with a countably infinite set of monadic predicate variables, the symbol $\mathbf{0}$ (for zero), the symbol $'$ (for successor), the symbol $<$ (for less than), $=$, the usual propositional connectives, parentheses, comma, and the existential quantifier and the universal quantifier. Previously, we have tried to maintain a careful distinction between the formal language and the symbols used to talk about a relational system. Since the reader has learned how to make the distinction when necessary, we will purposely ignore it from now on. We let $x, y, z, x_1, x_2, \ldots$ denote individual variables, and $F, F_1, F_2, \ldots$ denote monadic predicate variables and henceforth call them *set variables*. The formulas of $\mathsf{L}(S)$ are defined as follows: Any individual variable is a term, 0 is a term, and if t is a term, then t' is a term. (For $n > 1$, t followed by n primes will be abbreviated by $t^{(n)}$.) If t_1 and t_2 are terms and F is a set variable, then $F(t_1)$, $(t_1 = t_2)$, and $(t_1 < t_2)$ are atomic formulas. Every atomic formula is a formula of $\mathsf{L}(S)$; if A and B are formulas of $\mathsf{L}(S)$, then $\sim A$, $(A \supset B)$, $(A \vee B)$, $(A \wedge B)$, and $(A \equiv B)$ are formulas of $\mathsf{L}(S)$; if A is a formula of $\mathsf{L}(S)$, x is an individual variable, and F is a set variable, then $(\exists x)A$, $(\forall x)A$, $(\exists F)A$, and $(\forall F)A$ are formulas of $\mathsf{L}(S)$. We will refer to $(\exists x)$ and $(\forall x)$ as "number quantifiers" and to $(\exists F)$ and $(\forall F)$ as "set quantifiers." Of course, $(\forall x)A$ and $\sim(\exists x)\sim A$ are logically equivalent, as are $(\forall F)A$ and $\sim(\exists F) \sim A$. For $\sim(t_1 = t_2)$, write $(t_1 \neq t_2)$. Instead of writing $F(x)$ or $F(x^{(n)})$, we will write $x \in F$ or $x^{(n)} \in F$ and write $x^{(n)} \notin F$ for $\sim F(x^{(n)})$. We are only interested in the interpretation of these formulas in terms of $\mathfrak{S}$ and so we will be thinking of the individual variables as ranging over the natural numbers, the set variables as ranging over subsets of N, and the $'$ as the successor function. When we come to the proof of the decidability of the weak monadic second-order theory of successor (final subsection), we will allow the set variables to range only over the finite subsets of N. (It seems preferable to develop the theory in the more natural setting of all subsets of N and to make the restriction at the last minute.)

If the reader recalls how assignment, valuation, and truth were defined for first-order languages relative to a relational system, he can easily see that for the language $\mathsf{L}(S)$ and the relational system $\mathfrak{S}$, an assignment is a mapping from individual variables to natural numbers and from set variables to subsets of the natural numbers. Valuations are then defined by induction on the formation of formulas in the obvious way and a sentence true in $\mathfrak{S}$ is defined accordingly. In particular, given a formula A with free individual and set variables, A is satisfied in $\mathfrak{S}$ if the existential closure is true in $\mathfrak{S}$. Rather than develop this formally, we just give examples to guide the reader's intuition.

Examples (1) The sentence $(\exists F)(\exists x)(\forall y)(x \in F \wedge (y > x \supset y \notin F))$ says that there is a nonempty subset with a largest element, and therefore the sentence is true in $\mathfrak{S}$.

(2) The sentence $(\forall F)(\forall x)(\forall y)(x \in F \wedge y \in F \supset x < y \vee x = y \vee y < x)$ is true in $\mathfrak{S}$ because it says that every subset is linearly ordered.

(3) The sentence

$$(\forall F)(\forall x)(\forall y)(\exists z)((x \in F \wedge y \in F \wedge x < y) \supset (x < z \wedge z < y \wedge z \in F))$$

is false in $\mathfrak{S}$ because it says that every subset is densely ordered.

(4) The sentence $(\forall F_1)(\exists F_2)(\forall x)(x \in F_1 \wedge x \notin F_2) \vee (x \notin F_1 \wedge x \in F_2))$ is

true in $\mathfrak{S}$ because it says that every set has a complement.

(5) The formula

$$(\exists x)(x \in F \wedge x' \in F \wedge \cdots \wedge x^{(n)} \in F \wedge (\forall y)(y < x \vee y > x^{(n)} \supset y \notin F)$$

is satisfiable in $\mathfrak{S}$ because it is satisfied by any subset of N which is just $n + 1$ consecutive elements of N.

(6) The formula

$$(\exists x)(\exists y)(\exists z)(\forall u)\big((x \neq y \wedge x \neq z \wedge y \neq z \wedge x \in F \wedge y \in F \wedge z \in F \wedge u \in F) \supset (u = x \vee u = y \vee y = z)\big)$$

is satisfied by any subset of N containing exactly three elements.

(7) The formula $(\forall x)(x \in F \wedge x' \notin F)$ is not satisfied by any subset of N.

(8) The formula $(\forall x)(x \in F_1 \wedge x \in F_2 \supset x \in F_3)$ is satisfied by any three subsets of N provided one of them contains the intersection of the other two.

Exercise 14.37 Write a formula with one free set variable which is satisfied by every finite subset of N and no others. (Recall that there is no first-order formula describing just the finite subsets of a universe.)

Notice that the sentence $(\exists F)A$ is true in $\mathfrak{S}$ if and only if the set of sets satisfying A is not empty.

Lemma 14.6 The following are true in $\mathfrak{S}$:

(i) $(\forall x)(\forall F)A \equiv (\forall F)(\forall x)A$

(ii) $(\exists x)(\exists F)A \equiv (\exists F)(\exists x)A$

(iii) $(\exists x)(\forall F_1)A \equiv (\exists F_2)(\forall F_1)(\forall x)\big((\exists z)(z \in F_2) \wedge (x \in F_2 \supset A)\big)$

(iv) $(\forall x)(\exists F_1)A \equiv (\forall F_2)(\exists F_1)(\exists x)\big((\exists z)(z \in F_2) \supset (x \in F_2 \wedge A)\big)$

PROOF Since we have no formal definitions, the proof is necessarily informal. We do (iii) in terms of some fixed but arbitrary assignment to the free variables of A. Suppose the right side is true in $\mathfrak{S}$. That means there is some nonempty subset of N, call it S_2, such that any $n \in S_2$ and every subset S_1 of N satisfy A; since S_2 is nonempty, we can conclude that there is an n such that for all $S \subseteq N$, n and S satisfy A. That is to say, $(\exists x)(\forall F_1)A$ is true. Suppose the left side is true. Thus, there is some $n \in N$, call it n^*, such that for all $S \subseteq N$, n^* and S satisfy A. But then $\{n^*\}$ is a nonempty subset of N and it has the property that for every subset $S \subseteq N$ and every $n \in N$, if $n \in \{n^*\}$, then n and S satisfy A.

Exercise 14.38 Finish the proof of the lemma. ▮

Exercise 14.39 Prove that each of the following is true in $\mathfrak{S}$:

(i) $(x = y) \equiv (\forall F)(x \in F \equiv y \in F)$
(ii) $(x = 0) \equiv {\sim}(\exists y)(y' = x)$
(iii) $(x < y) \equiv (\exists F)\big((\forall z)(z' \in F \supset z \in F) \wedge x \in F \wedge y \notin F\big)$

Exercise 14.40 Give intuitive arguments to show that the following formulas are true in $\mathfrak{S}$:

(a) $(A \supset (\exists F)B) \equiv (\exists F)(A \supset B)$, where F has no free occurrences in A.
(b) $((\exists F)A \supset B) \equiv (\forall F)(A \supset B)$, where F has no free occurrences in B.
(c) $(A \supset (\forall F)B) \equiv (\forall F)(A \supset B)$, where F has no free occurrences in A.
(d) $((\forall F)A \supset B) \equiv (\exists F)(A \supset B)$, where F has no free occurrences in B.

Lemma 14.7 For any formula A (in $\mathsf{F}(S)$), there is formula A^* (in $\mathsf{F}(S)$) such that A^* is in prenex normal form with all set quantifiers preceding number quantifiers and $A \equiv A^*$ is true in $\mathfrak{S}$.

Exercise 14.41 Prove the lemma.

Formulas of the form $x\eta F_i$, $x'\eta F_i$, $x^{(n)}\eta F_i$, for any $n > 1$ and η either $\in$ or $\notin$, will be called *elementary formulas*. (We will continue to use η in this way.) A formula of the form

$$(\exists x)(x^{(m_1)}\eta_1 F_{i_1} \wedge \cdots \wedge x^{(m_k)}\eta_k F_{i_k})$$

where $m_1 \leq m_2 \leq \cdots \leq m_k$ and the F_{i_j} are not necessarily distinct, is to be called a *principal formula*. A *unit formula* is an elementary formula, a principal formula, or a principal formula preceded by a negation symbol. Any unit formula is a *basic formula* and if A and B are basic formulas, then so are $A \vee B$, $A \wedge B$, $A \supset B$, and ${\sim}A$. A formula is in *basic disjunctive normal form* (or *basic conjunctive normal form*) if it is a disjunction of conjunctions of unit formulas (or a conjunction of disjunctions of unit formulas). By Theorem 13.3, every basic formula is logically equivalent to a formula in basic disjunctive normal form and to a formula in basic conjunctive normal form.

Lemma 14.8 Suppose that A is in basic disjunctive normal form. Then, $(\exists x)A$ and $(\forall x)A$ are each logically equivalent to formulas in basic disjunctive normal form.

PROOF Suppose that $(\exists x)A$ is of the form $(\exists x)(D_1 \vee \cdots \vee D_k)$. This formula is logically equivalent to $(\exists x)D_1 \vee \cdots \vee (\exists x)D_k$. For any j, consider $(\exists x)D_j$. It is of the form $(\exists x)(C_1 \wedge \cdots \wedge C_k)$, where each C_i is a unit formula, and we assume that x has free occurrences in $C_1, \ldots, C_m$ but has no free occurrence in $C_{m+1}, \ldots, C_t$. (It follows that $C_1, \ldots, C_m$ are elementary formulas.) This is logically equivalent to $(\exists x)(C_1 \wedge \cdots \wedge C_m) \wedge C_{m+1} \wedge \cdots \wedge C_t$. Thus, $(\exists x)(C_1 \wedge \cdots \wedge C_m)$ is a principal formula and each C_j, $m < j \leq t$, is either a principal formula or an elementary formula in which x does not occur. The proof for $(\forall x)A$ is left to the reader. ▮

Lemma 14.9 For every formula A (in $\mathsf{F}(S)$) having no free individual variables, there is a formula A^* (in $\mathsf{F}(S)$) such that $A \equiv A^*$ is true in $\mathfrak{S}$, A and A^* have the same free set variables, and A^* is of the form $Q(P)$, where Q is a string of set quantifiers and P is a disjunction of conjunctions of principal formulas and principal formulas preceded by a negation sign.

PROOF By exercise 14.39, we can assume that 0, $=$, and $<$ do not occur in A. Further, by Lemma 14.7, we can also assume that A is in prenex normal form with set quantifiers coming first. Thus, we can assume that A is of the form $Q_1 Q_2 A_0$, where Q_1 is a string of set quantifiers, Q_2 is a string of number quantifiers, and that A_0 is a disjunction of conjunctions of elementary formulas. Suppose Q_2 is $Q_3(\exists x)$. Then A is of the form $Q_1 Q_3(\exists x)A_0$; by the previous lemma, that is logically equivalent to $Q_1 Q_3 A_1$, where A_1 is in basic disjunctive normal form. Continuing inductively, we obtain $Q_1 A^*$, where A^* is in basic disjunctive normal form. Since A has no free individual variables, A^* is a disjunction of conjunctions of principal formulas or principal formulas preceded by negation signs. ▮

Finite Automata

A finite automaton can be thought of as a Turing machine with the following restrictions: (i) it never prints, (ii) when reading a blank square of tape, the machine does not move and does not change internal state, and (iii) when reading a nonblank square of tape, the machine always moves to the right and may change internal state. In terms of the functions Q_T, S_T, and D_T defined in section 5.3 for Turing machines, if M is a finite automaton, we have, for s_j not $*$, $Q_M(q_i, s_j) = q_t$, $S_M(q_i, s_j) = s_j$, and $D_M(q_i, s_j) = R$. Since the last two do not contribute anything in this case, we forget about them. The function Q_M is called the *transition function* and henceforth will be denoted by t, t', t_1, etc.

More formally, we say that a *finite automaton is given by*: a finite set of symbols $\Sigma = \{*, a_1, \ldots, a_n\}$; a finite set of internal states $Z = \{q_0, q_1, \ldots, q_{m-1}\}$, where q_0 is the initial state; a subset F of Z, called the *final* (*accepting*) *states*; and a transition function $t : Z \times \Sigma \to Z$ which meets the requirement that $t(q_i, *) = q_i$ for all i, $0 \leq i \leq m - 1$. A word of length 0 will be denoted by Λ and called the *empty word*. Generally when we write "$\tau \equiv a_{h_1} \ldots a_{h_t}$" we assume that no a_{h_i} is $*$. If $\tau \equiv a_{h_1} a_{h_2} \cdots a_{h_t}$, then the behavior of M on τ is: at time 0,

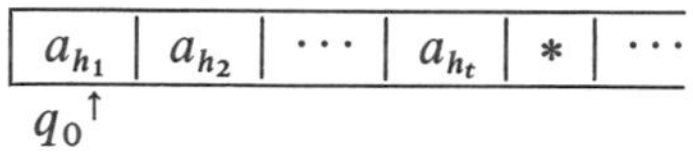

a_{h_1}	a_{h_2}	$\cdots$	a_{h_t}	$*$	$\cdots$

$q_0 \uparrow$ (under a_{h_1})

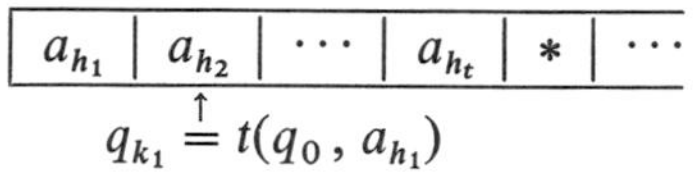

at time 1,

a_{h_1}	a_{h_2}	$\cdots$	a_{h_t}	$*$	$\cdots$

$\uparrow$ (under a_{h_2}) $q_{k_1} \doteq t(q_0, a_{h_1})$

at time t,

$\vdots$

a_{h_1}	a_{h_2}	$\cdots$	a_{h_t}	$*$	$\cdots$

$\uparrow$ (under $*$) $q_{k_t} \doteq t(q_{k_t - 1}, a_{h_t})$

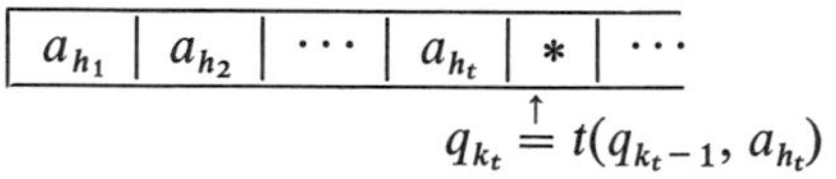

Since M does not move when reading $*$, the machine halts a time t. If, in the illustration, $q_{k_t} \in F$, then we say M *accepts* τ, and if $q_{k_t} \notin F$, then M *rejects* τ, or does not accept τ. The transition function t can be used to define a map $\hat{t} : Z \times \{\text{words on } \Sigma\} \to Z$. The map $\hat{t}$ is defined by induction on the lengths of words τ as follows: if $|\tau| = 0$, then τ is Λ and $\hat{t}(q_i, \Lambda) = t(q_i, *) = q_i$, and if $|\tau| \geq 0$ and $a_j \in \Sigma$, then $\hat{t}(q_i, \tau a_j) = t(\hat{t}(\tau), a_j)$. The set $\{\tau \mid \hat{t}(q_0, \tau) \in F\}$ is the set of *all words (tapes) accepted by M* and it is denoted by $\mathscr{T}_M$. A set of words $\mathscr{W}$ on an alphabet Σ is said to be *finite-automaton acceptable* if there is a finite automaton M such that $\mathscr{T}_M = \mathscr{W}$.

For any Σ and Z, let $\pi_1, \pi_2, \ldots$ denote members of $\Sigma \times Z$. Define projections $S : \Sigma \times Z \to \Sigma$ and $Q : \Sigma \times Z \to Z$ in the usual way: if π is (a_i, q_j), then $S(\pi) = a_i$ and $Q(\pi) = q_j$. Suppose M is a finite automaton with alphabet $\Sigma = \{*, a_1, \ldots, a_n\}$, set of internal states $Z = \{q_0, \ldots, q_{m-1}\}$, final states F, and transition function t, and let τ be a word on Σ. Suppose $\tau \equiv a_{h_1} a_{h_2} \cdots a_{h_t}$ and is written on a tape. If M is in state q_k and reads the leftmost symbol of τ, it will go through a certain sequence of states $\langle q_k, q_{k_1}, \ldots, q_{k_t} \rangle$ until it reaches a blank square of tape. That is, $t(q_k, a_{h_1}) = q_{k_1}, \ldots, t(q_{k_t - 1}, a_{h_t}) = q_{k_t}$. This gives the following sequence of pairs in $\Sigma \times Z$: $\langle (a_{h_1}, q_{k_1}), \ldots, (a_{h_t}, q_{k_t}) \rangle$, where $q_{k_1} = t(q_k, a_{h_1})$, and for each j, $1 \leq j < t$, $q_{k_{j+1}} = t(q_{k_j}, a_{h_{j+1}})$. With this in mind, given M and τ, for each k, $0 \leq k \leq m - 1$, let $\tau^{M,k}$ denote the sequence $\langle \pi_1, \ldots, \pi_k \rangle$ which satisfies the following conditions: (i) $S(\pi_1) S(\pi_2) \cdots S(\pi_t) \equiv \tau$, (ii) $Q(\pi_1) = t(q_k, S(\pi_1))$, and (iii) for all j, $1 < j \leq t$, $Q(\pi_j) = t(Q(\pi_{j-1}), S(\pi_j))$. Then, in particular, if $\tau \in \mathscr{T}_M$, the sequence $\tau^{M,0}$ satisfies a fourth condition, (iv) $Q(\pi_t) \in F$.

Exercise 14.42 Show that for any finite set $\mathscr{W}$ of words on an alphabet Σ there is a finite automaton M for which $\mathscr{T}_M = \mathscr{W}$.

Exercise 14.43 Show that for any alphabet Σ, the set of all words on Σ is finite-automaton acceptable.

Let M be a finite automaton with alphabet Σ, states $Z = \{q_0, \ldots, q_{m-1}\}$, final states F, and transition function t. We say that M *has a circuit from* q_i *to* q_j if there is a word τ on Σ such that $\hat{t}(q_i, \tau) = q_j$. Notice the following: (i) i and j may be equal, and (ii) for each i and each word τ on Σ, there is a j such that τ is a circuit from q_i to q_j. Define the set

$$C_{i,j} = \{\tau \mid \hat{t}(q_i, \tau) = q_j\}$$

It follows that

$$\mathscr{T}_M = \bigcup_{q_j \in F} C_{0,j}$$

A very obvious and very useful observation is the following.

The Pigeonhole Principle If there are k pigeonholes and m pigeons and every pigeon lives in a pigeonhole, then if $m > k$, there are some pigeonholes with more than one pigeon in residence.

Lemma 14.10 Let M be a finite automaton with m states. If $\tau \in C_{i,j}$ and $|\tau| = l > m$, then there is a k, $0 \le k \le m - 1$, and words ρ_1, ρ_2, and ρ_3 such that $\tau = \rho_1\rho_2\rho_3$ and $\rho_1 \in C_{1,k}$, $\rho_2 \in C_{k,k}$, and $\rho_3 \in C_{k,j}$.

Lemma 14.11 If a finite automaton M with m states accepts a word of length greater than m, then (a) M accepts a word of length less than or equal to m and (b) M accepts infinitely many words.

Exercise 14.44 Prove the lemmas.

Theorem 14.14 For a finite automaton M with m states, $\mathscr{T}_M \neq \varnothing$ if and only if there is a word of length $\le m$ which belongs to $\mathscr{T}_M$.

Theorem 14.15 The problem of determining, for a finite automaton M, whether or not $\mathscr{T}_M$ is empty is decidable.

Exercise 14.45 Prove the theorems.

The finite automata that we have been studying are also known as *deterministic finite automata* because the transition function t is from $Z \times \Sigma$ into Z and hence to each (q_i, a_j) there is a unique q_k such that $t(q_i, a_j) = q_k$. The definition of *nondeterministic finite automata* is the same as for deterministic finite automata except for the transition function. For nondeterministic finite automata, the transition function t_n is a mapping from the set

$Z \times \Sigma$ into the set of all nonempty subsets of Z. Let $\mathscr{P}^-(Z)$ denote the set of all nonempty subsets of Z. As before, a certain subset F of Z is designated as the set of final states. We define the mapping $\hat{t}_n : Z \times \{\text{words on } \Sigma\} \to \mathscr{P}^-(Z)$ by $\hat{t}_n(q_i, \Lambda) = \{q_i\}$ and $\hat{t}_n(q_i, \tau a_j) = \{t_n(q_k, a_j) \mid q_k \in \hat{t}_n(q_i, \tau)\}$. A nondeterministic finite automaton M_n accepts τ if $\hat{t}_n(q_0, \tau) \cap F \neq \varnothing$, and so $\mathscr{T}_{M_n} = \{\tau \mid \hat{t}_n(q_0, \tau) \cap F \neq \varnothing\}$. For any nondeterministic finite automaton M_n and word τ on its alphabet, we can define a tree $\mathscr{T}(M_n, \tau)$. (Refer to section 13.4 for the definition of tree. Given M_n and τ, a tree could be defined in various ways, but we choose to define it here so that branches look like sequences $\langle \pi_1, \ldots, \pi_t \rangle$ that were defined earlier.) All the vertices of $\mathscr{T}(M_n, \tau)$, except v_0, belong to $\Sigma \times Z$ and $v_0 = q_0$. $V_1 = \{(a_1, q_{1'}), \ldots, (a_1, q_{k'})\}$ for $\{q_{1'}, \ldots, q_{k'}\} = t_n(q_0, a_1)$. For $i > 0$, if $(a_i, q_j) \in V_i$, then $\text{arc}((a_i, q_j), (a_{i+1}, q_k))$, where $q_k \in t_n(q_j, a_{n+1})$. Then,

$$V_{i+1} = \{(a_{i+1}, q_k) \mid \text{for some } (a_i, q_j) \in V_i, \quad \text{arc}((a_i, q_j), (a_{i+1}, q_k))\}$$

If $|\tau| = t$, then the lengths of all the branches of $\mathscr{T}(M_n, \tau)$ based in V_0 are t. Further, $\tau \in \mathscr{T}_{M_n}$ if and only if there is some vertex $(a_t, q_f) \in V_t$ such that $q_f \in F$. A vertex $v \in V_i$ of a tree is a *branch point* if there is more than one vertex $v' \in V_{i+1}$ such that $\text{arc}(v, v')$. It follows from the definition of $\mathscr{T}(M_n, \tau)$ that a vertex (a_i, q_j) is a branch point if and only if $\overline{\overline{t_n(q_j, a_{i+1})}} > 1$. Except for v_0, let the vertices of $\mathscr{T}(M_n, \tau)$ be denoted by π_1, π_2, etc. Then any branch with base in V_1 is of the form $\langle \pi_1, \ldots, \pi_t \rangle$, where $Q(\pi_{i+1}) \in t_n(Q(\pi_i), S(\pi_{i+1}))$. Notice that the tree associated with a deterministic finite automaton M_d and word τ is just the single branch $\langle q_0, \pi_1, \ldots, \pi_t \rangle$, where $\langle \pi_1, \ldots, \pi_t \rangle = \tau^{M_d, 0}$ defined earlier.

Theorem 14.16 For every nondeterministic finite automaton M_n, there is a deterministic finite automaton M_d such that $\mathscr{T}_{M_d} = \mathscr{T}_{M_n}$.

PROOF The idea of the proof is to design a machine M_d so that for a word τ, the tree $\mathscr{T}(M_d, \tau)$ is the tree $\mathscr{T}(M_n, \tau)$ with all the branches squeezed together into one thick branch. Suppose M_n has alphabet Σ and set of states $Z = \{q_0, \ldots, q_{m-1}\}$, transition function $t_n : Z \times \Sigma \to \mathscr{P}^-(Z)$, initial state q_0, and set of final states F_n. Let $z_0, \ldots, z_{2^m - 2}$ denote the nonempty subsets of Z [i.e., elements of $\mathscr{P}^-(Z)$] and, in particular, z_0 is to denote $\{q_0\}$. We will say that a nondeterministic finite automaton, say M_n, has a circuit from i to z_j if there is a word τ such that $\hat{t}_n(q_i, \tau) = z_j$. Let

$$C^n_{i, z_j} = \{\tau \mid \hat{t}_n(q_i, \tau) = z_j\}$$

We will say that M_n has a pseudocircuit from 0 to k if there is a τ such that $q_k \in \hat{t}_n(q_0, \tau)$. Let

$$P^n_{0, k} = \{\tau \mid q_k \in \hat{t}_n(q_0, \tau)\}$$

Claim that

$$\mathscr{T}_{M_n} = \bigcup_{q_k \in F_n} P^n_{0,k}$$

Since, clearly,

$$\mathscr{T}_{M_n} = \bigcup_{z_j \cap F_n \neq \varnothing} C^n_{0,z_j}$$

we will show that

$$\bigcup_{q_k \in F_n} P^n_{0,k} = \bigcup_{z_j \cap F_n \neq \varnothing} C^n_{0,z_j} \tag{*}$$

If

$$\tau \in \bigcup_{q_k \in F_n} P^n_{0,k}$$

then $q_k \in F_n$ and $q_k \in \hat{t}_n(q_0, \tau) = z_j$ for some z_j, i.e., $\tau \in C^n_{0,z_j}$ for some z_j such that $z_j \cap F_n \neq \varnothing$. Suppose

$$\tau \in \bigcup_{z_j \cap F_n \neq \varnothing} C^n_{0,z_j}$$

Then, $\hat{t}_n(q_0, \tau) = z_j$, $z_j \cap F_n \neq \varnothing$, and so there is some $q_k \in z_j \cap F_n$. For that k, $\tau \in P^n_{0,k}$ and so

$$\tau \in \bigcup_{q_k \in F_n} P^n_{0,k}$$

Define the deterministic finite automaton M_d as follows. The alphabet is just Σ. The set of states is $Z' = \{z_0', \ldots, z'_{2^m-2}\}$, the initial state is z_0'. The set of final states F_d, is

$$\bigcup_{z_i \cap F_n \neq \varnothing} z_i'$$

The transition function $t_d : \Sigma \times Z' \to Z'$ is defined by

$$t_d(z_i', a_h) = z_j' \qquad \text{where} \quad z_j = \bigcup_{q_k \in z_i} t_n(q_k, a_h)$$

Circuits for deterministic finite automata have already been defined and for clarity we add the superscript, thus

$$C^d_{i,j} = \{\tau \mid \hat{t}_d(z_i', \tau) = z_j'\}$$

So

$$\mathscr{T}_{M_d} = \bigcup_{z_j' \in F_d} C^d_{0,j}$$

We will say M_d has a subcircuit from 0 to k if there is a word such that $\hat{t}_d(z_0', \tau) = z_j'$ and $q_k \in z_j$. Let

$$S^d_{0,k} = \{\tau \mid \hat{t}_d(z_0', \tau) = z_j' \quad \text{and} \quad q_k \in z_j\}$$

Since it is clear that

$$\bigcup_{q_k \in F_n} P^n_{0,k} = \bigcup_{q_k \in F_n} S^d_{0,k}$$

in order to show that $\mathcal{T}_{M_n} = \mathcal{T}_{M_d}$, we only need to show that

$$\bigcup_{z_j' \in F_d} C^d_{0,j} = \bigcup_{q_k \in F_n} S^d_{0,k}$$

By the definition of F_d, $z_j' \in F_d$ if and only if $z_j \cap F_n \neq \varnothing$. So we only need show that

$$\bigcup_{z_j \cap F_n \neq \varnothing} C^d_{0,j} = \bigcup_{q_k \in F_n} S^d_{0,k}$$

But this follows easily by an argument very similar to the one used in proving (*) above. ▮

Next, we look into the possibility of characterizing the sets that are finite-automaton acceptable. That is, is there some property whose definition does not depend on finite automata and which is true of the sets acceptable by finite automata and no others? Since a word on an alphabet Σ is formed by concatenating elements (i.e., letters) of Σ, we might look into extending that definition so as to "concatenate" the elements (i.e., words) of a set of words. First, for $\mathcal{U}$ and $\mathcal{V}$, each a set of words, we define the *product of* $\mathcal{U}$ *and* $\mathcal{V}$,

$$\mathcal{U} \cdot \mathcal{V} = \{\tau \mid \tau \equiv \alpha\beta, \alpha \in \mathcal{U}, \beta \in \mathcal{V}\}$$

For example, if $\mathcal{U} = \{\text{cat, house}\}$, $\mathcal{V} = \{\text{rat, mouse}\}$, then $\mathcal{U} \cdot \mathcal{V} = \{$catrat, catmouse, houserat, housemouse$\}$. Given a set of words $\mathcal{U}$ on Σ, we define, inductively, $\mathcal{U}^0 = \Lambda$, the empty word, $\mathcal{U}^{n+1} = \mathcal{U}^n \cdot \mathcal{U}$. Define the *closure of* $\mathcal{U}$ as

$$\mathcal{U}^* = \bigcup_{n \in N} \mathcal{U}^n$$

If one looks at each word in $\mathcal{U}$ as if it were a single symbol, then the elements of $\mathcal{U}^*$ are seen to be all "the words on $\mathcal{U}$." Notice that by the definition of closure the set of all words on Σ is Σ^*. (This notation is widely used in the literature.) For an alphabet Σ, we define the set of all *Σ-regular sets* (of words on Σ) by: (i) $\{a\}$ is a Σ-regular set for each $a \in \Sigma$, and (ii) if $\mathcal{U}$ and $\mathcal{V}$ are Σ-regular sets, then $\mathcal{U} \cup \mathcal{V}$, $\mathcal{U} \cdot \mathcal{V}$, and $\mathcal{U}^*$ are Σ-regular sets. A set of words is *regular* if it is Σ-regular for some Σ.

Exercise 14.46 Let Σ be an alphabet. (a) Show that the set of all words on Σ is regular. (b) Show that any finite set of words on Σ is regular.

Exercise 14.47 Suppose we have two sets of symbols Σ and Σ', where the symbols in Σ' are words on the symbols in Σ; e.g., $\Sigma = \{a, b\}$ and

$\Sigma' = \{bab, abba\}$. Show that if a set of words $\mathscr{R}$ is Σ'-regular, then it is Σ-regular but that the converse need not be true.

Let Σ_1 and Σ_2 be two alphabets, not necessarily distinct. If a mapping $\hat{\rho}$ from the set of all subsets of words on Σ_1 into the set of all subsets of words on Σ_2 has the property that for any sets $\mathscr{U}$ and $\mathscr{V}$ on Σ_1, $\hat{\rho}(\mathscr{U} \cdot \mathscr{V}) = \hat{\rho}(\mathscr{U}) \cdot \hat{\rho}(\mathscr{V})$, then $\hat{\rho}$ is a *regular homomorphism.*

Exercise 14.48 Show that if $\mathscr{U}$ is a Σ_1-regular set and $\hat{\rho}$ is a regular homomorphism, then $\hat{\rho}(\mathscr{U})$ is a Σ_2-regular set for the appropriate Σ_2. (Hint: By induction on the formation of regular sets.)

Exercise 14.49 Let Σ_1 and Σ_2 be two alphabets such that $\bar{\bar{\Sigma}}_1 \le \bar{\bar{\Sigma}}_2$ and let ρ map Σ_1 onto Σ_2. Define $\hat{\rho} : \{\text{words on } \Sigma_1\} \to \{\text{words on } \Sigma_2\}$ by induction on the length of the word: $|\tau| = 0$ implies $\tau = \Lambda$ and $\hat{\rho}(\Lambda) = \Lambda$, and, for any $a_i \in \Sigma_1$ and τ on Σ_1, $\hat{\rho}(\tau a_i) = \hat{\rho}(\tau)\rho(a_i)$. Show that $\hat{\rho}$ is a regular homomorphism.

The next two theorems establish that it is the property of being regular that characterizes the finite-automaton-acceptable sets.

Theorem 14.17 For each finite automaton M, the set of words $\mathscr{T}_M$ accepted by M is regular.

PROOF Suppose M is some finite automaton and any a_i or τ mentioned in the proof refer to symbols in or words on the alphabet of M. For any i, j, and $\tau \in C_{i,j}$, consider $\tau^{M,i} = \langle \pi_1, \ldots, \pi_k \rangle$. Define q_k to be the *maximum interior state of* $\tau^{M,i}$ if there is an n, $1 \le n < t$, such that $Q(\pi_n) = q_k$ and for all j, $1 \le j < t$, if $Q(\pi_j) = q_m$, then $m \le k$. Define $\lambda(\tau^{M,i}) = k$ if k is the maximum interior state of $\tau^{M,i}$. [Notice that neither the state (entering state) that the machine is in as it reads the leftmost symbol of the word τ nor the state it goes into (exiting state) upon reading the rightmost symbol of τ is considered in computing $\lambda(\tau^{M,i})$.] For example, if $\tau^{M,2} = \langle (a_1, q_5), (a_3, q_4), (a_2, q_6) \rangle$, then $\lambda(\tau^{M,2}) = 5$. For each k, $0 \le k \le m - 1$, define

$$C_{i,j}^k = \{\tau \mid \tau \in C_{i,j} \quad \text{and} \quad \lambda(\tau^{M,i}) \le k\}$$

and let

$$C_{i,j}^{-1} = \{a_k \mid t(q_i, a_k) = q_j\}$$

(A member of $C_{i,j}^{-1}$ has no interior state.) Notice that

$$\bigcup_{0 \le i, j \le m} C_{i,j}^{-1}$$

is finite, $C_{i,j}^{k-1} \subseteq C_{i,j}^{k}$ and therefore $C_{i,j} = C_{i,j}^{m-1}$. It is clear that if $\tau \in C_{i,j}^{k}$ and $\lambda(\tau^{M,i}) = k$, then there are words $\rho_0, \rho_1, \ldots, \rho_{r-1}, \rho_r$ such that $\tau \equiv \rho_0 \rho_1 \cdots \rho_{r-1} \rho_r$ and the following conditions are satisfied: $\rho_0 \in C_{i,k}$ and $\lambda(\rho_0^{M,i}) < k$; for each h, $0 < h < r$, $\rho_h \in C_{k,k}$ and $\lambda(\rho_h^{M,k}) < k$; and $\rho_r \in C_{k,j}$ and $\lambda(\rho_r^{M,k}) < k$.

Exercise 14.50 Prove that for all k, $0 \leq k \leq m-1$,

$$C_{i,j}^{k} = C_{i,j}^{k-1} \cup C_{i,k}^{k-1} \cdot (C_{k,k}^{k-1})^* \cdot C_{k,j}^{k-1}$$

Thus, we have, inductively, for all i, j, $0 \leq i, j \leq m-1$,

$$C_{i,j}^{0} = C_{i,j}^{-1} \cup C_{i,0}^{-1} \cdot (C_{0,0}^{-1})^* \cdot C_{0,j}^{-1} \quad \text{and is therefore regular}$$

$$\vdots$$

$$C_{i,j}^{m-1} = C_{i,j}^{m-2} \cup C_{i,m-1}^{m-2} \cdot (C_{m-1,m-1}^{m-2})^* \cdot C_{m-1,j}^{m-2} \quad \text{and is therefore regular.}$$

But $C_{i,j} = C_{i,j}^{m-1}$. Therefore, in particular, for each $q_j \in F$, $C_{0,j}$ is regular. Thus, finally, $\mathscr{T}_M = \bigcup_{q_j \in F} C_{0,j}$ is regular. ▮

Theorem 14.18 If $\mathscr{R}$ is a regular set, then $\mathscr{R}$ is finite-automaton acceptable.

PROOF Suppose Σ is an alphabet such that $\mathscr{R}$ is Σ-regular. The proof is by induction on the formation of Σ-regular sets. By exercise 14.42, there is a finite automaton that accepts just the elements of Σ. There are three cases for the induction step: $\mathscr{U} \cdot \mathscr{V}$, $\mathscr{U}^*$, and $\mathscr{U} \cup \mathscr{V}$. The most difficult, $\mathscr{U}^*$, will be shown here, with the others left as exercises. Suppose $\mathscr{U}$ is accepted by a finite automaton M with alphabet $\Sigma = \{*, a_1, \ldots, a_n\}$, set of internal states $Z = \{q_0, \ldots, q_{m-1}\}$, final states F, and transition function t. We must show that there is a finite automaton M^* which accepts $\mathscr{U}^*$. In fact, what we will show is that there is a nondeterministic finite automaton M_n which accepts $\mathscr{U}^*$. It then follows by Theorem 14.17 that there is a deterministic finite automaton which accepts $\mathscr{U}^*$. M_n is to have alphabet Σ, set of states $Z_n = \{q_{00}\} \cup Z$; q_{00} is the initial state and $\{q_{00}\} \cup F$ is the set of final states. The transition function t_n is defined as follows:

$$t_n(q_{00}, *) = q_{00}$$

$$t_n(q_{00}, a_h) = t(q_0, a_h) \qquad \text{for every} \quad a_h \in \Sigma$$

$$t_n(q_i, a_h) = \begin{cases} \{t(q_i, a_h)\} & \text{if} \quad q_i \notin F \\ \{t(q_i, a_h), t(q_0, a_h)\} & \text{if} \quad q_i \in F \end{cases} \qquad \text{for all } a_h \in \Sigma \text{ and } q_i \in Z$$

$\mathscr{U}^* = \mathscr{U}^0 \cup \mathscr{U}^1 \cup \cdots \cup \mathscr{U}^n \cup \cdots$, where $\mathscr{U}^0 = \Lambda$. M_n accepts Λ since $t_n(q_{00}, *) = q_{00}$, which is in the set of final states of M_n. We assume that $*$ does not occur in any word τ in the rest of the proof. So there are two things to be shown: (1) If $\tau_1, \ldots, \tau_p \in \mathscr{T}_M$, then $\tau_1 \cdots \tau_p \in \mathscr{T}_{M_n}$. (2) If $\tau \in \mathscr{T}_{M_n}$, then there are words $\tau_1, \ldots, \tau_p$ such that $\tau \equiv \tau_1 \cdots \tau_p$ and $\tau_1, \ldots, \tau_p$ all belong to $\mathscr{T}_M$. We begin by showing (2). Let $\tau \in \mathscr{T}_{M_n}$ and consider the tree $\mathscr{T}(M_n, \tau)$ In particular, consider any branch $\langle \pi_1, \ldots, \pi_t \rangle$ with base in V_1 and $Q(\pi_t) \in F$. Keep that branch fixed for the discussion. Let $\pi_{i_1}, \ldots, \pi_{i_p}$ be the branch points of that branch. We will show that $\langle i_1, \ldots, i_p \rangle$ has a subsequence $\langle i_1', \ldots, i_p' \rangle$ such that

$$S(\pi_1) \ldots S(\pi_{i_1'}), \quad S(\pi_{i_1'+1}) \ldots S(\pi_{i_2'}), \quad \ldots, \quad S(\pi_{i_p'+1}) \ldots S(\pi_t)$$

all belong to $\mathscr{T}_M$ and

$$S(\pi_1) \cdots S(\pi_{i_1'}) S(\pi_{i_1'+1}) \cdots S(\pi_{i_2'}) \cdots S(\pi_{i_p'+1}) \cdots S(\pi_t) \equiv \tau$$

Notice that for M_n, $\pi_{i_j} = (a_{i_j}, q_h)$ is a branch point if and only if $q_h \in F$. Furthermore, if that is the case, then there are just two arcs from π_{i_j}: $\mathrm{arc}((a_{i_j}, q_h), (a_{i_j+1}, q_k))$, where $t(q_h, a_{i_j+1}) = q_k$ and $\mathrm{arc}((a_{i_j}, q_h), (a_{i_j+1}, q_l))$, where $t(q_0, a_{i_j+1}) = q_l$. The sequence $\langle i_1', \ldots, i_p' \rangle$ of the branch in question is determined as follows: i_j belongs in $\langle i_1', \ldots, i_p' \rangle$ if and only if (a_{i_j+1}, q_l) belongs to the branch. Consider $\langle \pi_1, \ldots, \pi_{i_{1'}} \rangle$, $\langle \pi_{i_1'+1}, \ldots, \pi_{i_2'} \rangle$, $\ldots$, $\langle \pi_{i_p'+1}, \ldots, \pi_t \rangle$. $Q(\pi_1) = t(q_0, S(\pi_1))$ and $Z(\pi_{i_1'}) \in F$, $Q(\pi_{i_1'+1}) = t(q_0, S(\pi_{i_1'+1}))$ and $Z(\pi_{i_2'}) \in F, \ldots,$ and $Q(\pi_{i_p'+1}) = t(q_0, S(\pi_{i_p'+1}))$ and $Z(\pi_t) \in F$. Therefore, $S(\pi_1) \cdots S(\pi_{i_1'})$, $S(\pi_{i_1'+1}) \cdots S(\pi_{i_2'}), \ldots, S(\pi_{i_p'+1}) \cdots S(\pi_t)$ all belong to $\mathscr{T}_M$.

Exercise 14.51 Show that if $\tau_1, \ldots, \tau_p$ all belong to $\mathscr{T}_M$, then

$$\tau_1 \cdots \tau_p \in \mathscr{T}_{M_n}$$

The exercise completes the proof that $\mathscr{U}$ finite-automaton acceptable implies that $\mathscr{U}^*$ is finite-automaton acceptable.

Exercise 14.52 Show that if $\mathscr{U}$ and $\mathscr{V}$ are finite-automaton acceptable, then $\mathscr{U} \cdot \mathscr{V}$ and $\mathscr{U} \cup \mathscr{V}$ are finite-automaton acceptable. (Hint: for $\mathscr{U} \cdot \mathscr{V}$, if $\mathscr{U}$ is accepted by M_1 with the set of states Z_1 and $\mathscr{V}$ is accepted by M_2 with set of states Z_2, think about a finite automaton with states $Z_1 \times Z_2$.)

This completes the induction, so the theorem is proved. ▌

Exercise 14.53 (a) Show that if $\mathscr{U}$ is a finite-automaton-acceptable set of words, than the complement of $\mathscr{U}$, $\overline{\mathscr{U}} = \{\text{all words on } \Sigma\} - \mathscr{U}$, is finite-automaton acceptable. (b) It follows that if $\mathscr{U}$ and $\mathscr{V}$ are finite-automaton acceptable, then $\mathscr{U} \cap \mathscr{V}$ is finite-automaton acceptable.

Exercise 14.54 Suppose Σ and Σ' are two alphabets and $\hat{\rho}$ is a regular homomorphism from the set of all words on Σ to the set of all words on Σ'.

For a word τ' on Σ', $\hat{\rho}^{-1}(\tau') = \{\tau \mid \tau$ on Σ, $\hat{\rho}(\tau) = \tau'\}$. Show that if for each τ', in the set of all words on Σ', $\hat{\rho}^{-1}(\tau')$ is finite, and if $\mathscr{R}$ is a Σ'-regular set, then $\hat{\rho}^{-1}(\mathscr{R})$ is a Σ-regular set.

The Decidability of the Weak Monadic Second-Order Theory of Successor

For every k, define the set

$$B_k = \{\omega \mid \omega \text{ is a word on } \{0, 1\} \quad \text{and} \quad |\omega| = k\}$$

So, if $\omega \in B_k$, then $\omega = \mathrm{b}_1 \mathrm{b}_2 \cdots \mathrm{b}_k$, where each b_i is 0 or 1. We will sometimes think of each ω as a binary number reading least significance to most significance from left to right, that is, if $\omega \equiv \mathrm{b}_1 \mathrm{b}_2 \cdots \mathrm{b}_k$, then the binary number is $\mathrm{b}_1 \cdot 2^0 + \mathrm{b}_2 \cdot 2^1 + \cdots + \mathrm{b}_k \cdot 2^{k-1}$. For example, $B_4 = \{0000, 1000, 0100, 1100, 0010, \ldots, 1111\}$ and we think of 0000 as representing the number represented in decimal by 0, of 1010 as 5, of 1111 as 15, etc. An infinite sequence of elements of B_k will be called an *infinite k-sequence* and be denoted by $\sigma_{k,\infty}$. Thus, $\sigma_{k,\infty} = \langle \omega_0, \omega_1, \ldots \rangle$, where each $\omega_i \in B_k$. Let σ_k denote finite k-sequences. We use "k-sequence" to mean either finite or infinite when we do not wish to be specific. For every k, let Σ_k be a set of 2^k distinct symbols $\{a_0, a_1, \ldots, a_{2^k-1}\}$ and assume that if $i \neq j$, then $\Sigma_i \cap \Sigma_j = \varnothing$. An *infinite k-word on* Σ_k is the same as a word on Σ_k except that infinitely many symbols occur. We will call words on Σ_k *finite k-words*, and denote them by τ_k, and let $\tau_{k,\infty}$ denote infinite k-words on Σ_k. We use "k-words" to mean either finite k-words or infinite k-words when we do not wish to be specific. We wish to be able to switch back and forth between k-sequences on B_k and k-words on Σ_k. In order to do so, we make the obvious correspondence between Σ_k and B_k and use it to make a correspondence between k-sequences and k-words. Define the map $v_k : \Sigma_k \to B_k$ by $v_k(a_i) = \mathrm{b}_1 \cdots \mathrm{b}_k$, where $i = \mathrm{b}_1 \cdot 2^0 + \mathrm{b}_2 \cdot 2^1 + \cdots + \mathrm{b}_k \cdot 2^{k-1}$. For example, $v_k(a_0) = 0^k$ and $v_k(a_5) = 1010^{k-3}$. Since v_k is clearly one-to-one, we also have $v_k^{-1} : B_k \to \Sigma_k$. For a given k, we can extend the definition of v_k, in the obvious way, to define a mapping $\hat{v}_k$ from k-words to k-sequences. Thus, if $\tau_{k,\infty} \equiv a_{i_0} a_{i_1} \cdots$, then $\hat{v}_k(\tau_{k,\infty}) = \langle v_k(a_{i_0}), v_k(a_{i_1}), \ldots, \rangle$; similarly for $\hat{v}_k(\tau_k)$. Conversely, if $\sigma_{k,\infty} = \langle \omega_0, \omega_1, \ldots \rangle$, then $\hat{v}_k^{-1}(\sigma_{k,\infty}) = v_k^{-1}(\omega_0) v_k^{-1}(\omega_1) \ldots$, and similarly for $v_k(\sigma_k)$.

Define the map $\rho_{-1} : B_k \to B_{k-1}$ by $\rho_{-1}(\mathrm{b}_1 \cdots \mathrm{b}_{k-1} \mathrm{b}_k) = \mathrm{b}_1 \cdots \mathrm{b}_{k-1}$. So, for example, if $k = 5$ and $\mathrm{b}_1 \cdots b_{k-1} \mathrm{b}_k$ is 11001, then $\rho_{-1}(11001) = 1100$. Then ρ_{-1} can be used in the obvious way to define a map $\hat{\rho}_{-1} : \{k\text{-sequences}\} \to \{(k-1)\text{-sequences}\}$. That is, if $\sigma_{k,\infty} = \langle \omega_0, \omega_1, \ldots \rangle$, then $\hat{\rho}_{-1}(\sigma_{k,\infty}) = \langle \rho_{-1}(\omega_0), \rho_{-1}(\omega_1), \ldots \rangle$; similarly for $\hat{\rho}_{-1}(\sigma_k)$. The corresponding maps $\bar{\rho}_{-1} : \Sigma_k \to \Sigma_{k-1}$ and $\hat{\bar{\rho}}_{-1} : \{\text{words on } \Sigma_k\} \to \{\text{words on } \Sigma_{k-1}\}$ are defined as one would expect: $\bar{\rho}_{-1}(a_i) = v_{k-1}^{-1}(\rho_{-1}(v_k(a_i)))$, for each $a_i \in \Sigma_k$, and if $\tau_k \equiv a_{i_0} a_{i_1} \cdots a_{i_t}$, then $\hat{\bar{\rho}}_{-1}(\tau_k) \equiv \bar{\rho}_{-1}(a_{i_0}) \bar{\rho}_{-1}(a_{i_1}) \cdots \bar{\rho}_{-1}(a_{i_t})$, similarly for

$\hat{\bar{\rho}}_{-1}(\tau_{k,\infty})$. For any $m < k$, we can define, inductively, $\rho_{-m} : B_k \to B_{k-m}$ by

$$\rho_{-m}(\mathrm{b}_1 \cdots \mathrm{b}_k) = \underbrace{\rho_{-1}(\cdots \rho_{-1}}_{m \text{ times}}(\mathrm{b}_1 \cdots \mathrm{b}_k))$$

and define $\hat{\rho}_{-m}$, $\bar{\rho}_{-m}$, and $\hat{\bar{\rho}}_{-m}$ accordingly. This is illustrated by

$$\begin{array}{ccc} \Sigma_k & \xrightarrow{\nu_k} & B_k \\ {\scriptstyle \bar{\rho}_{-m}} \downarrow & & \downarrow {\scriptstyle \rho_{-m}} \\ \Sigma_{k-m} & \xleftarrow{\nu_{k-m}^{-1}} & B_{k-m} \end{array}$$

Notice that for $\mathrm{b}_1 \cdots \mathrm{b}_k \in B_k$, $\rho_{-1}^{-1}(\mathrm{b}_1 \cdots \mathrm{b}_k) = \{\mathrm{b}_1 \cdots \mathrm{b}_k 0, \mathrm{b}_1 \cdots \mathrm{b}_k 1\}$. Define $\rho_{+1} : B_k \to \{\text{pairs of elements in } B_{k+1}\}$ by $\rho_{+1}(\mathrm{b}_1 \cdots \mathrm{b}_k) = \rho_{-1}^{-1}(\mathrm{b}_1 \cdots \mathrm{b}_k)$. Similarly, we can define $\bar{\rho}_{+1} : \Sigma_k \to \{\text{pairs of symbols in } \Sigma_{k+1}\}$. If $a_i \in \Sigma_k$, then define $\bar{\rho}_{+1}(a_i)$ to be $\nu_{k+1}^{-1}(\rho_{+1}(\nu_k(a_i)))$. The computation is as follows: $\nu_k(a_i) = \mathrm{b}_1 \cdots \mathrm{b}_k$, where $\mathrm{b}_1 \cdot 2^0 + \cdots + \mathrm{b}_k \cdot 2^{k-1} = i$, then $\rho_{+1}(\nu_k(a_i)) = \{\mathrm{b}_1 \cdots \mathrm{b}_k 0, \mathrm{b}_1 \cdots \mathrm{b}_k 1\}$ and finally, $\nu_{k+1}^{-1}(\rho_{+1}(\nu_k(a_i))) = \{d_i, d_{i+2^k}\}$, where $d_i, d_{i+2^k} \in \Sigma_{k+1}$. ρ_{+m} and $\bar{\rho}_{+m}$ can be defined inductively. The corresponding maps $\hat{\rho}_{+m}$ on k-sequences and $\hat{\bar{\rho}}_{+m}$ on k-words are then defined in the usual way. Considering elements of B_k as words on $\{0, 1\}$, we notice that the mappings ρ_{-m} and ρ_{+m} chop off or add to the right ends of the words. We now perform similar operations on the left ends of the words on $\{0, 1\}$ which are elements of B_k. Define $\lambda_{-1} : B_k \to B_{k-1}$ by $\lambda_{-1}(\mathrm{b}_1 \mathrm{b}_2 \cdots \mathrm{b}_k) = \mathrm{b}_2 \cdots \mathrm{b}_k$ and $\lambda_{+1} : B_k \to \{\text{pairs of elements in } B_{k+1}\}$ by $\lambda_{+1}(\mathrm{b}_1 \cdots \mathrm{b}_k) = \{0\mathrm{b}_1 \cdots \mathrm{b}_k, 1\mathrm{b}_1 \cdots \mathrm{b}_k\}$. Then, $\bar{\lambda}_{-1} : \Sigma_k \to \Sigma_{k-1}$ by $\bar{\lambda}_{-1}(a_i) = \nu_{k-1}^{-1}(\lambda_{-1}(\nu_k(a_i)))$. Thus, for $a_i \in \Sigma_k$, $\bar{\lambda}_{-1}(a_i) = c_{[i/2]} \in \Sigma_{k-1}$. (Notice that if $\mathrm{b}_1 \cdot 2^0 + \mathrm{b}_2 \cdot 2^1 + \cdots + \mathrm{b}_k \cdot 2^{k-1} = i$, then $\mathrm{b}_2 \cdot 2^0 + \cdots + \mathrm{b}_k \cdot 2^{k-2} = [i/2]$.) $\bar{\lambda}_{+1} : \Sigma_k \to \{\text{pairs of symbols in } \Sigma_{k+1}\}$ by $\bar{\lambda}_{+1}(a_i) = \nu_{k+1}^{-1}(\lambda_{+1}(\nu_k(a_i)))$. Thus, $\nu_k(a_i) = \mathrm{b}_1 \cdots \mathrm{b}_k$ such that $\mathrm{b}_1 \cdot 2^0 + \cdots + \mathrm{b}_k \cdot 2^{k-1} = i$, then, $\lambda_{+1}(\nu_k(a_i)) = \{0\mathrm{b}_1 \cdots \mathrm{b}_k, 1\mathrm{b}_1 \cdots \mathrm{b}_k\}$, and finally, $\nu_{k+1}^{-1}(\lambda_{+1}(\nu_k(a_i))) = \{d_{2i}, d_{2i+1}\}$. λ_{+m}, $\bar{\lambda}_{+m}$, $\hat{\lambda}_{+m}$, and $\hat{\bar{\lambda}}_{+m}$ can be defined in the usual way.

Exercise 14.55 (a) Verify that $\hat{\bar{\rho}}_{-m}$ and λ_{-m} are regular homomorphisms. (b) Show that if $\mathscr{R}$ is a Σ_k-regular set, then both $\hat{\bar{\rho}}_{+m}(\mathscr{R})$ and $\hat{\bar{\lambda}}_{+m}(\mathscr{R})$ are Σ_{k+m}-regular sets. (Hint: Refer to exercise 14.54.)

Let $F_1, \ldots, F_k$ be a sequence of sets of natural numbers. We say that the *infinite k-sequence* σ_k *describes* $F_1, \ldots, F_k$, where $\sigma_{k,\infty} = \langle \omega_0, \omega_1, \ldots \rangle$, and for each n, $\omega_n = \mathrm{b}_1 \mathrm{b}_2 \cdots \mathrm{b}_k$, such that

$$\mathrm{b}_1 = \begin{cases} 1 & \text{if } n \in F_1 \\ 0 & \text{if } n \notin F_1 \end{cases}, \ldots, \qquad \mathrm{b}_k = \begin{cases} 1 & \text{if } n \in F_k \\ 0 & \text{if } \mathrm{n} \notin F_k \end{cases}$$

Thus, for every sequence of sets $F_1, \ldots, F_k$, there is an infinite k-sequence describing them; and conversely, to every infinite k-sequence $\sigma_{k,\infty}$, there is a sequence of sets which is described by $\sigma_{k,\infty}$. This idea can be extended to infinite words on Σ_k by means of ν_k and ν_k^{-1}. Thus, we say that the *infinite k-word $\tau_{k,\infty}$ describes $F_1, \ldots, F_k$* if $\hat{\nu}_k(\tau_{k,\infty})$ describes $F_1, \ldots, F_k$. It follows that for every sequence of k sets there is an infinite k-word describing it and for every infinite k-word there is a sequence of k sets described by it.

For each formula A of $\mathsf{L}(S)$ with k free set variables (and no free individual variables), define the set of infinite k-sequences $\mathscr{S}_k(A)$ by

$$\mathscr{S}_k(A) = \{\sigma_{k,\infty} \mid \text{the sequence of } k \text{ sets described by } \sigma_{k,\infty} \text{ satisfies } A\}$$

The corresponding set of infinite k-words is

$$\mathscr{W}_k(A) = \{\tau_{k,\infty} \mid \text{the sequence of } k \text{ sets described by } \tau_{k,\infty} \text{ satisfies } A\}$$

Exercise 14.56 Suppose A is a sentence on the language $\mathsf{L}(S)$ of the form $(\exists F_1)B$, where F_1 has a free occurrence in B. Show that A is true in $\mathfrak{S}$ (see the first subsection of this section) if and only if $\mathscr{S}_1(B) \neq \varnothing$.

Notice that if $F_1, \ldots, F_k$ are all finite sets, then the infinite k-sequence that describes them is "eventually zero." That is, if $F_1, \ldots, F_k$ are finite sets and $\sigma_{k,\infty} = \langle \omega_0, \omega_1, \ldots \rangle$ describes them, then there exists an n^* such that $\omega_{n^*} \not\equiv 0^k$ but for all $n > n^*$, $\omega_n \equiv 0^k$. Similarly, if $F_1, \ldots, F_k$ are finite sets and $\tau_{k,\infty}$ describes them, then there is an n^* such that $a_{i_{n^*}} \neq a_0$ and for all $n > n^*$, $a_{i_n} = a_0$. We map the set of infinite k-words that describe finite sets into the set of k-words in the obvious way: if $\tau_{k,\infty} \equiv a_{i_1} \cdots a_{i_{n^*}} a_0 a_0 \cdots$ and $a_{i_{n^*}} \neq a_0$, then $\tau_k \equiv a_{i_1} \cdots a_{i_{n^*}}$. Thus, every sequence of k finite sets is described by a finite k-word and every finite k-word describes some sequence of k finite sets. (We leave it to the reader to set up a formal definition using all the appropriate mappings.) For each formula A with k free set variables, define $\widetilde{\mathscr{W}}_k(A)$ by

$$\widetilde{\mathscr{W}}_k(A) = \{\tau_k \mid \text{the sequence of } k \text{ finite sets described by } \tau_k \text{ satisfies } A\}$$

We now have a correlation among (1) finite sets, (2) monadic second-order formulas, and (3) sets of words. This correlation will be used to establish a relation between formulas and regular sets.

Theorem 14.19 † For any formula A (of the language $\mathsf{L}(S)$) with k free set variables (and no free individual variables), the set $\widetilde{\mathscr{W}}_k(A)$ is regular.

PROOF By Lemma 14.9, we may assume that A is of the form $Q(P)$, where Q is a string of set quantifiers and P is a disjunction of principal formulas and principal formulas preceded by a negation symbol. Thus, the proof is by induction on the formation of such formulas.

† This is Theorem 5.3(a) of Elgot (1961, p. 27).

For the base step, A is a principal formula, i.e., of the form

$$(\exists x)(x^{(m_1)}\eta_1 F_{i_1} \wedge \cdots \wedge x^{(m_t)}\eta_t F_{i_t})$$

where $m_1 \leq \cdots \leq m_t$. We may assume that $F_1, \ldots, F_k$ are the distinct free set variables occurring in A and that $0 \leq n_1 < \cdots < n_p$ are the distinct superscripts on x's occurring in A.

Example If A is $(\exists x)(x^{(2)} \in F_1 \wedge x^{(2)} \notin F_2 \wedge x^{(5)} \in F_1 \wedge x^{(6)} \in F_3)$, then F_1, F_2, F_3 are the free set variables and 2, 5, 6 and distinct superscripts.

For each n_j, define a set S_{n_j} of elements in B_k as follows: $\mathrm{b}_1 \cdots \mathrm{b}_k \in S_{n_j}$ if and only if for each F_i such that $x^{(n_j)}\eta F_i$ occurs in A, $\mathrm{b}_i = 1$ if η is $\in$ and $\mathrm{b}_i = 0$ if η is $\notin$. Thus, continuing the example, we have $S_2 = \{100, 101\}$, $S_5 = \{100, 110, 101, 111\}$, and $S_6 = \{011, 101, 011, 111\}$. For each j, $1 \leq j \leq p$, we also have the set $v_k^{-1}(S_{n_j})$ of elements of Σ_k. We use these to define a certain set $\mathscr{R}_A$ of words on Σ_k:

$$\tau \in \mathscr{R}_A \quad \text{if and only if} \quad \tau \equiv \tau_1 a_{i_1} \tau_2 a_{i_2} \cdots \tau_p a_{i_p} \tau_{p+1}$$

satisfying the following conditions: (i) for each j, $1 \leq j \leq p$, $a_{i_j} \in v_k^{-1}(S_{n_j})$, (ii) $|\tau_1| \geq n_1$ and $|\tau_j| = n_j - n_{j-1} - 1$ for $1 < j \leq p$.

Exercise 14.57 Show that $\mathscr{R}_A = \mathscr{W}_k(A)$.

It is easy to see that $\mathscr{R}_A$ is finite-automaton acceptable. Notice that for each $\tau_1 a_{i_1} \tau_2 \cdots \tau_p a_{i_p} \tau_{p+1} \in \mathscr{R}_A$, $a_{i_1} \tau_2 \cdots \tau_p a_{i_p}$ has length $(n_p - n_1 + 1)$ and τ_1 and τ_{p+1} can be any words on Σ_k. Since n_1 and n_p are fixed, the set of all words $a_{i_1} \tau_2 \cdots \tau_p a_{i_p}$ is finite; call it $\mathscr{C}$. Let $\mathscr{U}$ be the set of all words on Σ_k. Then $\mathscr{R}_A$ is just $\mathscr{U} \cdot \mathscr{C} \cdot \mathscr{U}$. By exercises 14.42 and 14.43, $\mathscr{C}$ and $\mathscr{U}$ are finite-automaton acceptable. By exercise 14.53, $\mathscr{U} \cdot \mathscr{C} \cdot \mathscr{U}$ is finite-automaton acceptable, i.e., $\mathscr{R}_A$ is regular.

There are four cases for the induction step.

1. A is $B \wedge C$. Suppose $F_1, \ldots, F_s$ are the distinct free set variables in B and $F_{r+1}, \ldots, F_k$ are the distinct free set variables in C, where $r \leq s$ and the total number of distinct free set variables in A is k. By the induction hypothesis, $\mathscr{W}_s(B)$ and $\mathscr{W}_{k-r}(C)$ are regular. By exercise 14.55, $\hat{\rho}_{+(k-s)}(\mathscr{W}_s(B))$ is a Σ_k-regular set, and $\hat{\lambda}_{+r}(\mathscr{W}_{(k-r)}(C))$ is a Σ_k-regular set. Then, by exercise 14.53, $\hat{\rho}_{+(k-s)}(\mathscr{W}_s(B)) \cap \hat{\lambda}_{+r}(\mathscr{W}_{(k-r)}(C))$ is a Σ_k-regular set.

Exercise 14.58 Show that

$$\mathscr{W}_k(A) = \hat{\rho}_{+(k-s)}(\mathscr{W}_s(B)) \cap \hat{\lambda}_{+r}(\mathscr{W}_{(k-r)}(C))$$

Exercise 14.59 Show that $\mathscr{W}_k(A)$ for A of the form $\sim B$ and of the form $B \vee C$ is regular, assuming that the theorem holds for B and C.

The exercise gives us cases 2 and 3. For case 4, A is $(\exists F_k)B$, where B contains free set variables $F_1, \ldots, F_k$, and by the induction hypothesis, $\mathscr{W}_k(B)$ is regular.

Exercise 14.60 (a) Show that $\mathscr{W}_{k-1}((\exists F_k)B) = \{\hat{\rho}_{-1}(\tau_k) \mid \tau_k \in \mathscr{W}_k(B)\}$. (b) Conclude that, since $\hat{\rho}_{-1}$ is a regular homomorphism, $\mathscr{W}_{k-1}(A)$ is regular. ▌

Theorem 14.20 † The set of sentences of L(S) that are true about the finite subsets of the natural numbers is a recursive set.

PROOF By Lemma 14.7, we may assume that any sentence is in the form $(\exists F)A$ or $\sim(\exists F)A$. (1) $(\exists F)A$ is true for the finite subsets of N if and only if the set of finite subsets of N satisfying A is nonempty. Given A, we can construct $\mathscr{W}_1(A)$, which is the set of words which describe the finite sets satisfying A; thus, $(\exists F)A$ is true if and only if $\mathscr{W}_1(A) \neq \varnothing$. Since $\mathscr{W}_1(A)$ is regular, we can construct a finite automaton M_A such that $\mathscr{T}_{M_A} = \mathscr{W}_1(A)$. Since the problem of deciding of an arbitrary finite automaton whether or not the set of words which it accepts is empty is decidable, we can decide whether or not $\mathscr{W}_1(A) \neq \varnothing$ and thus whether or not $(\exists F)A$ is true. (2) The sentence $\sim(\exists F)A$ is true if and only if there are no finite sets which satisfy A. Thus, $\sim(\exists F)A$ is true if and only if $\mathscr{W}_1(A) = \varnothing$. ▌

Exercise 14.61 It is not really necessary to introduce the definition of regular set and prove Theorems 14.17 and 14.18 in order to prove Theorem 14.19. Reformulate the problem without using regular sets and prove whatever needs to be proved in order to prove Theorem 14.20.

In Elgot (1961), several corollaries are proved each of which is established by a correspondance between the set of finite subsets of N and some other set. For example, a one-to-one correspondance $\theta : \{\text{finite subsets of } N\} \to N$ can be defined by $\theta(\{n_1, \ldots, n_k\}) = 2^{n_1} + \cdots + 2^{n_k}$. Given two finite subsets of N, S_1 and S_2, an obvious question to ask is, can we describe the set $\theta^{-1}(\theta(S_1) + \theta(S_2))$? Or, more immediately, is there a formula in L(S) with three free set variables F_1, F_2, F_3 which is satisfied by three finite subsets N—S_1, S_2, and S_3—if and only if $\theta(S_3) = \theta(S_1) + \theta(S_2)$? If we carry out the addition of two numbers in binary representation, we see that, really, three sequences of 0's and 1's are involved: the two for the numbers being added and the set of "carries." Consider the formula (of L(S))

† This is the corollary to Theorem 5.3(a) in Elgot (1961, p. 31). It is also proved in Büchi (1960, p. 78), and it has been proved but not published by Ehrenfeucht.

$$(\exists C)(0 \notin C \wedge (0 \in F_3 \equiv 0 \in F_1 \circledvee 0 \in F_2)$$
$$\wedge (\forall x)(x' \in C \equiv ((x \in F_1 \wedge x \in F_2) \vee (x \in F_1 \wedge x \in C)$$
$$\vee (x \in F_2 \wedge x \in C)) \wedge x' \in F_3 \equiv ((x' \in C \wedge x' \in F_1$$
$$\wedge x' \in F_2) \vee (x' \in F_1 \circledvee x' \in F_2 \circledvee x' \in C)))) \qquad (*)$$

($A \circledvee B$ is an abbreviation for $((A \vee B) \wedge \sim(A \wedge B))$, i.e., the exclusive or.) Verify that if $S_1, S_2, S_3 \subseteq N$ satisfy (*), then $\theta(S_1) + \theta(S_2) = \theta(S_3)$. Recall the language $\mathsf{L}(n)$ of section 14.5. The formula (*) of $L(S)$ is satisfied by the finite subsets S_1, S_2, and S_3 of N, if and only if the formula $x + y = z$ of $\mathsf{L}(n)$ is satisfied by the natural numbers $\theta(S_1)$, $\theta(S_2)$, and $\theta(S_3)$. By this device, one can define the mapping $*$: {formulas of $\mathsf{L}(n)$} $\rightarrow$ {formulas of $\mathsf{L}(S)$}, by induction on the formation of formulas in $\mathsf{L}(n)$, so that A is true in $\mathfrak{N}$ if and only if A^* is true about the finite subsets of N. Thus, the decision problem for the set of sentences of $\mathsf{L}(n)$ true in $\mathfrak{N}$ is reduced to the decision problem for the set of sentences of $\mathsf{L}(S)$ true about the finite subsets of N. Since the latter is decidable by Theorem 14.20, the former is decidable without Theorem 14.11.

We give next a brief survey of some of the papers related to the material presented in this section. The principal paper in which finite automata were formulated is Kleene (1956). He develops them by way of an investigation of models for nerve nets. Although obscure in a few places, it is nevertheless a fascinating paper. Some parts of it are reformulated in Copi *et al.* (1958). The first systematic treatment of finite automata (including the material of the previous subsection) can be found in Rabin and Scott (1959); the reader is encouraged to read it. Along lines very similar to Elgot (1961) are Büchi (1960, 1962, and 1965). That the monadic second order theory of successor is decidable is shown in Büchi (1962). In Siefkes (1970), Büchi's methods are examined in detail. In Läuchli (1968), the decidability of the weak second-order theory of linear ordering is proved. Rabin (1967) is an interesting survey paper about finite automata and related topics, including those mentioned here, and it has a good bibliography. The last section of that paper is an introduction to the notion of tree-automata, which are seen to be a natural extension of the ordinary finite automata. In Elgot and Rabin (1966), various extensions of the monadic second-order theory of successor are shown to be decidable or undecidable. In that paper, the interest changes from relational systems having the natural numbers as the universe to relational systems which have words as the universe and functions and relations on words. What kind of formal language is appropriate for discussing "subword," "initial subword," "concatenation," "replacing one subword by another" as in a derivation, etc.? [See Thatcher (1966)]. Notice, for example, that if $\mathscr{U}$ is the set of words on {0, 1}, there are two natural mappings from

$\mathscr{U}$ to $\mathscr{U}$, one that concatenates 0 to the right end of a word and one that concatenates 1 to the right end of a word. These can be thought of as "successor functions" in a treelike ordering of words. For an alphabet with k symbols, there is a similar notion of k successor functions; such theories have come to be known as "multiple successor arithmetics." Thatcher (1966) shows that the first-order theory of multiple successor arithmetic with a binary predicate constant for "subword" is undecidable. In Thatcher and Wright (1968) and Doner (1965), it is shown that the weak monadic second-order theory of multiple successor arithmetic is decidable. Rabin (1969) proves that the standard theorems for finite automata also hold for tree-automata and then uses tree-automata to establish the decidability of the monadic second-order theory of multiple successor arithmetic. From this follow the decidability of the monadic second-order theory of linear ordering, the decidability of the theory of Boolean algebras with quantification over ideals, as well as the decidability of some questions in topology and game theory. Looking back at the Kleene (1956) paper, one can see the great vision of his investigations and how the development of his ideas leads naturally to these later results.

14.7 CONCLUDING REMARKS

What it means to say that a relation M (or function f) of a relational system $\mathfrak{U}$ is defined in a first-order language $\mathsf{L}(*)$ was discussed briefly in section 10.5. We would like to extend that notion to include second-order languages. Since a relational system for a second-order language has not been precisely defined, the new definition, though compatible with the old one, is a bit less precise. If U is a set and f is an n-ary function on U to U and L is a first- or second-order language, we say f *is definable in* L if there is a formula A of L, with $n+1$ free individual variables, such that A is satisfied by $u_1, \ldots, u_n, u_{n+1} \in U$ if and only if $f(u_1, \ldots, u_n) = u_{n+1}$. For example, let $\mathsf{L}(R)$ be the second-order language that allows quantification over binary predicate variables and has the symbols 0 and $'$, then addition is defined by the following formula:

$$(\forall P)\big((0, x) \in P \wedge (\forall u)(\forall v)((u, v) \in P \supset (u', v') \in P) \supset (y, z) \in P\big) \qquad (*)$$

The reader can verify for himself that $n_1, n_2, n_3 \in N$ satisfy $(*)$ if and only if $n_1 + n_2 = n_3$.

As we know, any theory of the natural numbers is undecidable in which addition and multiplication are definable. So, one way of looking at decidability problems is in terms of definability problems. There is a natural pair of questions for any given language L:

1. Are the natural numbers expressible and addition and multiplication definable in L?

2. Is the set of sentences of L true about the natural numbers a recursive set?

Of course, a positive answer to one question implies a negative answer to the other, and conversely. For example, by Presburger's theorem, the theory of addition of natural numbers on the first-order language (L(n)) with just 0, successor, and addition is decidable and so it follows that multiplication is not definable in that language. Similarly, since the second-order language (L(S)) with successor and monadic predicate variables is decidable [Büchi (1962)], it follows that not both addition and multiplication are definable in that language. In both these cases, we use decidability to imply undefinability. Next, we consider an example of definability implying undecidability.

Theorem 14.21 The set of sentences of L(R) that are true about the natural numbers is undecidable.

PROOF By (*) above, we know that addition is definable in L(R). We will see that divisibility $|$ is definable in L(R) and that multiplication $\cdot$ is definable (in a first-order language) by means of $|$ and $'$ and is therefore definable in L(R). Claim that

$x|y$ if and only if

$$(\forall F)\big(0 \in F \wedge (\forall z)\big(z \in F \supset (z + x) \in F \supset y \in F\big)\big) \qquad (**)$$

Suppose that $x, y \in N$ and $x|y$. Then, y is a multiple of x, i.e.,

$$y = \underbrace{x + \cdots + x}_{k\text{-times}}$$

for some k. We see that x and y also satisfy (**) because for any $F \subseteq N$, if $0 \in F$ and x plus any element of F belongs to F, then $(0 + x) \in F$ and all multiples of x belong to F and so $y \in F$. On the other hand, suppose that $x, y \in N$ and satisfy (**). In general, if an element satisfies all sets with such-and-such property, then that element satisfies any particular such set. Consider the set F_x consisting of just the multiples of x. Certainly $0 \in F_x$ and if $z \in F_x$, i.e., if z is a multiple of x, then $(z + x) \in F$. Then, $y \in F_x$ implies that y is a multiple of x. Next, we will show that multiplication is definable in any first-order language with symbols 0, $|$, and $'$. For $x, y \in N$, we will write $x \perp y$ if x and y are relatively prime and $x \cdot y$ for the least common multiple of x and y. (Refer to section 6.3 for the definitions.) Notice that if $x \perp y$, then $x \cdot y = x \cdot y$. $x \perp y$ and $x \cdot y = z$ are (first-order) definable by means of 0, $'$, and $|$ as follows:

$$x \perp y \quad \text{if and only if} \quad (\forall z)(z \mid x \wedge z \mid y \supset z = 0')$$

$$x \cdot y = z \quad \text{if and only if} \quad (x \mid z \wedge y \mid z \wedge (\forall u)(x \mid u \wedge y \mid u \supset z \mid u))$$

Claim that $x \cdot y = z$ if and only if

$$\begin{aligned}
\big((x = 0 \vee y = 0) \wedge z = 0\big) &\vee (\forall u)(x \mid u \wedge y \mid u \wedge z \mid u) \\
&\vee (\forall u)(\forall v)(\forall w)(x \perp u \wedge y \perp v \wedge z \perp u \wedge z \perp v \\
&\quad \wedge u \perp v \wedge w \mid (x \cdot u)' \wedge w \mid (y \cdot v)' \\
&\quad \supset (\exists t)\big(w \mid t \wedge t' = z \cdot (u \cdot v)\big) \qquad (***)
\end{aligned}$$

There are two special cases. If x or y is 0, then $x \cdot y = z$ if and only if x, y, and z satisfy (***). If $x = y = z = 1$, then $x \cdot y = z$ and $(\forall u)(x \mid u \wedge y \mid u \wedge z \mid u)$. For the general case, suppose $x \cdot y = z$ and consider (***). $x \perp y$ implies that $x \cdot u = x \cdot u$, etc. Thus, $w \mid (x \cdot u)'$ and $w \mid (y \cdot v)'$ imply that $x \cdot u \equiv -1 (\text{mod } w)$ and $y \cdot v \equiv -1 (\text{mod } w)$ so, multiplying these, we have $(x \cdot y) \cdot (u \cdot v) \equiv 1 (\text{mod } w)$. Since $x \cdot y = z$, $z \cdot (u \cdot v) \equiv 1 (\text{mod } w)$, i.e., there is a k such that $k \cdot w + 1 = z \cdot (u \cdot v)$. Thus, $k \cdot w$ is the number required by $(\exists t)$. On the other hand, suppose (***) is satisfied by $x, y, z \in N$. Keeping those numbers fixed, we will see that $x \cdot y = z$ as follows. As before, for any u, v, and w, if $x \cdot u \equiv -1 (\text{mod } w)$ and $y \cdot v \equiv -1 (\text{mod } w)$, then $z \cdot (u \cdot v) \equiv 1 (\text{mod } w)$. By multiplying the first two and substituting the third, we have $(x \cdot y) \cdot (u \cdot v) \equiv z \cdot (u \cdot v) (\text{mod } w)$ and consequently $x \cdot y \equiv z (\text{mod } w)$. Notice that there are arbitrarily large w for which this holds true. Since x, y, and z are fixed, if $w > \max(x \cdot y, z)$ then $x \cdot y \equiv z (\text{mod } w)$ implies that $x \cdot y = z$. Since $L(R)$ has symbols 0 and $'$, and $+$ and $\cdot$ are definable in $L(R)$, the recursive functions are n-representable in the set of sentences of $\mathsf{L}(R)$ true about the natural numbers. (Recall the proof of Theorem 14.7.) Thus, by Corollary 1 to Theorem 14.3, that set is undecidable. ▮

Let $\mathsf{L}(S^+)$ be the language $\mathsf{L}(S)$ with the additional symbol $+$. That is, $\mathsf{L}(S^+)$ is the second-order language with 0, $'$, $+$, and monadic predicate variables only.

Corollary 1 The set of sentences of $\mathsf{L}(S^+)$ that are true about the natural numbers is undecidable.

Corollary 2 Addition is not definable in $\mathsf{L}(S)$.

Exercise 14.62 Prove the corollaries.

So far in this chapter we have established some formal theories which parallel what we do with the natural numbers in everyday mathematics, and studied some properties of those theories. If $\mathsf{L}(x)$ is a language, let $\mathscr{T}(x)$ denote the set of sentences of $\mathsf{L}(x)$ true in the appropriate relational system with universe N. The results of the chapter are summarized as follows.

1. $\mathsf{L}(n)$ is the first-order language with equality, 0, $'$, $+$. $\mathfrak{N}_+ = \langle N; ', +; = \rangle$. The theory $\mathscr{T}(n)$ (previously denoted by Ps) is decidable, so multiplication is not definable in $\mathsf{L}(n)$.

2. $\mathsf{L}(N)$ is the language $\mathsf{L}(n)$ with the additional symbol $\cdot$. $\mathfrak{N} = \langle N; ', +, \cdot; = \rangle$. Recall that the theories R, Q, and P are the sets of theorems derivable from certain axioms. $\mathsf{R} \subset \mathsf{Q} \subset \mathsf{P} \subset \mathscr{T}(N)$, ($\mathscr{T}(N)$ was previously denoted by $\mathscr{N}$), and $\mathsf{Ps} = \mathscr{T}(n) \subset \mathsf{P}$. The theory R and all of its extensions are undecidable and so R is essentially undecidable, as, of course, are its extensions.

3. $\mathsf{L}(S)$ is the second-order language with 0, $'$, and only monadic predicate variables. $\mathscr{T}(S)$ is decidable. Divisibility, addition, and multiplication are not definable in $\mathsf{L}(S)$.

4. $\mathsf{L}(R)$ is the extension of $\mathsf{L}(S)$ obtained by adding binary predicate variables. $\mathscr{T}(S) \subset \mathscr{T}(R)$. Divisibility, addition, and multiplication are definable in $\mathsf{L}(R)$. $\mathscr{T}(R)$ is undecidable. There is a natural translation of $\mathscr{T}(N)$ into $\mathscr{T}(R)$ as implicitly indicated in the proof of Theorem 14.21.

5. $\mathsf{L}(S^+)$ is the extension of $\mathsf{L}(S)$ obtained by adding the symbol $+$. $\mathscr{T}(S) \subset \mathscr{T}(S^+)$. Divisibility and multiplication are definable in $\mathsf{L}(S^+)$. $\mathscr{T}(S^+)$ is undecidable. There is a natural translation of $\mathscr{T}(N)$ into $\mathscr{T}(S^+)$.

6. Any theory about the natural numbers which is an extension of $\mathscr{T}(N)$, $\mathscr{T}(R)$, or $\mathscr{T}(S^+)$ is undecidable.

In this list, each example of an undecidable theory is an essentially undecidable theory. It is natural to wonder if there are any interesting undecidable theories which have interesting decidable extensions. Let $\mathsf{L}(g)$ be the first-order language with $=$, a binary function symbol $*$, and a constant symbol **e**, and let G denote the first order theory with nonlogical axioms.

$$(\forall \mathbf{x})(\forall \mathbf{y})(\forall \mathbf{z})((\mathbf{x}*(\mathbf{y}*\mathbf{z})) = ((\mathbf{x}*\mathbf{y})*\mathbf{z}))$$
$$(\forall \mathbf{y})(\mathbf{e}*\mathbf{y} = \mathbf{y})$$
$$(\forall \mathbf{x})(\exists \mathbf{y})(\mathbf{y}*\mathbf{x} = \mathbf{e})$$

Every model of G is a group, by the definition of group. Let AG denote the set of theorems derivable from $\mathsf{G} \cup \{(\forall \mathbf{x})(\forall \mathbf{y})(\mathbf{x}*\mathbf{y} = \mathbf{y}*\mathbf{x})\}$. Every model of AG is an Abelian group. Of course, $\mathsf{G} \subset \mathsf{AG}$. G, which is the set of all sentences of $\mathsf{L}(g)$ true about all groups, is called *elementary group theory*. Similarly, AG is the *elementary theory of Abelian groups*. Using methods similar to those used in the prove of Theorem 14.21 (given the general name "method of interpretation"), Tarski has shown that elementary group theory is undecidable. [See Tarski, *et al.* (1953, Part III).] On the other hand, Szmielew (1955) has shown, using elimination of quantifiers, that the elementary theory of Abelian groups is decidable. [See also Ershov *et al.* (1965, pp. 62–65).] Thus, elementary group theory is undecidable but not essentially undecidable. The same relationship holds between elementary field theory (see the latter part of chapter XII for the definition) and the elementary theory of algebraically closed fields. In J. Robinson (1949), a method of interpretation is used to show that elementary field theory is undecidable. But Tarski (1949) used an elimination-of-quantifier argument to show that the elementary theories of algebraically closed and real closed fields arc decidable. [See also Ershov *et al.* (1965, pp. 49–52).] Clearly, these two methods are very useful; the reader can learn more about them in Ackermann (1954), Tarski *et al.* (1953) and Shoenfield (1967, pp. 57–65, 82–88). But, as one might imagine, these are not the only methods available. In A. Robinson (1965), the reader can learn more about completeness, which, as we know, is related to decidability. Many methods, including those already mentioned, are presented in Ershov *et al.* (1965), which the reader is urged to read. It is an excellent article surveying decidability and undecidability results and the methods used to obtain them. It also includes an extensive bibliography as well as tables showing which theories were then known to be decidable, which undecidable. The reader who wishes to learn about decidable and undecidable theories will have made a good beginning by reading the papers mentioned above.

REFERENCES

Ackermann, W. (1954). *Solvable Cases of the Decision Problem.* North-Holland Publ., Amsterdam.

Allen, J., and Luckham, D. (1970). An interactive theorem-proving program. In *Machine Intelligence* (D. Michie and B. Meltzer, eds.), Vol. 5, pp. 321–336. Am. Elsevier, New York.

Anderson, R., and Bledsoe, W. W. (1970). A linear format for resolution with merging and a new technique for establishing completeness. *J. Assoc. Comput. Mach.* **17**, 525–534.

Arbib, M. A. (1966). Speed-up theorems and incompleteness theorems. In *Automata Theory* (E. R. Caianiello, ed.), pp. 6–24. Academic Press, New York.

Arbib, M. A. (1969). *Theories of Abstract Automata.* Prentice-Hall, Englewood Cliffs, New Jersey.

Axt, P. (1963). Enumerations and the Grzegorczyk hierarchy. *Z. Math. Logik Grundlagen Math.* **9**, 53–65.

Blum, M. (1967a). A machine-independent theory of the complexity of recursive functions. *J. Assoc. Comput. Mach.* **14**, 322–336.

Blum, M. (1967b). On the size of machines. *Information and Control* **11**, 257–265.

Blum, M. (1969). On effective procedures for speeding up algorithms. *Assoc. Comput. Mach. Symp. Theory Comput., Marina del Ray, Cal., 1969*, pp. 43–53. Assoc. for Comput. Mach., New York.

Boone, W. W. (1959). The word problem. *Ann. of Math.* **70**, 207–265.

Boone, W. W., Haken, W., and Poénaru, V. (1968). On recursively unsolvable problems in topology and their classification. In *Contributions to Mathematical Logic* (H. A. Schmidt, K. Schütte, and H. J. Thiele, eds.), pp. 37–74. North-Holland Publ., Amsterdam.

Büchi, J. R. (1960). Weak second order arithmetic and finite automata. *Z. Math. Logik Grundlagen Math.* **6**, 66–92.

Büchi, J. R. (1962). On a decision method in restricted second order arithmetic. *Proc. Internat. Congr. Logic, Methodology and Philos. Sci. 1960* (E. Nagel, P. Suppes, and A. Tarski, eds.) pp. 1–11. Stanford Univ. Press, Stanford, California.

Büchi, J. R. (1965). Transfinite automata recursions and weak second order theory of ordinals. *Proc. Internat. Congr. Logic, Methodology and Philos. Sci. 1964* (Y. Bar-Hillel, ed.), pp. 3–23. North-Holland Publ., Amsterdam.

Cantor, D. G. (1962). On the ambiguity problem of Backus systems. *J. Assoc. Comput. Mach.* **9**, 477–479.

Carnap, R. (1947). *Meaning and Necessity.* Univ. of Chicago Press, Chicago, Illinois.

Chang, C. L. (1970). The unit proof and the input proof in theorem proving. *J. Assoc. Comput. Mach.* **17**, 698–707.

Chomsky, N. (1959). On certain formal properties of grammars. *Information and Control* **2**, 137–167.

Chomsky, N. (1968). *Syntactic Structures*. Mouton, The Hague.

Chomsky, N., and Shutzenberger, M. P. (1963). The algebraic theory of context-free languages. In *Computer Programming and Formal Systems* (P. Braffort and D. Hirschberg, eds.), pp. 118–161. North-Holland Publ., Amsterdam.

Church, A. (1936). An unsolvable problem of elementary number theory. *Amer. J. Math.* **58**, 345–363.

Church, A. (1956). *Introduction to Mathematical Logic*. Princeton Univ. Princeton, New Jersey.

Cleave, J. P. (1963). A hierarchy of primitive recursive functions. *Z. Math. Logik Grundlagen Math.* **9**, 331–345.

Cobham, A. (1965). The intrinsic computational difficulty of functions. *Proc. Intern. Congr. Logic, Methodology and Philos. Sci.* 1964 (Y. Bar-Hillel, ed.), pp. 24–30. North Holland Publ., Amsterdam.

Cocke, J., and Minsky, M. (1964). Universality of tag systems with $D = 2$. *J. Assoc. Comput. Mach.* **11**, 15–20.

Copi, I. M., Elgot, C. C., and Wright, J. B. (1958). Realization of events by logical nets. *J. Assoc. Comput. Mach.* **5**, 181–196.

Davis, M. (1953). Arithmetical problems and recursively enumerable predicates. *J. Symbolic Logic* **18**, 33–41.

Davis, M. (1958). *Computability and Unsolvability*. McGraw-Hill, New York.

Davis, M. (1963). Eliminating the irrelevant from mechanical proofs. *Proc. Symp. Appl. Math., Chicago and Atlantic City, 1963*, **15**, pp. 15–30.

Davis, M., ed. (1965). *The Undecidable*. Raven Press, Hewlitt, New York.

Davis, M., Putnam, H., and Robinson, J. (1960). The decision problem for exponential Diophantine equations. *Ann. of Math.* **74**, 425–436.

Doner, J. E. (1965). Decidability of the weak second-order theory of two successors. *Notices Amer. Math. Soc.* **12**, (1965) Abstr. 65T-468, p. 819.

Elgot, C. C. (1961). Decision problems of finite automata design and related arithmetics. *Trans. Amer. Math. Soc.* **98**, 21–51.

Elgot, C. C. and Rabin, M. O. (1966). Decidability and undecidability of extensions of second order theory of successor. *J. Symbolic Logic* **31**, 169–181.

Ershov, Y. L., Lavrov, I. A., Taimanov, A. D., and Taitslin, M. A. (1965). Elementary theories. *Russian Math. Surveys* **20**, 35–105.

Floyd, R. W. (1964). *New Proofs of Old Theorems in Logic and Formal Linguistics*. Comput. Assoc., Wakefield, Massachusetts.

Frege, G. (1892). Über Sinn und Bedeutung. *Z. Phil. und phil. Kritik*. **100**, 25–50. [An Engl. transl. appears in *Phil. Rev.* (1948), **57**, 207–230.]

Friedberg, R. M. (1957). Two recursively enumerable sets of incomparable degrees of unsolvability. *Proc. Nat. Acad. Sci. U.S.A.* **43**, 236–238.

Friedman, J. (1963). A semi-decision procedure for the functional calculus. *J. Assoc. Comput. Mach.* **10**, 1–24.

Ginsburg, S. (1966). *The Mathematical Theory of Context-Free Languages*. McGraw-Hill, New York.

Gödel, K. (1930). Die Vollständigkeit der Axiome des logischen Funktionenkalkuls. *Monatsh. Math. Phys.* **37**, 349–360.

Gödel, K. (1931). Über formal unentscheidbare Sätze de Principia Mathematica und verwandter System I. *Monatsh. Math. Phys.* **38**, 173–198 [an Engl. transl. appears in Davis (1965)].

Gross, M., and Lentin, A. (1970). *Introduction to Formal Grammars*. Springer, Berlin.

Grzegorczyk, A. (1953). Some classes of recursive functions. *Rozprawy Mat.* **4.** Instytut Mathematyczne Polskiej Akademic Nauk, Warsaw.
Grzegorczyk, A. (1961). *Fonctions Recursives.* Gauthier-Villars, Paris.
Hartmanis, J., and Stearns, R. E. (1964). Computational complexity of recursive sequences. *Proc. Ann. Symp. Switching Circuit Theory Logical Design, 5th, Princeton, 1964*, pp. 82–90. I.E.E.E., New York.
Hartmanis, J., and Stearns, R. E. (1965). On the computational complexity of algorithms. *Trans. Amer. Math. Soc.* **117**, 285–306.
Hartmanis, J. (1968). Computational complexity of one-tape Turing machines. *J. Assoc. Comput. Mach.* **15**, 325–339.
Helm, J. P., and Young, P. R. (1971). On size versus efficiency for programs admitting speed-ups. *J. Symbolic Logic* **36**, 21–27.
Henkin, L. (1949). The completeness of the first-order functional calculus. *J. Symbolic Logic* **14**, 159–166.
Herbrand, J. (1930). Recherches sur la théorie de la démonstration. *Travaux de la Societé des Sciences et des Lettres de Varsovie, Classe III science mathématiques et physiques*, **33.** Instytut Mathematyczne Polskiej Akademic Nauk, Warsaw.
Hermes, H. (1965). *Enumerability, Decidability, Computability.* Springer, Berlin.
Herstein, I. N. (1964). *Topics in Algebra.* Ginn (Blaisdell), Boston.
Hilbert, D., and Bernays, P. (1934). *Grundlagen der Mathematik* Vol. I. Springer, Berlin (reprinted in 1968).
Hopcroft, J. E., and Ullman, J. D. (1968). Relations between time and tape complexities. *J. Assoc. Comput. Mach.* **15**, 414–427.
Hopcroft, J. E., and Ullman, J. D. (1969a). *Formal Languages and Their Relation to Automata.* Addison-Wesley, Reading, Massachusetts.
Hopcroft, J. E., and Ullman, J. D. (1969b). Some results on tape-bounded Turing machines. *J. Assoc. Comput. Mach.* **16**, 168–177.
Kleene, S. C. (1952). *Introduction to Metamathematics.* Van Nostrand, Reinhold, Princeton, New Jersey.
Kleene, S. C. (1956). Representation of events in nerve nets and finite automata. In *Automata Studies* (C. E. Shannon, and J. McCarthy, eds.), pp. 3–42. Princeton Univ. Press, Princeton, New Jersey.
Kleene, S. C. (1967). *Mathematical Logic.* Wiley, New York.
Kneale, W. and Kneale, M. (1962). *The Development of Logic.* Oxford Univ. Press (Clarendon), London and New York.
Kurosh, A. G. (1956). *The Theory of Groups.* Chelsea, New York.
Läuchli, H. (1968). A decision procedure for the weak second order theory of linear order. In *Contributions to Mathematical Logic* (H. A. Schmidt, K. Schutte, and H.-J. Thiele, eds.), pp. 189–197. North-Holland Publ., Amsterdam.
Loveland, D. (1970a). A simplified format for model elimination theorem-proving procedure. *J. Assoc. Comput. Mach.* **17**, 349–363.
Loveland, D. (1970b). A linear format for resultion. *Lecture Notes in Mathematics*, Vol. 125, pp. 147–162, Springer, Berlin.
Magnus, W. (1932). Das Identitäts problem für Gruppen mit einer definierendem Relation. *Math. Ann.* **106**, 295–307.
Magnus, W., Karrass, A., and Solitar, D. (1966). *Combinatorial Group Theory.* Wiley (Interscience), New York.
Markov, A. A. (1958). Insolubility of the problem of homeomorphy (in Russian). *Proc. Intern. Congr. Math., Edinburgh 1958*, pp. 300–306. Cambridge Univ. Press, London and New York.
Matijasevic, J. V. (1970). Enumerable sets are Diophantine. *Soviet Math. Dokl.* **11**, 354–358.

Mendelson, E. (1964). *Introduction to Mathematical Logic*. Van Nostrand, Reinhold, Princeton, New Jersey.

Minsky, M. (1967). *Computation: Finite and Infinite Machines*. Prentice-Hall, Englewood Cliffs, New Jersey.

Muchnik, A. A. (1956). Negative answer to the problem of reducibility of the theory of algorithms (in Russian). *Dokl. Akad. Nauk SSSR* **108**, 194–197.

Myhill, J. (1960). Linear-bounded automata. WADC Tech. Rep. 57–624. Wright-Patterson Air Force Base, Ohio.

Myhill, J. (1964). *Math. Reviews* **28**, 2045.

Novikov, P. S. (1958). On the algorithmic unsolvability of the word problem in group theory. *Amer. Math. Soc. Transl.* [2] **9**, 1–120.

Péter, R. (1967). *Recursive Functions*, 3rd ed. Academic Press, New York.

Post, E. (1936). Finite combinatory processes. Formulation I. *J. Symbolic Logic* **1,** 103–105 [reprinted in Davis (1965)].

Post, E. L. (1943). Formal reductions of the general combinatorial decision problem. *Amer. J. Math.* **65**, 197–215.

Post, E. (1944). Recursively enumerable sets of positive integers and their decision problems. *Bull. Amer. Math. Soc.* **50**, 284–316 [reprinted in Davis (1965)].

Post, E. (1946). A variant of a recursively unsolvable problem. *Bull. Amer. Math. Soc.* **52,** 264–268.

Post, E. (1947). Recursive unsolvability of a problem of Thue. *J. Symbolic Logic* **12**, 1–11 [reprinted in Davis (1965)].

Prawitz, D. (1970). A proof procedure with matrix reduction. *Lecture Notes in Mathematics*, Vol. 125, pp. 207–214. Springer, Berlin.

Presburger, M. (1929). Über die Vollständigkeit eines gewissen Systems der Arithmetik ganzer Zahlen, in welchem die Addition als einzige Operation hervortritt. *Congr. Math. des Pays Slaves, Warsaw, 1929*, **1**, pp. 92–101.

Quine, W. van O. (1953). *From a Logical Point of View*. Harper, New York.

Rabin, M. O. (1958). Recursive unsolvability of group theoretic properties. *Ann. of Math.* **67**, 172–194.

Rabin, M. O. (1960). Degree of difficulty of computing functions and a partial ordering of recursive sets. Tech. Rep. No. 2. Hebrew Univ. Jerusalem, Israel.

Rabin, M. O. (1963). Real-time computation. *Israel J. Math.* **1**, 203–211.

Rabin, M. O. (1967). Mathematical theory of automata. *Proc. Symp. Applied. Math.: Math. Aspects of Comput. Sci. New York City, 1967*, **19**, pp. 153–175. Am. Math. Soc., Providence, Rhode Island.

Rabin, M. O. (1969). Decidability of second-order theories and automata on infinite trees. *Trans. Amer. Math. Soc.* **141**, 1–35.

Rabin, M. O., and Scott, D. (1959). Finite automata and their decision problems. *IBM J. Res. Develop.* **3**, 115–125.

Ritchie, R. W. (1963). Classes of predictably computable functions. *Trans. Amer. Math. Soc.* **106**, 139–173.

Robinson, A. (1956). *Complete Theories*. North-Holland Publ., Amsterdam.

Robinson, A. (1963). *Introduction to Model Theory and to the Metamathematics of Algebra*. North-Holland Publ., Amsterdam.

Robinson, G., and Wos, L. (1969). Paramodulation and theorem-proving in first order theories with equality. In *Machine Intelligence* (D. Michie and B. Meltzer, eds.), Vol. 4, pp. 135–150. Am. Elsevier, New York.

Robinson, J. (1949). Definability and decision problems in arithmetic. *J. Symbolic Logic* **14**, 98–114.

Robinson, J. (1950). General recursive functions. *Proc. Amer. Math. Soc.* **1**, 703–718.

Robinson, J. (1952). Existential definability in arithmetic. *Trans. Amer. Math. Soc.* **72**, 437–449.
Robinson, J. A., (1965a). A machine oriented logic based on the resolution principle. *J. Assoc. Comput. Mach.* **12**, 23–41.
Robinson, J. A. (1965b). Automatic deduction with hyper-resolution. *Internat. J. Comput. Math.* **1**, 227–234.
Robinson, J. A. (1967). A review of automatic theorem-proving. *Proc. Symp. Appl. Math. New York City, 1967*, **19**, Am. Math. Soc., Providence, Rhode Island. pp. 1–18.
Robinson, R. M. (1947). Primitive recursive functions. *Bull. Amer. Math. Soc.* **53**, 925–942.
Robinson, R. M. (1958). Restricted set-theoretical definitions in arithmetic. *Proc. Amer. Math. Soc.* **9**, 238–242.
Rogers, H. (1967). *Theory of Recursive Functions and Effective Computability*. McGraw-Hill, New York.
Rosser, J. B. (1936). Extensions of some theorems of Gödel and Church. *J. Symbolic Logic* **1**, 87–91.
Rotman, J. J. (1965). *The Theory of Groups: An Introduction*. Allyn & Bacon, Rockleigh, New Jersey.
Russell, B. (1940). *An Inquiry into Meaning and Truth*. Humanities Press, New York.
Sacks, G. E. (1963). *Degrees of Unsolvability*. Princeton Univ. Press, Princeton, New Jersey.
Shepherdson, J. C., and Sturgis, H. E. (1963). Computability of recursive functions. *J. Assoc. Comput. Mach.* **10**, 217–255.
Shoenfield, J. R. (1967). *Mathematical Logic*. Addison-Wesley, Reading, Massachusetts.
Siefkes, D. (1970). Büchi's monadic second order successor arithmetic. *Lecture Notes in Mathematics*, Vol. 120. Springer, Berlin.
Slagle, J. (1967). Automatic theorem proving with renamable and semantic resolution. *J. Assoc. Comput. Mach.* **14**, 698–709.
Smullyan, R. M. (1961). *Theory of Formal Systems*. Princeton Univ. Press, Princeton, New Jersey.
Szmielew, W. (1955). Elementary properties of Abelian groups. *Fund, Math*, **41**, 203–271.
Tarski, A. (1949). Arithmetical classes and types of mathematical systems, mathematical aspects of arithmetical classes and types, arithmetical classes and types of Boolean algebras, arithmetical classes and types of algebraically closed and real-closed fields. *Bull. Amer. Math. Soc.* **55**, 63–64.
Tarski, A. (1956). Some observations on the concepts of ω-consistency and ω-completeness. In *Logic, Semantics, Meta-mathematics.*, pp. 279–295. Oxford Univ. Press (Clarendon), London and New York.
Tarski, A., Mostowski, A., and Robinson, R. M. (1953). *Undecidable Theories*. North-Holland Publ., Amsterdam.
Thatcher, J. W. (1966). Decision problems for multiple successor arithmetics. *J. Symbolic Logic* **31**, 182–190.
Thatcher, J. W., and J. B. Wright, (1968). Generalized finite automata theory with an application to a decision problem of second-order logic. *Math. Systems Theory* **2**, 57–81.
Turing, A. M. (1936–1937). On computable numbers, with an application to the Entscheidungsproblem. *Proc. London Math. Soc.* [2], **42**, 230–265 [reprinted in Davis (1965)].
Turing, A. M. (1950). Computing machinery and intelligence. *Mind*, **54**, 433–460.
Wang, H. (1957). A variant to Turing's theory of computing machines. *J. Assoc. Comput. Mach.* **4**, 63–92.
Wittgenstein, L. (1961). *Tractatus Logico-Philosophicus*. (D. F. Pears and B. F. McGuiness, translators). Humanities Press, New York.
Yamada, H. (1962). Real-time computation and recursive functions not real-time computable. *IRE Trans. Electron. Comput.* **EC-11**, 753–760.

Young, P. R. (1969a). Speed-ups by changing the order in which sets are enumerated. *Assoc. Comput. Mach. Theory of Comput. Marina del Rey. Cal., 1969*, pp. 89–92. Assoc. Comput. Mach., New York.

Young, P. R. (1969b). Toward a theory of enumerations. *J. Assoc. Comput. Mach.* **16**, 228–248.

INDEX

The symbols and theorem numbers in brackets refer to those used in the text. Page numbers in boldface refer to those pages on which definitions of terms or statements of theorems are given. The letter n following a page number indicates that the entry is cited in a footnote to that page

E

F

G

H

Q

R

S